Modelling of Concrete Performance

Modelling of Concrete Performance

Hydration, microstructure formation and mass transport

Koichi Maekawa, Rajesh Chaube and Toshiharu Kishi

ROUTLEDGE
London and New York

First published 1999 by E & FN Spon, an imprint of Routledge
11 New Fetter Lane, London EC4P 4EE

Simultaneously published in the USA and Canada
by Routledge
29 West 35th Street, New York, NY 10001

Typeset in Times by Mathematical Composition Setters Ltd, Salisbury
Printed and bound in Great Britain by Biddles Ltd, Guildford and King's Lynn

British Library Cataloguing in Publication Data
A catalogue record for this book is available from the British Library

Library of Congress Cataloging in Publication Data
Modelling of concrete performance: hydration, microstructure formation, and mass transport/Koichi Maekawa, Rajesh Chaube, and Toshiharu Kishi.
p. cm
Includes bibliographical references and index.
1. Concrete—Evaluation. 2. Concrete—Curing—Mathematical models.
3. Concrete cracking. I. Chaube, Rajesh, 1955– .
II. Kishi, Toshiharu, 1955– . III. Title.
TA440.M243 1999
620. 1'36—dc21 98–31036
CIP

ISBN 0-419-24200-7

Contents

Preface

This book offers a theoretical and computational platform on which structural concrete performance is evaluated and concrete quality at early age is examined in both space and time under differing environmental actions. Here, concrete is treated as a composite material consisting of growing microscale pores, which govern the basic mechanical and physical features of concrete with respect to long-term durability. On this basis, hygrothermal physics of concrete has a central position in this book, based on which the scientific knowledge and engineering simulation of concretes and their performances are presented.

The authors examine three thermodynamic features of early aged concrete. One feature is the hydration model associated with heat generation and water consumption of cementitious powders. Its mathematical expression under arbitrary environments in the micropores with respect to temperature, pore water and its solution is presented. The second feature is the hysteresis state equilibrium of liquid water, vapour and other idealized gases in the microscale pores. The moisture migration and percolation in space are linked with the thermodynamic equilibrium. The third feature is the geometrical treatment of the formation of micropores built by cement and pozzolanic hydrates. The micropore structure primarily controls the mechanical performance of hardening concrete. However, the use of physical chemistry is also included in this book. Some work to link the physical–chemical system of concrete with the deformational field associated with mechanical stresses is submitted as a bridge between material science and structural engineering for reinforced concrete.

It is well known that long-term durability of concrete in building and infrastructures can be achieved if material quality, concreting works, structural detailing and dimensioning is appropriately performed. For systematically resolving problems related to structural concrete durability, performance-based design is one of the most promising methods of unified structural and durability design. Within this framework, it is indispensable to accurately verify attained concrete performances and to simulate structural behaviours in time and space under external actions given as design conditions. It is also recognized that the quality of concrete in structures and defects induced at early age are governing factors for long-term performance of concrete. Young concrete performance matured after birth, is the authors' main concern in this book. The technology discussed is expected to be an initial input to help ensure long-term concrete performances and transient behaviours of structural concrete in the future.

In another respect, this book can be thought of as an intermediate report from 1981 of the concrete engineering research unit at the University of Tokyo. Their proposal of a 'lifespan simulator of reinforced concrete' follows the book entitled *Nonlinear Analysis and Constitutive Models of Reinforced Concrete*, published in 1991.

In 1981, I started my research on the temperature prediction of massive concrete and thermal crack risk evaluation. This study became the start of this book. Soon after, I encountered difficulty concerning the adiabatic temperature rise model for concrete heat generation, because it provides a singular solution of the hydration process of cement in concrete under the unrealistic condition of perfect thermal isolation. For rational and accurate prediction of thermal stresses, it became clear that the heat of hydration of cement must be predicted for realistic conditions of temperature and moisture inside structures. As a matter of fact, the temperature inside a structure is in turn associated with the heat of hydration itself.

The author's study of the hygrochemical system of hardening concrete began again in 1989 with the fact that hydration of cement cannot be defined as a unique material property or an intrinsic quality, but has to be predicted as a engineering behaviour coupled with the thermodynamic fields developing in structures. Furthermore, it was clearly recognized that a much higher precision of measuring heat released under the adiabatic temperature rise condition is crucial for constructing chemical reaction based modelling for growing concrete performance. Fortunately, the enhanced adiabatic temperature rise testing machine was developed by Dr Y. Suzuki of Sumitomo Cement, who had worked with me since 1981 and contributed to the increasing accuracy of measurements of the energy release rate of cement in concrete.

From 1990 to 1992 I spent two valuable years in the Asian Institute of Technology in Thailand on secondment from the Japanese Government. The importance of the coupling of moisture consumption, migration and state equilibrium was perceived and both experimental and theoretical studies were initiated. The blueprint was made in this period. At the beginning of 1990s, three outstanding researchers joined the project at the University of Tokyo after my return from Thailand.

Dr Toshiharu Kishi, one of the co-authors of this book, chiefly investigated heat of hydration modelling of cementitious powder consisting of multimineral compounds and created a linkage gate with liquid water/vapour/ entropy migration and micropore formation. His work was important for every subsequent stage.

Dr Rajesh P. Chaube is the main developer of the computer code DuCOM which is on the internet and regarded as a practical result of this book. He is one of the brightest young scholars at the University and has greatly contributed to the understanding of pore structural formation, its mathematical idealization and associated percolation. His outstanding capability in science and computer technology was actively applied with the massive parallel computer system in the Intelligent Modelling Laboratory (IML) at the University of Tokyo.

Dr Takumi Shimomura, a current faculty member of Nagaoka University of Science and Technology, collaborated in the study of the drying shrinkage model based upon micropore geometry and hygrothermal state equilibrium. Dr Tetsuya Ishida, Research Associate of the University of Tokyo, is actively extending the applicability of concrete performance modelling up to full-scale reinforced concrete under external forces and environmental actions. It was useful to have had excellent younger staff to steer the project on lifespan simulation. Their work serves as a basis of the current project to integrate DuCOM with the nonlinear structural mechanics of damaging reinforced concrete.

On behalf of all the co-authors, I acknowledge the valuable guidance and suggestions given by Professor Hajime Okamura in the Department of Civil Engineering, University of Tokyo. He has been the authors' research chief since 1980. His intellect and engineering insight have been a keystone of the authors' activities.

Let me repeat again that the aim of this book is to propose a scheme of tracing the physical–chemical quality and conditions in concrete structures of three-dimensional extent and to begin discussion of performance based durability design. The authors of this book regard this text as an incomplete description of technical and scientific elements which need to be further investigated. We are just beginning the study of the long-term performance of concrete structures and its prediction. But, we hope that this book will help the study of rational performance based design of concrete structures and serve students and practitioners who are responsible for objectively ensuring required performances, planning of concrete construction and management of recycle use of cementitious construction materials in the next millennium.

Koichi Maekawa
University of Tokyo
April, 1998

Abstract

Strength and microstructure development of cementitious materials are generally the crucial factors that influence several deterioration mechanisms controlling the durability of concrete. While individual processes leading to deterioration have been studied extensively, an integrated approach to predict quantitatively the durability defining factors of concrete is lacking. To accomplish this task, the major governing phenomena of interest must be studied in detail. For example, during the hardening phase of concrete, the risk of cracking due to thermal as well as drying shrinkage deformation needs to be investigated. Therefore, heat generation problems due to hydration should be carefully examined. In the hardened state, transport of external agents into the internal microstructure of concrete governs the deterioration phenomena. In this regard, transport processes in concrete with special emphasis on moisture transport need to be investigated. Furthermore, economic considerations of project cost and efficiency optimizations may require that a construction schedule minimizes the curing periods of concrete as far as possible. To discover the impact of curing conditions on concrete durability performance, we must study the parameters due to which the hydration and strength development processes might be affected.

Consideration of all the requirements in a durability evaluation system calls for an integrated approach. The aim of this book is to attempt the establishment of such an integrated rational computational framework for the durability analysis of concrete structures. This is done by studying the durability of concrete, starting from the stages of early age development of concrete, with regard to its microstructure and strength development. The method involves dynamic coupling of cement hydration, moisture transport and microstructure formation phenomena.

The microstructure development model assumes a monosized particle dispersion and uses a particle expansion model based upon the degree of hydration. Furthermore, assuming a linear bulk porosity variation in the expanding cluster of a particle, important microstructural parameters such as the surface area and bulk porosity distribution parameters, are computed with time. The moisture transport model is based on the multi-component division of concrete space porosity and is applicable to any arbitrary combination of drying and wetting. Moisture transport related characteristics are directly evaluated from the microstructure of the matrix. Since long-term durability of concrete is primarily influenced by its mass transport characteristics, special attention has been given to the relationships of transport characteristics to the microstructure by taking into account the composite nature of concrete.

The hydration model is based upon a multicomponent division of cement and pozzolanic powder materials and obtaining the rate of heat generation for each of these components, based on modified Arrhenius' law. Mutual interactions among the reacting constituents during hydration and their dependence on the availability of water in the concrete micropore is also quantitatively formulated. Simultaneously studying these three basic phenomena at any point enables the tracing of microstructure development with increase in the degree of hydration for arbitrary temperature and moisture content. The computational models of structure formation, moisture transport and hydration are then integrated into a finite-element simulation program called DuCOM. The simulation method typically starts from the casting stage of concrete and computes the development of several properties such as strength, porosity and microstructure, etc., along with the temperature and pore moisture content history with time.

This coupled computational system is verified by applying it to sets of experimental data related to different physical phenomena. For example, the effect of different curing conditions on the strength gain, moisture loss and the pore structure development for specimens of different mix proportions and curing conditions can be quantitatively investigated. Similarly, it is applied to predict the moisture loss and gain behaviour under arbitrary drying-wetting conditions. Furthermore, simulated concrete performances are linked with volumetric deformation of concrete composite driven by water loss, hydration and their combination.

Chapter 1

Introduction

- Survey of the literature
- Scope and aim of this book
- A proposed strategy of concrete performance evaluation and layout of this book

1.1 General

Serviceability and safety of concrete structures have been the main concern of structural engineers and currently required performance is generally assessed by conducting structural analyses based on nonlinear solid mechanics. Another issue of great engineering importance is how to retain performance during the expected life of buildings and infrastructures. The length of time serviceability and safety of structures remain is generally referred to as *durability*, which has also been a long-term topic of interest to engineers with extensive study in the past. The serviceability limit of concrete structures is primarily governed by the extent of damage resulting from daily service loads and various deterioration processes, which might be active throughout the structure's life. These deterioration processes can be physical, chemical or mechanical, and their net effect is to weaken the integrity and tightness of the complex internal microstructural building blocks of concrete. Figure 1.1 shows a hypothetical concrete structure exposed to some of the typical and most common types of deterioration processes which may be active over the lifespan of structures. These deterioration processes may be due to weathering, occurrence of extreme temperatures, abrasion, electrolytic action, repeated fatigue loads and possible attack by natural or industrial liquids and gases assisted by moisture ingress.

Internal deterioration of concrete may result from the alkali–aggregate reaction, volume change occurring due to the difference of thermal properties of cement paste and aggregate and, most important of all, permeability of concrete [1]. Thermal stresses induced in concrete due to cement hydration and pozzolanic activity might lead to thermal cracks whereas stresses induced by the loss of moisture and its redistribution in the

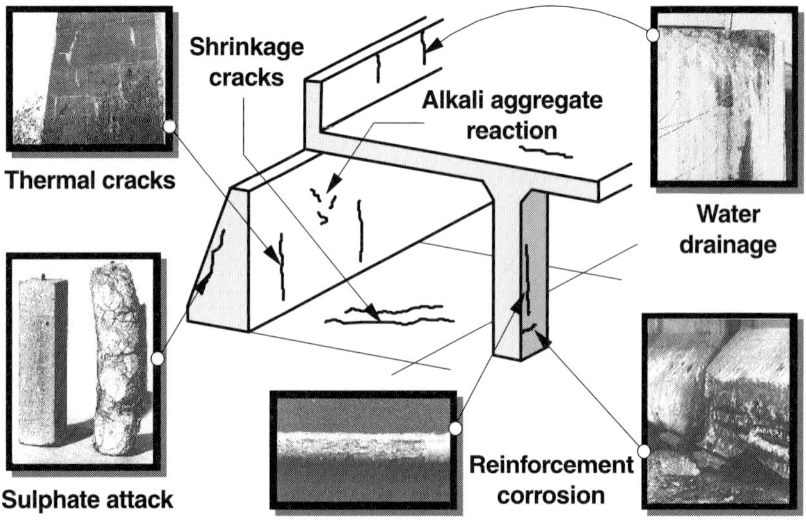

Figure 1.1 Deterioration mechanisms of concrete structures [3].

period of curing and service life may lead to finer drying shrinkage cracks. In the case of cement-rich mixed concrete, autogenous shrinkage associated with self-desiccation might cause internal microdefects, too. Such flaws which might be produced in the very early stages of concrete microstructural development and hardening obviously limit the serviceability life-span of the structure and require expensive maintenance schedules. It is common experience that the concrete performance achieved at early age affects the subsequent long-term serviceability and safety of concrete structures, but the concrete quality achieved inside structures has never been well evaluated in design. To check the life-long overall performance of concrete structures, modelling of concrete performance at early age, the main topic of this book, is crucially important.

In many of the mechanisms of deterioration, moisture transport in concrete is involved. For example, one of the main hazards to a concrete structure comes from the risk of corrosion of its steel reinforcing bars. Corrosion of steel bars depends partially on the chloride concentration at the bar, which needs to reach a critical level in order for corrosion to proceed. Chloride ions mainly reach the reinforcing bars through water-filled pores and/or cracks, gaps and joints. Since corrosion leads to the increase in the volume of steel, cracking and spalling of the concrete cover may follow, thus further deteriorating the structures [2]. Sulphate attack also depends on the sulphate ion transport through the water-filled pores. The movement of water in the pore void structure determines the freeze–

thaw resistance of concrete. Furthermore, imperviousness of concrete to any liquid is important in consideration of various forms of liquid or hazardous materials' retaining structures. In fact, the Comité Euro-International du Béton (Euro-International Committee for Concrete) (CEB) design guide for durable concrete structures states [3]

> 'It can be seen that the combined transportation of heat, moisture and chemicals, both within the concrete mass and in exchange with the surroundings (the microclimate), and the parameters controlling these transport mechanisms, constitute the principal elements of durability. The presence of water or moisture is the single most important factor controlling the various deterioration processes, apart from mechanical deterioration.'

Moisture transport processes and the impermeability of concrete have therefore traditionally occupied a very important place in the consideration of durability. Moreover, many of the early age defects caused by shrinkage, thermal dilatation and creep, etc., are related to the moisture present in concrete and thus require a quantitative estimate of the pore water distribution in the concrete in time and space domains [4]. The early age development phenomenon in real-life concrete structures is quite complex. It involves a simultaneous occurrence of microstructure growth and bonding of crystals associated with dissolution of cement grains and depositions of hydration products. Since hydration is an exothermic reaction, it is accompanied by liberation of large amount of heat. Moreover, the rate of hydration reaction is dependent on the amount of water available in the reacting mixture for reactant diffusion and dissolution. Therefore, an exchange of moisture with the environment and its internal redistribution within the hydrating concrete may further influence the overall hardening process. The damage due to associated risks of thermal cracking during both the hardening and cooling stages of concrete may be influenced accordingly and would affect the serviceability and/or safety of concrete structures over the long term life-span. It is obvious, therefore, that to predict this damage, various mechanisms related to the hydration and development of internal structure of concrete must also be understood.

1.2 A brief survey of the literature

1.2.1 Microstructure formation and cement hydration

The hydration phenomenon of cement in concrete has been extensively studied in the past since it is the sole source of inherent temperature rise during hardening and also leads to the formation of the hardened cement matrix. The importance of thermal effects in early age concrete has been recognized since Tetmayer [5] did early temperature measurements of concrete. Thermal problems have been discussed primarily in the light of optimization of the pouring sequence and the stripping of formwork to minimize the risk of cracking in massive concrete structures, such as dams. Earlier studies in this regard were primarily qualitative and limited to recommendations and guidelines in codes of practice. Recently, the thermal cracking risks have been associated with the structural functions and durability of concrete and a more analytical approach to deal with the problem has been proposed. One of the methods involves the prediction of temperature fields in concrete with time and is generally based on the differential equation

$$\frac{\partial T}{\partial t} = \mathrm{div}(D_T \nabla T) + Q \tag{1.1}$$

where D_T is the coefficient of thermal diffusivity, T is the temperature at any point and Q is the rate of heat generation. In general, the function Q is the heart of this analytical treatment and, accordingly, numerous attempts have been made to formulate this function, ranging from purely empirical formulae to sophisticated multicomponent models that take into account the hydration process of each mineral compounds. Probably, Rastrup [6] was the first to introduce the concept of *degree of hydration* and he recognized the influence of temperature on the rate of cement reaction. Several researchers have taken into account the sensitivity of temperature on hydration, i.e. the effect of curing temperature on the rate of hydration and subsequent temperature rise. Suzuki *et al.* [7] proposed a lumped model to predict the rate of heat liberation of cement using Arrhenius' law of chemical reaction. A hydration model of cement proposed by Tomozawa [8] considers the rate of penetration of reaction front into a cement particle, based upon diffusion of moisture across the product layer. It assumes a spherical shape for the cement particles and that diffusion across the product layer is based on the Fick's law of diffusion. Furthermore, the initial resistance to mass transport from the surface layer is the rate controlling process which gradually turns to a diffusion controlled process. The simplicity of the model makes it highly versatile, though some of the empirical coefficients must be obtained experimentally and fitted for different types of cement.

Generally, the overall rate of hydration of cement has been recognized to be a summation of the rate of hydration of its individual components. Moreover, the differences in the rate of reactions of individual constituents of cement, such as C_3S and C_2S, have been recognized for a long time. The studies in this regard have primarily been experimental and, for several reasons, a large scatter in the experimental results exists and as such some interactions among the hydrating constituents of cement might exist. The proponents of the *independent hydration concept* have disregarded the mutual interactions among the individual constituents of cement [9, 10]. It appears that this concept has found a widespread acceptance and forms the basis for the application of least squares analysis of hydration data. This approach enables the evaluation of the total degree of hydration of cement as a weighted summation of the independent rate of hydration of its constituents. Another concept put forward to account for the hydration rate of cement is the *equal fractional rates concept* which states that hydration proceeds at a more or less uniform rate and affects all of the cement compounds of a cement particle more or less simultaneously. In reality, however, one or both of these concepts might prove incorrect. Nevertheless, in a polymineral system, it appears that the interactions of a physical and chemical nature lead to mechanisms due to which the hydration of individual constituents proceeds neither at equal fractional rates nor independently from each other. In this regard, a formulation which takes into account these interactions on chemical and physical grounds is highly desired. This is more important now than ever before, since the wide availability of cheap computational power has enabled the application of refined material models that require extensive computations. These models can be easily adopted in the sophisticated finite-element analysis methods that are widely used at present for the prediction of temperature rise and the associated thermal risks of concrete structures.

Although the heat generation associated with hydration has been studied extensively, the research into the associated microstructure development process has been largely qualitative in nature. Several researchers have described the shapes and stereological distributions of the hydrated products that are formed with time. Numerous scanning electron microscope (SEM) studies have described the morphological features of the products of hydration, such as calcium silicate hydrate (CSH) gel and $Ca(OH)_2$ crystals, and the influence of parameters, such as mix proportions, curing temperature and admixtures on their development [11, 12]. While a huge database exists that qualitatively describes the features of the maturing microstructure of cement, a quantitative description has been lacking. Van Breugel's [10] work is pioneering in this regard since it attempts to trace the development of strength and degree of hydration from a quantitative microstructure development model based upon cement particle growth concepts.

1.2.2 Mass transport

The basic constituents of concrete, i.e. aggregate and cement paste, both contain pores. In addition, concrete might contain voids caused by incomplete compaction and/or by bleeding. These voids may occupy from a fraction of one up to a few percent of the whole volume of concrete. A higher percentage of these voids represents a highly honeycombed concrete of poor quality and low mechanical strength [1]. In selfcompacting high-performance concrete, aggregates are fully enveloped by the cement paste and the porous transition layers created close to the solid aggregates' surface are hardly connected. Therefore, it is the permeability of cement paste that has the greatest impact on transport properties of the concrete composite [13]. For high water-to-powder ratio mixes and high aggregate content concretes, interfacial transition zones may have the effect of increasing the permeation characteristics of concrete [14]. In general, for most of the concretes encountered in practice now, it is the quality of the hardened cement-paste matrix that determines the permeability of the concrete. Several analytical researches on transport properties of concrete can be primarily classified into two groups:

1. microstructural aspects dealing with the size, shape and characteristics of different kinds of pore systems present in concrete;
2. mechanism and mathematical formulation of fluid movement into the concrete and mathematical simulation models to represent the physical characteristics of concrete as a whole.

Microstructure influence on mass transport

It is often said that it is the total open porosity and porosity distributions of the hardened matrix that influence the amount and rate of moisture transport in concrete respectively. The exposure history of concrete to the environment can also have a profound effect on the overall moisture transport behaviour. For example, a porous medium generally exhibits higher moisture content and moisture conductivity during the drying phase compared to the wetting phase.

Microstructural characteristics of various kinds of concrete and their relationship to the influence of different additives and curing environments have also been widely investigated [11, 12]. For example, it has been reported that addition of fines, such as limestone powders and silica fume, leads to a denser pore structure whereas high curing temperatures lead to a coarser porosity distribution. The transition zone lying between aggregates and the hardened cement paste (hcp) matrix has also been studied extensively. It has been reported that a distinct transition zone of several micrometres thickness exists which probably has a different mineralogical composition and higher porosity than the bulk cement paste [10, 15]. Garboczi [2] has done computer simulations to estimate the volumetric aggregate contents at which the transition zones form a continuous percolating path across the specimen. It was concluded that the aggregate grading and thickness of the interfacial or transition zone define the percolation characteristics of concrete. In general, a high water-to-binder ratio mortar mix containing more than 45% by volume of aggregate (sand) has a high probability of transition zone percolation.

Clearly, this is not desirable since it effectively leads to a more porous concrete of lower durability performance when exposed to moisture or hazardous chemical environments. For self-compacting concrete [13], the requirement for flowability and segregation resistance consequently imposes a lower aggregate and water content restriction on the mix design, leading to a concrete where transition zones are either nonexistent or nonpercolating. Such systems have greatly reduced the burden on site supervision by eliminating the human factor in placement to a large extent. Further rationalization of construction systems, durability requirements and economic principles may dictate the use of new materials of consistent and uniform quality so that the consistency of a high-quality concrete might be guaranteed. There are also some innovative applications of using high-porosity lightweight aggregates. These aggregates could serve as the internal reservoirs of water in concrete and would help in the autogenous curing of concrete even when good external curing conditions could not be provided on the site [16].

Analysis of moisture transport, which is a part of rational concrete performance evaluation systems, especially in specialized concretes, requires adequate modelling of interactions among its various components, such as aggregates and matrix. A treatment which takes into account the moisture transport interactions between the components of concrete enables the prediction of the limits in which the traditional analytical techniques that treat concrete as a uniform homogenous porous medium would be valid or could be extrapolated. Furthermore, a multicomponent treatment of transport processes gives a better physical understanding of the overall moisture transport and associated phenomena, such as early age drying shrinkage and strength development of concrete.

Transport mechanisms

Transport of liquids and gases takes place in the pores, microcracks and voids of the concrete. Under usual conditions, a moisture potential gradient in the porous medium is the main driving force of the transport process. At low relative humidity, the main mechanism for moisture transport is through the diffusion process of vapours. However, at high relative humidity, capillary condensed pores constitute the bulk of the saturated pore system. In the presence of a saturated face (or wetting of concrete surface), capillary suction becomes important. For such a case, diffusion of gases is totally impeded. This process is usually quite rapid compared to the drying process wherein the main mechanism is considered to be diffusion. Thus, regarding basic transport mechanisms, it can be summarized that the transport of vapours and inert gases takes place by diffusion and by filtration under applied pressure gradient. Transport of liquid occurs due to diffusion, capillary absorption and by filtration. It must be noted here that a kind of sealing effect of even well-hydrated concrete under prolonged moisture exposure has been observed [17] leading to significantly reduced moisture suction and transport.

Usually, the mathematical formulation used to express above mechanisms have been lumped into either Darcy's law or simply Fick's law of diffusion. In all cases, permeation characteristics of concrete can be defined by basically two transport coefficients, permeability and diffusivity. The latter is defined by the constant D in Fick's first law as the flux in moles (or kg) per time per unit cross-sectional area for an applied unit concentration or saturation gradient. The dimension of D is length squared per unit time. Permeability is based upon Darcy's law. Either Fick's second law of diffusion or Darcy's law have been used thus far in the mathematical formulation of the problem [1, 4, 18]. Fick's second law of diffusion for a particular phase can be derived by inserting Fick's first law into the phase mass conservation equation. Stated mathematically, if θ denotes the volumetric water content and $D(\theta)$ denotes 'diffusivity', a material property, then rate of change of θ within a control volume in the one-dimensional case can be given by the following equation:

$$\frac{\partial \theta}{\partial t} = \frac{\partial}{\partial x}\left[D(\theta)\frac{\partial \theta}{\partial x}\right] \tag{1.2}$$

Partial derivatives are used because θ varies in both the x and t domain. Much of the effort in the past has been devoted to the determination of the 'material' property, $D(\theta)$ [19]. It has to be noted that such a formulation lumps all the information related to the pore structure and transport mechanisms into one parameter, D (diffusivity). Thus the multiphase transport dynamics of the porous medium and their relationship to physical characteristics are not adequately represented by equation (1.2). Transport properties of concrete have generally been analysed by treating concrete as a isotropic, homogeneous and uniform porous medium [2, 3, 4, 18]. In this approach, the physical characteristics of various components of concrete, such as aggregates, the cement-paste matrix, aggregate cement paste interfaces and bleeding paths, are lumped into a single representative porous medium.

While this approach can treat analytically most of the cases of practical interest for concrete made of normal mixes and low porosity aggregates, it fails to consider rationally the moisture transport behaviour in special concretes, for example those which contain high porosity lightweight aggregates. A different analytical approach is also needed for the cases where, due to high aggregate contents, a distinct aggregate–cement matrix interface might be present and affect the overall mass transport. Since consideration of interface effects and suction characteristics, such as isotherms, are out of the question in such a lumped treatment described above, it cannot describe the hysteresis effects observed experimentally. Such a treatment does not reflect the overall physics of transport processes either microscopically or macroscopically. Moreover, the time-dependent sealing phenomenon of concrete under prolonged moisture exposure has not yet been investigated and analytically formulated.

1.3 Scope and objectives

The ultimate objective of the methodology described in this book is to find a so-called *life-span simulator* of structural concrete based on the time-dependent microscopic modelling of varying concrete performance under both environmental and mechanical actions. This is to be achieved in a manner that is similar to the established methods of simulators of structural behaviour that are used for ensuring the required serviceability, safety and seismic performance of buildings and infrastructures under static and dynamic mechanical actions [20]. Here, it has to be recognized that at present structural mechanics can simulate plasticity and damage of building materials rooted in large-scale defects, such as steel yielding and buckling, concrete cracking and crushing. However, the varying mechanical characteristics of materials in different environments are regarded as input information for structural mechanics and are not inherently predicted within the realm of solid mechanics of concrete and reinforced concrete.

It could be noticed from the foregoing survey that a wealth of information exists dealing with the various aspects of formation and development of concrete microstructure and strength. The transport of agents, such as moisture and gases through the internal network of concrete microstructure and macroscopic flaws such as cracks, has also been studied. To achieve the goal of an integrated method of predicting concrete performance for durability assessment, all of these phenomena need to be combined into a unified system. Furthermore, since many of these phenomena would be interdependent (for example, hydration is dependent on the availability of moisture which in turn is again dependent on transport process), the interrelationships of these processes must be clearly and carefully sorted out in such a unified coupled system. To identify such relationships and rationalize the overall system, we need to re-examine and re-investigate the individual subsystems involved. The fundamental rigour adopted at the level of modelling individual components would help us to take the results obtained at micro levels to the more practical macro level of applications. Clearly, this goal appears to be ambitious and far away at present. Thus, it is important to clearly state the scope of the investigations pursued in this book.

The primary aim of this book is to describe a scheme integrating the research done at several levels of young concrete thermodynamics and material engineering for assessing concrete performance. Here, modelling of concrete performance covers microstructure, hydration, temperature, moisture content, state of equilibrium, mechanical strength and volumetric change. The integration is to be sought in terms of a computational tool that can take into account the development of microstructure and strength of any concrete exposed to any arbitrary environment. Furthermore, it needs to predict the long-term serviceability and safety of concrete structures in terms of predicting the structural concrete performance together with the transport of various external agents into the concrete microstructure. Of course, to achieve this goal the framework must be benchmarked and validated by experimental results. The scope of this book is restricted to the establishment of a general computational framework only, that can deal with the durability problems of concrete engineering ranging from early age thermal cracking and desiccation-based volumetric change to long-term creep, carbonation or chloride attack, etc. The framework discussed here is based on simple and rational physical material models of concrete. No attempt has been made to actually formulate the model for deterioration mechanisms, such as carbonation or chloride attack, but it is understood that such subsystems can be successfully added to the overall computational system later. For example, although the system can compute the temperature rise, development of microstructure and moisture transport within concrete during the hardening stages, to compute the associated thermal or drying shrinkage cracks requires a corresponding constitutive model for concrete deformation.

The status of the developments described in this book in the overall framework for the life-span simulator of concrete structural performances is shown in Figure 1.2. For assessing the serviceability and safety required in the design period, we have to recognize the different scales on which concrete performances should be described. That is to say, to recognize the microscale with CSH gel and capillary pores that the authors intend to cover in this book, and the macroscale volumes involving cracks, which are predicted in the stress and deformational fields of structural mechanics. The structural mechanics simulation of nonlinearity has a close similarity to the varying concrete performance rooted in the microscale chemical physics of capillary tension, thermal expansion, strength development and stiffness of constituent concrete and steel. Since the macromechanical system is built up from the micromaterial ones, it is quite natural to see a strong influence of the microscale concrete performance on the macrostructural behaviour, or of hygrothermal physics on dynamic structural mechanics.

However, the inverse stream of influence exists especially when concrete structures are much deteriorated. For example, steel corrosion provoked by the chloride penetration through concrete may produce structural cracking.

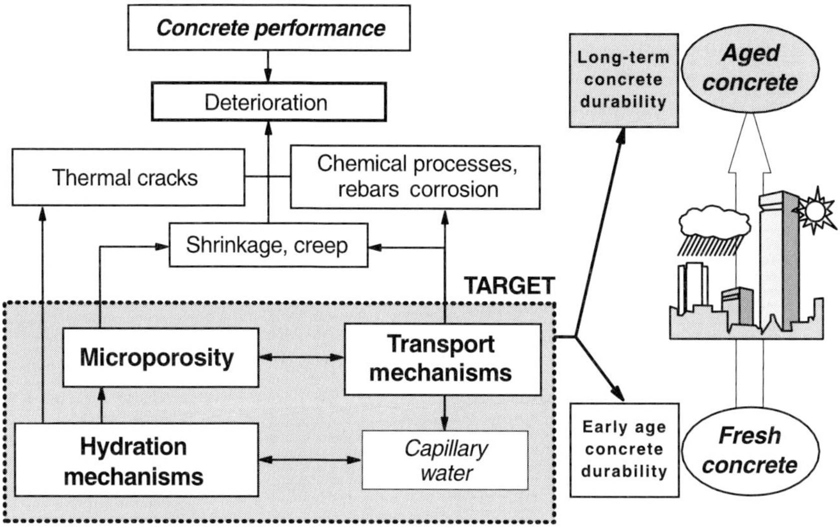

Figure 1.2 Framework for rational evaluation of durability.

These cracks, which can be predicted by nonlinear mechanics, make mass transport of moisture much easier and, in turn, the corrosion is accelerated by rapid concentration of hazardous agents through these defects. As shown in Figure 1.3, macroscopic structural mechanics and microscopic

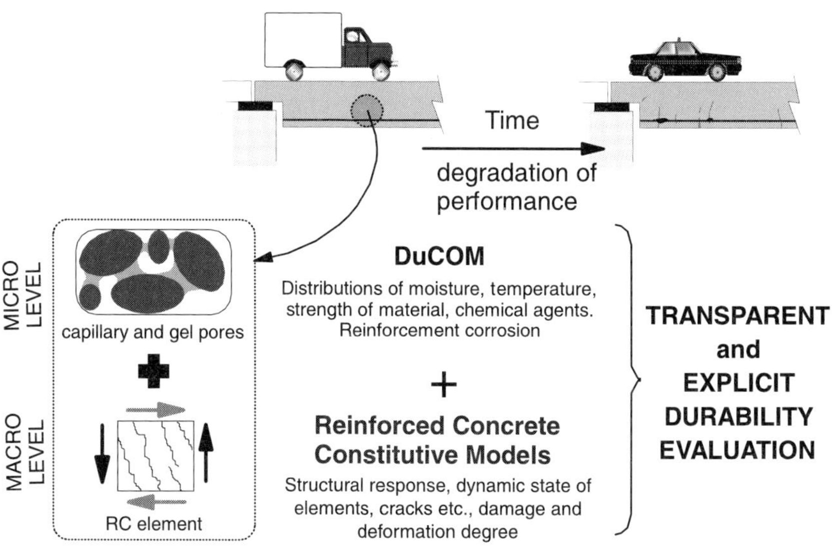

Figure 1.3 Simulation of serviceability and safety of a structure over its lifetime.

hygrothermal chemical physics mutually interact. The role of modelling concrete performance on the microscale is the target of this book and is achieved through a computational simulation model called DuCOM (Durability of COncrete Model) that can be in fact thought of as a substitute for numerous experiments that deal with the observations of the effects of different material and environmental conditions on the durability performance of concrete structures. In view of the development of this simulation code, several goals targeted by this book are summarized as below.

- Understand and analytically formulate the microstructure development of cementitious materials. The model should take into account the effect of mix proportions, admixtures and initial powder characteristics. Moreover, the model should be extensible.
- Reorganize and reformulate the hydration phenomenon of cement, since it is very basic to the development of microstructure and generation of heat. Obviously, the model should be able to consider the chemical characteristics and interactions among the constituents of components of cement and take into account the availability or depletion of water in the concrete microstructure.
- Develop an analytical formulation to describe the basic transport behaviour of liquid water as well as vapour in the concrete microstructure. The model should be extensible and applicable to arbitrary conditions of drying and wetting. Moreover, it should consider the effect of dynamic changes in the microstructure of concrete on transport behaviour.
- Develop an integrated computer simulation model that combines the micropore structure development, hydration and mass transport processes into a unified code structure. The simulation model should be practical, so that it can be used to analyse structures of practical interest. The primary input (*extrinsic* parameters) needed in the model should be kept to a minimum, though the structure of the model should be extensible enough to allow for any variations in the definition of *intrinsic* or internal material parameters.
- At all stages of the development, efforts will be made to develop physically meaningful and rational material models. These can be simplified and reused as practical models for real life applications.

In the development of the simulation model as well as the submodels, an indication of their approximate range of the accuracy and reliability is also given. At this stage, the structural serviceability and safety can be simply assessed by running already developed nonlinear mechanical and structural analysis programs in which the achieved concrete performances predicted by DuCOM are input information for each finite element. This one-way stream of information on concrete performance is necessary and sufficient in most engineering practice. But provided that the mutual interactions between macrodefects and mass transport are predominant, the DuCOM system discussed in this book should be unified with the structural mechanics simulator as shown in Figure 1.3. Coupling the nonlinear mechanics of the computer program coded COM3, which the authors have worked with, with the DuCOM discussed here gives the model under current investigation.

1.4 Research strategy and outline of this book

The research described in this book is in fact an assimilation and extension of the concepts already known to the concrete research community. However, instead of the usual empirical approach, a more rigorous and fundamental approach based on physical reasoning has been adopted. At the same time, the investigations are guided more by the engineering applications of the research than by pure scientific knowledge. Extensive use has been made of the vast amount of data already present in the literature by its reinterpretation and by using it to verify and benchmark the simulation material models. However, key benchmarking of the material submodels has been done from a select set of experimental data. The layout of this book is as follows.

Chapter 2 is an investigation into the current practices that quantify and evaluate the durability of concrete structures. A proposal is made which calls for an integrated approach for quantitative evaluation of various parameters related to the serviceability of concrete structures. A sample case study using the proposed methods of achieving concrete performances related to durability is demonstrated. In this scheme, the early age development process of concrete performance is shown to be a phenomenon of prime importance.

The next three chapters basically deal with a description of the physical material models that characterize the basic physical phenomena responsible for long-term durability of concrete. They are, namely, structure formation and hardening of concrete, hydration and temperature rise and the mass transport behaviour in concrete as a composite porous medium.

Chapter 3 addresses the hardening stage of concrete. Salient features of the phenomena occurring during the first few hours of hardening are explained in terms of the morphological changes in the microstructure. Accordingly, a simple model of microstructure development is proposed that is based upon the concept of cement particle growth. The concepts of inner and outer products of hydration are introduced and, based upon a few measurable microscale properties of the hydration products, macroscopic properties, such as the porosity and surface area of hardened cement paste, are obtained using a few simple mathematical manipulations. Key to this microstructure development model is a multicomponent concept for the heat of hydration of cement based on Arrhenius' law of chemical reaction. This hydration model takes into account the heat generation rates of the individual components of cement and pozzolanic nonorganic minerals and the mutual interactions between them. The hydration phenomena and subsequent microstructure generation are found to be highly dependent on the availability of moisture in the internal pores of concrete.

Chapter 4 investigates the mass transport process in cementitious materials. A basic formulation is suggested that takes into account the bulk movements of liquid water as well as diffusive vapour movements. The fact that water in the cementitious microstructure can exist in various forms is recognized and accordingly, a multicomponent division of pore water has been done. Most important moisture characteristics, such as isotherms and conductivity, have been analytically derived for a known cementitious microstructure. These models can be applied under any arbitrary combination of drying and wetting. The complete analytical model of moisture transport can generally explain the moisture transport behaviour in mortars satisfactorily.

Chapter 5 discusses the implications of the fact that concrete is a composite material from the view point of transport processes in concrete. The aim of this chapter is to find the range and conditions under which the single and uniform porous medium assumption usually adopted for concrete can be used. This is done by examining the percolation behaviour of interfaces using computer simulations and analysing the moisture transport behaviour in concrete by treating it as a composite material. It is found that special concretes, such as ones containing very high porosity aggregates, need to be dealt with as a composite whereas, for usual concretes, the assumption of a conventional single porous medium might be valid.

The next chapter combines the theoretical frameworks of material models described in the previous three chapters into a computational simulation tool.

Chapter 6 describes the development and applications of a three-dimensional finite-element numerical model of simulating maturing concrete performance, codenamed DuCOM. This model has been achieved by simplifying the theoretical models of hydration, microstructure formation and moisture transport into simpler, computer-friendly models and tying them together into a coupled finite-element program of both diffusive heat and mass transfer applicable to three-dimensional space and time domains. The program is verified against a set of available experimental results and several simulation tests are done to show the effect of the water-to-cement ratio, curing conditions and admixture materials on strength and microstructure development.

Chapter 7 gives a detailed description of the multicomponent model for heat of hydration of Portland cement and pozzolanic additives. It is a fairly selfcontained chapter that describes the basics of Arrhenius' law based heat generation rates for various mineral components of Portland cements and pozzolans. The cause of overall heat generation and the dependence of hydration rate on temperature as well as the types of chemical mineral reactions are explored and formulated on a physico-chemical basis. The extension of basic modelling to blended powder materials and its applicability to several practical cases is illustrated. Also a scheme is shown whereby the thermodynamic approach to heat generation of cement can be used for thermal crack control design of concrete structures.

Chapter 8 and the Appendices conclude this book and give recommendations and suggestions for future investigations. Here, discussion on performance based design of structures is emphasized as a new scheme of engineers' decision making process in which modelling of concrete performance is thought to be essential.

Appendices are attached after the main text. Here the source code of the multicomponent heat generation and heat of hydration model discussed in Chapter 7 is published. This code can compute the temperature variation of massive concrete structures for evaluating thermal crack risks. The entire DuCOM model discussed in this book for concrete performance simulation is partially open to anyone through the Internet. The home page structure and the way of running the program and down-loading analytical results are described. As an example of unifying DuCOM with the mechanical structural problems of engineering encountered in practice, the computation of volume change caused by moisture migration, hydration and micropore structural formation (usually called *drying and autogenous shrinkage*) is presented. The deformation simulated here is free of external restraint. But the linkage of micropore physics with the mechanical stress field is a long-term aim to unify structural mechanics and material sciences.

References

[1] Neville, A.M., *Properties of Concrete*, Elsevier Science, Amsterdam, 1991.

[2] Garboczi, E.J., Permeability, diffusivity and microstructural parameters: a critical review, *Cement and Concrete Research*, 1990, **20**, 591–601.

[3] CEB, *Durable Concrete Structures, CEB Design Guide*, Thomas Telford, UK, 1992, 3–7.

[4] Bazant, Z.P. and Najjar, L.J., Nonlinear water diffusion in nonsaturated concrete, *Matrx. et Constr.*, 1972, **5**(N25), 3–20.

[5] Tetmayer, T., *Deutsche Topfer-und Ziegler-Ztg.*, 1883, 234.

[6] Rastrup, E., Winter concreting, Proceedings of the RILEM Symposium, Copenhagen, RILEM, 1956.

[7] Suzuki, Y., Harada, Y., Maekawa K. and Tsuji, Y., Quantification of heat of hydration generation process of cement in concrete, *Concrete Library of JSCE*, 1990, No. 16, 111–24.

[8] Tomozawa, F., A study on strength development of concrete in view of hydration rate, *Proceedings of Japan Arch. Inst., Kanto Branch Conf.*, 1970, No. 41, 229–240.

[9] Metrier, D., Jawed, I., Sun, T.S., and Skalny, J., Surface studies of hydrated ß-C2S, *Cement and Concrete Research*, 1980, **10**(3), 425–432.
[10] van Breugel, K., Simulation of hydration and formation of structure in hardening cement-based materials, Ph.D thesis submitted to Delft Technological Institute, Netherlands, 1991.
[11] Gard, J.A. and Taylor, H.F.W., Calcium silicate hydrate(II) (C-S-H(II)), *Cement and Concrete Research*, 1976, **6**(5), 667–678.
[12] Jennings, H.M. and Xi., Y., Microstructurally based mechanisms for modelling shrinkage of cement paste at multiple levels, Proceedings of the 5th, International RILEM Symposium on Creep and Shrinkage of Concrete, Barcelona, E&FN Spon, London, 1993, 85–102.
[13] Okamura, H., Maekawa, K. and Ozawa, K., *High Performance Concrete*, Giho-do Press, Tokyo, 1993.
[14] Chaube, R.P. and Maekawa K., A study of the moisture transport process in concrete as a composite material, *Proceedings of Japanese Concrete Institute*, 1994, **16**(1), 895–900.
[15] Synder, K.A., *et. al.*, Interfacial zone percolation in cement aggregate composites, in *Interfaces in Cementitious Composites*, Maso J.C. (ed.), E&FN Spon, London, 1990.
[16] Silvia, W. and Reinhardt, H.W., A new generation of high performance concrete: concrete with autogenous curing, *Advanced Cement Based Materials*, 1997, No. 6, 59–68.
[17] Hearn, N., Detwiler, R.J. and Sframeli, C., Water permeability and microstructure of three old concrete, *Cement and Concrete Research*, 1994, **24**(4), 633–640.
[18] Hall, C., Water sorptivity of mortars and concretes: a review, *Magazine of Concrete Research*, 1989, **41**(147), 51–61.
[19] Aikita, H., Fujiwara, T. and Ozaka, Y., Water movement within mortar due to drying and wetting, *Proceedings of JSCE*, 1990, No. 420, 61–69.
[20] Okamura, H. and Maekawa K., *Nonlinear Analysis and Constitutive Models of Reinforced Concrete*, Giho-do Press, Tokyo, 1991.

Chapter 2

Assessment of achieved concrete performance and durability design

- Current practices on durability
- Practical durability evaluation proposal
- A case study of quantitative evaluations of durability parameters

2.1 Introduction

A 1990 National Research Council report in the US indicates that it will cost \$2–\$3 trillion over the next 20 years to repair all the US concrete structures that are undergoing deterioration due to corrosion or poor construction and maintenance practices. Considering that concrete is one of the most extensively used materials in the \$1.5 trillion per year world-wide construction industry, the extra costs of maintenance, repair or destruction of damaged concrete structures is enormous not only in financial terms but also in terms of the lost capacities of services and the added burden on the already overloaded environments. In such a situation, it is the duty of the concrete community to address the issue of producing a better concrete, i.e. more durable, less wasteful of our limited resources and gentler to the environment. More precisely, in order to realize adequate durability and long-term performance of concrete structures, we need to identify the quality parameters that can provide a rational and sound basis for the *in situ* quality control and specifications.

The question is: are we good enough in asking for specifications that would give us a better concrete or, alternatively, are we asking for the right specifications at all? At the current stage of development, certainly our answer is 'not enough' (Figure 2.1). For example, concrete is typically ordered by strength requirements, say 30 MPa or 50 MPa at a certain age. This is because the specimen strength is quantifiable and easily measured at the production stage for quality control and assurance. Similarly, it is logically possible to ask the supplier for a concrete whose performance would last for 500 years or 100 years. How do we ask such a question, and even if we get a concrete that is certified for 500 years of life, how can we be sure of its durability? What if the environmental conditions change with

Figure 2.1 Current status of durability design practice.

time? We need to establish a system that answers these questions and also helps us to ask better questions in our pursuit of making a higher-performance concrete. In this chapter, we will explore the requirements for material design and methods of evaluating time-dependent concrete performance that need to be enforced to produce durable concrete in the real world. The mechanics of how those requirements are actually implemented or enforced in the concrete design and technology community, however, generally require a socio-engineering and socio-economical outlook on the problem. Those who are interested in this issue are invited to turn to Chapter 8.

2.2 Current practices in durability evaluation

The design, dimensioning and detail of concrete structures is typically based upon structural performance considerations of safety and serviceability. In the case of urban infrastructure, requirements of landscape friendliness and aesthetics might also become essential for designers. This is primarily done not because the requirements on the structural performance criteria are more stringent, but because sound assessment methods for the durability of concrete structures do not exist. This situation prevails even though a lot of work on durability of concrete structures has been done and designers are aware of the severity of the problems. Designers can easily find a huge body of knowledge on how to make structural concrete durable in terms of the material, construction and structural detailing including the arrangement of reinforcement. In fact, certain codes of practice have described how material, construction and reinforcement, etc., should be managed as specified items. Here, it must be pointed out that these statements have been qualitative and plenty of codes used to be like 'engineering manuals and data books' in comparison with structural design codes for safety and serviceability requirements. According to Sakai [1], the ideal design system of concrete structures is shown in Figure 2.2. In such systems, environmental conditions are a prerequisite to design. These environments include the environment to which the structure is exposed as well as the amenity environment (aesthetics). Another important prerequisite is the maintenance according to which levels of performance requirements for durability need to be changed. However, the basic question remains of identifying the *interfaces* that link environment and maintenance prerequisites to the design subsystems. Hoff [2] notes that in order to include stringent durability requirements in codes and standards, two approaches are possible: performance and prescriptive requirements, perhaps a combination of these two can provide the best solution.

2.2.1 Performance requirements

The establishment of this criterion assumes that designers understand the nature of concrete and its interaction with both the environment and service loads. Using a knowledge of the environment and its definition, the service life of a concrete structure can then be predicted. Currently, methods to measure performance are approximate and only an educated guess can be

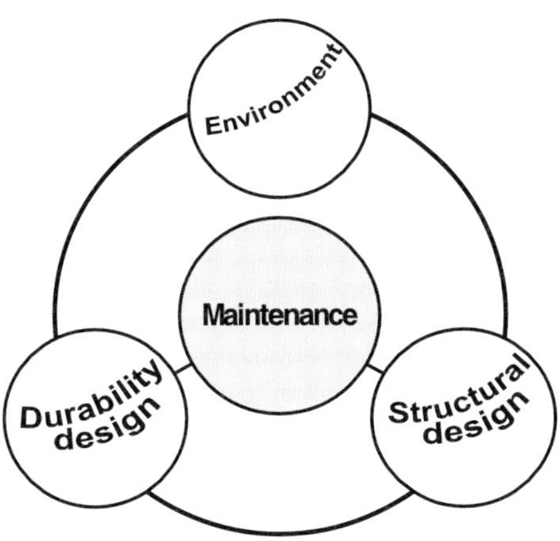

Figure 2.2 An ideal design system [1].

made at the performance criteria. The performance specification may provide flexibility and encourage economic mix proportioning so that the minimum amount of, or the cheapest, material that satisfies a performance standard can be used. The key question to be raised here is how do designers know in what way the performance of concrete is going to be affected by a certain material in certain usage conditions or how do they make an educated guess? To find answers to such questions, suitable material models need to be developed that link performance levels to the ingredient material compositions and the environment to which the concrete will be exposed. Needless to say, such prediction methods would be the key to success of performance requirement based design systems in objectively assessing the serviceability and safety of structures over their life-span. Also, in general, such prediction methods would be out of the realm of design codes, such as empirical formulae, and could efficiently utilize sophisticated computational methods.

2.2.2 *Prescriptive requirements*

Specifications that give the amount of each ingredient to be put into concrete used to be and still are common in design methodology. Typical limits that are currently used in the design process are the minimum amount of binder material, or the maximum amount of water that can be used. The goal of such requirements is to obtain a tough and impermeable structure. Such methods, however, tend to be conservative and lack versatility because the nature of ingredients and environmental conditions tend to be different at different places. However, prescriptive requirements tend to be easily understood by the users and in fact can be easily tested during construction, although the achieved performance of the prescribed items are seldom verified on site.

It appears that in the near future, the need to ensure transparency and rationalization performance requirements will dictate the direction of durability design methods for concrete structures. To meet this challenge, a significant amount of theoretical knowledge of cementitious material behaviour needs to be accumulated. Availability of cheap computational power would mean transforming that knowledge into virtual materials that could be tested under any arbitrary environmental conditions. At the very least, such methods could provide a basis for the development of more rational prescriptive methods that could be adopted by the bulk of the concrete community. The scheme of modelling, varying concrete performance over time, proposed in the following chapters in fact follows the performance requirement approach.

A durability design proposal [3] was first issued by the Japanese Society of Civil Engineers (JSCE) in 1989 and internationally introduced at the CEB plenary session in 1989. The original and crucial point of this proposal concerning durability engineering was that it numerically scored the overall durability performance by using some fictitious durability points, with which the durability of concrete structures could be objectively ranked. For the first time, this new concept made clear that the evaluation of the limit state for durability of concrete structures would be the core of development. It has to be noted here that the durability scoring procedure is at this moment in a primitive stage with simplified empirical formulae. In fact, the time dependent safety and serviceability of designed structures were not explicitly checked during the design life but only an indistinct concept called *durability* was used for the pragmatic marking system. But, this fuzzy and empirical formulation based upon engineering judgment proved to be closer to the facts in practice than semitheoretical equations were.

Currently, a computational approach which enables designers to predict ageing and deterioration of structural concrete in space and time domains is still under investigation. Further research and development along the lines of verifying the time-dependent performance of concrete is required to make durability design sophisticated and more versatile. With this background, the current authors propose a new scheme of checking the durability related limit states using a full computational approach based on coupled mass transport, hydration and the theory of micropore structure formation.

The JSCE durability assessment proposal [3] is summarized in Figure 2.3. The overall durability of structural concrete is numerically scored according to the following linear summation rule,

$$DI = F(\phi_1, \phi_2, \ldots, \phi_n) = \sum_{k=1}^{n} F_k(\phi_k) \tag{2.1}$$

where ϕ_k is the influencing factor on the durability performance, such as water-to-cement ratio, water content, slump and spacing of reinforcement, etc., and DI is defined as the durability index, a scalar quantity. In fact, the influencing factors interact nonlinearly but the linear summation hypothesis is accepted as a first approximation. After quantitatively and comparatively defining the weighting factor importance F_k of each influencing factor on DI, empirical formulae were primarily used since an integrated theoretical approach was never established.

The above equation is used only for checking the required durability performance, i.e. '50 year's maintenance free', condition as,

$$DI \geqslant DI_{req} \qquad DI_{req} = 100 \quad \text{for the usual environment} \tag{2.2}$$

where DI_{req} is the required value of the index, which is changed in accordance with the design environmental conditions. This approach enables the easy computation of the durability index, and the estimation of the durability performance of reinforced concrete. However, when the durability limit state is changed, the functions F and F_k in equation (2.1) and

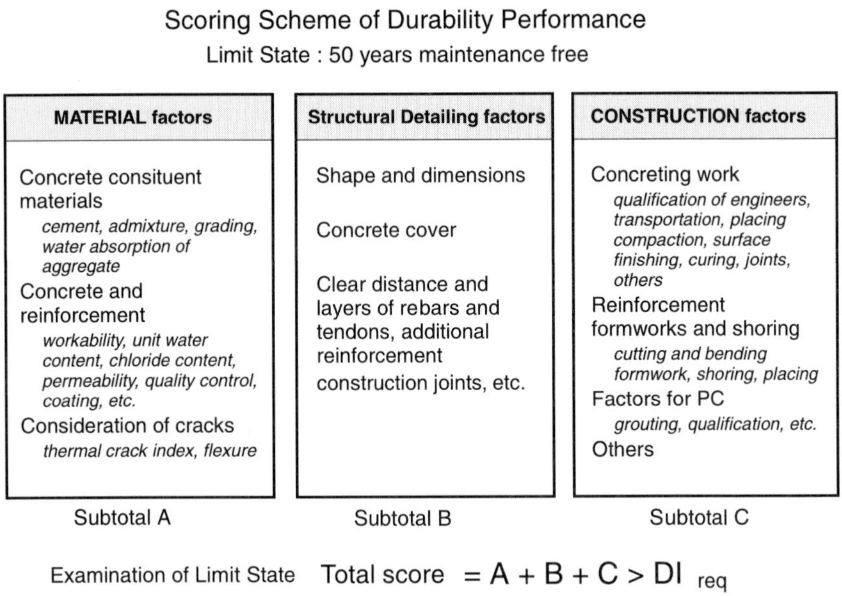

Figure 2.3 The durability assessment proposal of the Japan Society of Civil Engineers.

the requirement in equation (2.2) have to be modified for each limit state level. Since its versatility is limited, the durability examination method of equation (2.1) and equation (2.2) can be regarded as a tentative method. In this regard, it is crucial at present for designers to have a clear idea of the idealized method of ensuring durability performance of structural concrete in a future scheme of design.

Following the methods discussed above, it will be useful to consider the structural design methods with regard to the safety and serviceability limit states. The development of enhanced constitutive laws for concrete and reinforced concrete [4] have made it possible to predict numerically the structural response and the mechanical states of constituent materials in time and three-dimensional space under any external mechanical action, even though further research is needed for better precision. This generic structural analysis method which derives from microscopic modelling of concrete can be used to examine macroscopic structural performance with respect to the limit states to be examined. On similar lines, the objective of this study is to seek a so called life-span simulator of structural concrete based on the microscopic modelling of concrete, analogous to the generic structural analysis method which is well established in the field of structural engineering.

2.3 Proposal for concrete performance evaluation

The authors' idea of a versatile method of evaluating concrete performance is shown in Figure 2.4 and is similar to the dynamic nonlinear finite-element analysis technique of reinforced concrete, which is considered to be the most advanced way of examining structural and mechanical performances in safety and serviceability [4]. The emphasis in the proposed method is on identifying a versatile framework for material modelling that captures the fundamental parameters influencing the long-term durability characteristics of concrete at the microstructure level. For example, these models include, but are not limited to, heat generation, strength and microstructure development, permeability and sorption characteristics of concrete. Various deterioration mechanisms, such as corrosion or sulphate attack, that damage the internal structure and strength of concrete can be also be considered by fitting the physical laws describing these phenomena to the generic framework which will widen the scope of prediction of serviceability limits of concrete structures.

The generic framework should generally be able to describe the development and ageing process of concrete for any environmental history and the dynamic linkage with macroscopic damage such as cracking.

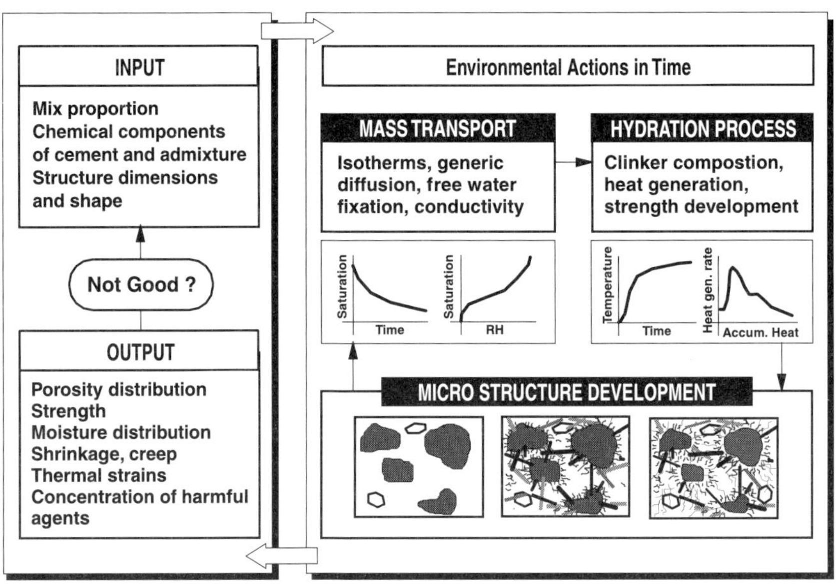

Figure 2.4 A general method of durability evaluation of concrete.

Moreover, it should correlate these with certain key parameters, such as permeability, of concrete for arbitrary types of fluid being transported. If these characteristics are well understood at the microstructure scale, they can be integrated over the domain of interest to predict the time-dependent performance of concrete over its life-span. It is believed that the physico-chemical characteristics of concrete at the microstructure level are fundamental in nature and a performance assessment system based on these characteristics can be seamlessly integrated into the well-developed system of examining structural and mechanical performances over the life of a structure.

The output of this general framework for assessing concrete performance which is the basis of estimating durability is:

1. micropore structure of concrete as expressed by pore size distribution;
2. degree of hydration of cement in concrete;
3. pore water content in concrete;
4. internal stresses related to shrinkage and external actions;
5. chemistry of the pore solutions, which is related to the corrosion of steel and the deterioration of the cement-paste matrix.

The above information is to be computed in three-dimensional space and time domains. The input information required by the model are:

1. mix proportions of concrete;
2. mineral composition of cement (or kind of cement);
3. geometrical shape and dimensions of structure;
4. initial temperature of concrete at placement;
5. environmental boundary conditions to which the structure is exposed.

It must be said that the generalized computational method of assessing concrete performance has not yet been established even though previous research has contributed significantly to the process. In the authors' opinion, now is the right time to start establishing the generalized computational method of examining durability performance of structural concrete. Concerted efforts are required to synthesize past scientific endeavour and start the studies necessary in line with the scheme shown in Figure 2.4. As a matter of fact, the output on concrete performance has to be obtained over the life-span so that structural safety and serviceability in time could be assessed. In this book, modelling at early ages accompanying micropore structure development is extensively focused as the first step of the project, and the achieved concrete quality at the beginning of its life generally governs the subsequent ageing and the time for which that quality is maintained in service.

2.3.1 Assumptions

A mathematical description of the nature of phenomena to be predicted in the proposed scheme is quite complex. In addition, several of these simultaneously occurring phenomena might be coupled to each other. Although significant knowledge exists that explains these phenomena, we are certainly nowhere near the stage of claiming that all the mechanisms related to the complex physico-chemical processes in cementitious materials are well understood. What should be our stand-point in such a situation? In this book, we have adopted a more practical approach to the problem. To a large extent, our endeavours are guided by the engineering implications of this research to the concrete community. At the same time, an attempt has been made to ensure that the models are rational at the material level by capturing the basic physics of the phenomena involved and incorporating their salient characteristics. Certainly, it is not claimed that the treatment adopted by the authors is completely comprehensive, but the treatment does provide some key indicators which can be traced, and based upon which an actual design system could be built upon.

The basic assumption in such a methodology concerns the overall predictability of the system. It is implicitly assumed that for a given mix-design, material specifications and environmental conditions, we will always end up with similar concrete. Obviously, this is not entirely true in real-life situations, since, random statistical variations apart, the methods of handling the material at fabrication stage and, most importantly, the human factor can lead to larger deviations in the properties of seemingly similar materials. It is not clear, at this stage of research, how these extraneous factors could be incorporated in the proposed method of durability evaluations. Tentatively, we have aimed the model towards intelligent construction materials, such as selfcompacting high-performance concrete [5], that require little or no human intervention and therefore eliminates to a large extent the uncertainties involved in the quality at construction stage.

2.3.2 Schematics of quantitative evaluation

The overall scheme for evaluation of the concrete performance attained does not consider different aspects of durability separately through unrelated empirical formulae, but takes an integrated approach. For example, instead of simply talking about the permeability and penetration depths of certain fluids in concrete, the question is raised about the life history of concrete in question. Considering its initial mix-design, the curing conditions and the environmental conditions, a parameter, such as permeability, is automatically obtained. In addition, any damage in the microstructure due to deterioration can be accounted for and dynamically updated.

The integration is primarily done by analytically examining the inter-relationships of hydration, moisture transport and pore-structure development processes, based upon fundamental physical material models related to each physical process. These processes and their inter-relationships are then translated into computer models that can be solved in a dynamically coupled manner. An outline of the overall scheme is provided next. In the framework, physical processes related to moisture transport are formulated at the micropore scale and integrated over a representative elementary volume (REV) to give macroscale mass transport behaviour. The hydration process is based on a multicomponent heat of hydration model for powder materials and is dependent on the free water available for hydration. Thus the average degree of hydration as well as the hydration of each clinker component can be obtained. The development of the pore structure at early ages is obtained using a pore structure development model based on the average degree of hydration. The predicted computational pore structures of concrete are used as a basis for moisture transport computations. In this way, applying a dynamic coupling of pore-structure development to the moisture transport and hydration models, the development of strength along with moisture content and temperature can be traced with the increase of degree of hydration for any arbitrary initial and boundary conditions. This methodology serves as a basis for the quantitative evaluation of the parameters relevant to the durability of structural concrete. The final formulation is however kept simple enough to be directly used in a regular finite-element (FE) based computational code. This integrated approach is highly computation intensive and is a job that could be easily delegated to a computer.

2.3.3 Implications of the proposal

The assessment of varying concrete performance considered here may mathematically cover the items treated in the durability design shown in Figure 2.3 as material and structural details, but cannot cover those items categorized as construction factors. In fact, we know construction has a significant influence on the durability performance of structural concrete. However, by utilizing self-compacting high-performance concrete (HPC), which can be placed inside formwork which has densely packed reinforcing bars, there is no need for any applied compaction [5], and the uncertainties regarding construction methods can be completely avoided in the assessment framework. Selfcompacting HPC was originally developed for achieving reliable structural concrete of high durability when high-quality workmanship cannot be expected at construction sites. In other words, HPC eliminates the uncertainties in compaction and construction quality, and makes the computational approach meaningful. The technology of self-compacting HPC will be the basis of the computational approach described above. The generalized way of simulating the durability of concrete can be applied to any limit state of durability performance evaluation.

Owing to the requirements for superfluidity and segregation resistance of selfcompacting HPC, a small amount of free water and a low water-to-powder ratio are specified. This often leads to a very tight texture of the micropores and durability much beyond the required levels when sufficient curing is provided. Therefore, under some conditions, the curing procedure can be obviated, making rapid construction possible. Here, let us consider one of the possible applications of a quality assessment method of concrete. It is essential to evaluate quantitatively the level of concrete quality which is actually achieved to ensure long-term durability as shown in Figure 2.5. Unfortunately, quantification of the curing of concrete and related effects has never been possible, but decisions still need to be taken to optimize the construction schedule. In such cases, the integrated system for assessing concrete performance can help in optimizing the resources in the planning and construction stages of concrete structures by quantitatively showing the effects of different curing conditions or schedules on concrete durability. Moreover, the degradation of durability can be traced not only for a few days or months but for the entire lifetime of concrete structures, once the environmental conditions to which concrete is exposed are known. An example of the versatility and power of the simulation approach is explained in the next section. Where the engineering problem of premature drying in a concrete structure is related to the micropore structure formation, strength

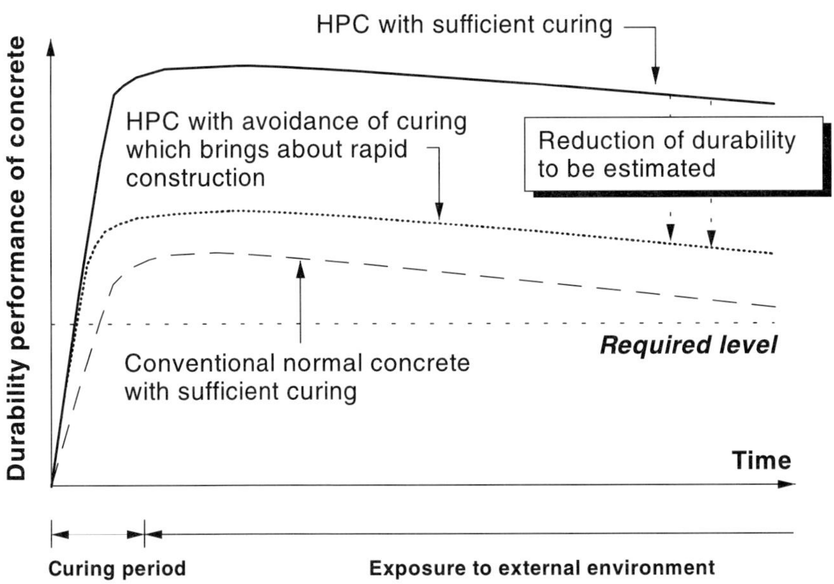

Figure 2.5 Degradation of durability performance of concrete.

development and hydration processes are investigated under various curing conditions.

This section demonstrates the application of a sample modelling integrated in a software program called DuCOM. The explanation of DuCOM (which is a computational model realized through computer simulation methods) and several associated material models is described in later chapters. The case study [6] taken up in this section involves studying the effect of various curing conditions on mortars prepared with different mix proportions. In this evaluation, the strengths at 7 and 28 days of concrete cylinders were measured for each set of mix proportions and curing conditions. The details of the experimental procedure and discussions can be found in Chapter 6 and reference [7]. The details of mix proportions and curing conditions are shown in the Table 2.1.

2.4 A case study of durability evaluation

Based upon the experimental test details of HPC mortars, several computations were performed using DuCOM, considering the similar cement type, mix proportions, temperatures and curing conditions to quantitatively evaluate the coupled systems of moisture transport and hydration. The finite-element discretization of the cylindrical specimen is shown in Figure 2.6. In using this evaluation method, apart from the bulk characterization of properties, such as strength, porosity or moisture content, information about different locations in the structure can also be obtained. Accordingly, zones having a high risk of damage can be identified well in advance and preventive measures can be taken during the design stage itself. Such solutions are not possible in the conventional treatment of durability design but experimental verification has been successful. A reasonable agreement between the computed and measured strength values for all the cases in Figure 2.6 shows the practicality and viability of the proposed method.

Table 2.1 Mix proportions and curing conditions [7]

Case	*W/C (%)*	*Air (%)*	*Mix unit weight (kg/m^3)* Water	Ordinary Portland cement	Medium heat cement	Lime	Slag	Sand	Gravel
MS	33.5	3.5	172	—	513	28	—	828	827
S6	55.8	3.5	172	—	308	17	200	828	827
OP	55.0	4.5	165	300	—	—	—	927	924

Case	*Curing condition*
SL	Sealed
16	16 hours stripped
2D	2 days stripped
WT	Submerged

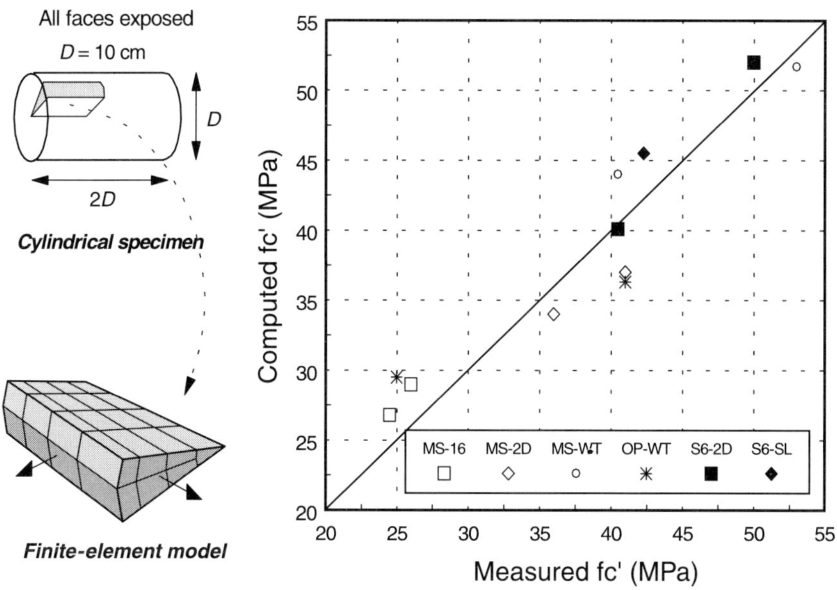

Figure 2.6 Predictions of strength for different curing conditions.

2.5 Summary

Quite often concrete structures fail in service due to damage done by deterioration and a loss of functions under environmental actions. This may be because the structures are designed with structural performance considerations in mind. A similar durability performance based designing method for concrete structures has not yet been put into practice. This is probably due to the difficulty in quantifying the adequate durability performance criteria and lack of assessing concrete performance. Codes are therefore forced to limit the durability design advice to general engineering notes on good construction practices, etc., that are prescriptive in nature. To consider rationally the durability evaluation system, we need to consider the performance-based specifications in detail. The examination of durability performance system introduced by the JSCE is one such system.

However, the JSCE system lacks the necessary versatility of application in that it cannot predict the service life of concrete manufactured with arbitrary materials and exposed to arbitrary environments. An integrated durability evaluation method is proposed that is based upon computer simulation of a physical mechanisms-based material model. The method is versatile and can predict the completed concrete performance in structures over their entire life-span. The approach is based upon combining structure formation, hydration and transport processes in concrete and computing their mutual interactions as well as the interactions with the environment. The application of one such method in terms of durability related concrete qualities is illustrated and shows reasonable agreement with the experiments. At present however the proposed method cannot take the job-site related human factors into account in durability considerations.

References

[1] Sakai, K., Concrete technology in the century of the environment, in K. Sakai (ed.), *Integrated Design and Environmental Issues in Concrete Technology*, E&FN SPON, London, 1996.

[2] Hoff, G.C., Toward rational design of concrete structures — integration of structural design and durability design, in K. Sakai (Ed.), *Integrated Design and Environmental Issues in Concrete Technology*, E&FN SPON, London, 1996.

[3] JSCE Subcommittee on Durability Design for Concrete Structures, Proposed recommendation on durability design for concrete structures, *Concrete Library JSCE*, 1989, No. 14, 35–71.

[4] Okamura, H. and Maekawa, K., *Nonlinear Analysis and Constitutive Models of Reinforced Concrete*, Giho-do Press, Tokyo, 1991.

[5] Okamura, H., Maekawa, K. and Ozawa, K., *High Performance Concrete*, Giho-do Press, Tokyo, 1993.

[6] Maekawa, K., Chaube, R.P. and Kishi, T., Coupled mass transport, hydration and structure formation theory for durability design of concrete structures, in K. Sakai (Ed.), *Integrated Design and Environmental Issues in Concrete Technology*, E&FN SPON, London, 1996.

[7] Shimomura, T. and Uno, Y., Study on properties of hardened high performance concrete stripped at early age, *Proceedings of the JSCE*, 1995, **26**(508), 15–22.

Chapter 3

Microstructure formation and hydration phenomena

- Concrete microstructure and its mathematical representation
- Cluster expansion based microstructure development theory
- Multicomponent cement heat of hydration model
- Strength development model of high-performance concrete mixes

3.1 Introduction

The calculation procedures that are routinely adopted in current engineering practice for the prediction of temperature fields, strengths and stresses in massive hardening concrete require knowledge of heat generation and strength development functions with time. Usually, these are experimentally obtained or specified in code practices for each concrete mix design and curing conditions to be analysed. Naturally, the process is tedious and little insight can be gained in the early age development phenomenon through such methods. Moreover, the simplified macromodels, such as isothermal, adiabatic or semi-adiabatic hydration curves obtained in controlled laboratory experiments, may not adequately describe the material behaviour under field conditions, due to dynamic variations in the temperature and moisture contents. For these reasons, there is an increasing demand for computer simulation methods that can provide a model to be used in macroscale calculations under arbitrary conditions of temperature and moisture.

The conventional approach to these problems has been to propose hydration models that usually predict the degree of hydration or amount of heat liberated with time under the assumption of fixed boundary conditions. These models are basically rate equations based upon several kinetic models and have served to generate input data to be used in structural programs devoted to macroscale static mechanics. The rationality of such methods is doubtful since the effects of temperature and moisture on the models of hydration cannot be dynamically considered in such a scheme. For a real-time integration, many of these models are too simplistic in nature and lack the versatility to explain the hydration behaviour under arbitrary conditions.

However, a formulation with less versatility can be exceptionally effective for huge concrete structures, such as dams. Here, almost all the volume of concrete can be presumed to be thermally isolated except for the surface zones, since the loss of heat from inside the structures to the external environment is much less than the heat generated inside the concrete. Then, the hydration process is nearly the same at any location, and the adiabatic temperature rise measured with high accuracy in the laboratory is expected to be correct for massive concrete.

A few digital computer models have also been proposed that simulate the development of cementitious microstructure due to hydration on a pixel-based model. The hydration products and various phases are represented as pixels of different nature and properties. The rate of their deposition and transport in the pore solution phase is governed either by a simple stoichiometric balance of various phases or is based upon some variation of the homogenous nucleation models [1]. These digital models are generally quite useful to give a qualitative insight into the nature of hydration products and the paste microstructure.

However, due to the digital nature of the simulations, these methods are limited in their resolution and require large computer resources. Though continuum models, such as the one proposed by van Breugel [2] and others [3], are sophisticated in nature and useful for research, some are not incorporated in the real time computational schemes that take into account the effect of temperature and moisture availability in a real-time basis over the entire structure domain. Without the dynamic incorporation of such models in the integrated scheme of coupled heat and mass transfer solutions, it is hard to predict macroscopic structural behaviour under arbitrary conditions of casting and curing.

It is believed that, since the development of microstructure, hydration and moisture transport are intrinsically coupled, an integrated approach might be required for macroscale computations of temperature and related stress models that make use of the sophisticated microscale material models directly. On these lines, it is the goal of this chapter to develop analytical models of microstructure development and cement hydration that are based upon fundamental physical material models pertaining to each physical process. At the same time, an important consideration is that the formulations should be simple enough to be incorporated in a real-time computational scheme of heat and moisture transport in a full-sized concrete structure from a practical view point. The development of a microstructure formation model is necessary since the mass transport characteristics are dependent on it.

The hydration process is based on a multimineral component heat of hydration model of powder materials and is dependent on the free water available for hydration [4]. The development of the pore structure at early ages is obtained using a pore structure development model based on the average degree of hydration. The predicted computational pore structures of concrete would be used as a basis for moisture transport computation. In this way, applying a dynamic coupling of pore-structure development to both the moisture transport and hydration models, the development of strength along with moisture content and temperature can be traced with the increase in the degree of hydration for any arbitrary initial and boundary conditions.

3.2 Microstructural development theory

Early age development phenomena of concrete structures have been extensively studied in terms of the development of macroscopic characteristics, such as permeability, strength, stiffness, etc., with time. Studies dealing with the micro-aspects of pore structure development in cementitious materials have been primarily confined to experimental realms. Moreover, little effort has been made to convert the fruits of these experimental studies into a computational system that could be used as a simulation tool for predicting concrete performance, especially related to durability of concrete structures. The primary motivation of this book is to create bridges between the various theories that deal with the prediction of moisture content, degree of hydration, strength, and so on. It may not be possible to account for all the complex details of pore structure development of real materials in such a model. However, the authors believe that a rational consideration of the salient characteristics of early age phenomena would be sufficient at the current stage to build a system that can be put to use in the real world.

3.2.1 Stereological description of the initial state of mix

The particle sizes of commercially available cement are spread over at least two orders of magnitude. Thus, manufactured pastes contain particles that range from colloidal sizes to particles that are about 150 μm in size. Just after mixing with water, many of the smaller particles dissolve completely, leaving room for the larger particles to grow as the hydration proceeds. In structure development, it appears that the larger particles play an important role in deciding the morphological features of the paste. It might be possible therefore to represent the entire particle distribution of powder materials by a representative volumetrically averaged particle size. As a starting point for our model, we assume that the powder material can be idealized as consisting of uniformly sized spherical particles. An important implication of this assumption is that the characteristics of the entire particle size distribution are lumped into a specific particle size. This may appear to be a gross approximation, however the initial colloidal dispersion characteristics of the paste are more important than the particle size distribution itself.

For usual W/P (water-to-powder) ratio pastes encountered in practice, the distance between powder particles is about the same order of magnitude as their average size. The theoretical arguments point to the fact that initially a completely homogenous and dispersed system should exist. However, there are observations that point to the agglomeration and flocculation of particles, which might result from the adherence of smaller powder particles to the larger ones. In such a case, we could consider the average size of the flocculated particles as the representative size of the powder material particle. In any case, the same size assumption is simple enough to deal with any initial configuration of the dispersed system although it is understood that a system which considered a particle size distribution as a starting point would be more theoretically complete. When the initial average size of the powder particles is known, it is assumed that the system comprising these colloidal dispersions of powder particles in the mix is homogenous. In a uniformly dispersed system of spherical particles of radius r_o, the average volumetric concentration G and the mean spacing between outer surfaces of two particles s is related by the following relationship.

$$G = \frac{G_o}{(1 + s/2r_o)^3} \tag{3.1}$$

where G_o is the maximum volumetric concentration or the limiting filling capacity that can be achieved. The theoretical maximum of G_o is $\pi/3\sqrt{2}$. However, in our model, G_o is computed from the following empirical relationships, considering a large spread in the size of particles, as

$$G_o = 0.79(BF/350)^{0.1} \qquad G_o \leqslant 0.91 \tag{3.2}$$

where BF represents the Blaine fineness index. For usual cements, r_o is taken as 10 μm and the average BF index is taken as 340 m^2/kg. If the W/P ratio of the paste is ω_o then the average spacing s between powder particles can be obtained as (using equation (3.1))

$$s = 2r_o[\{G_o(1 + \rho_p\omega_o)\}^{1/3} - 1] \tag{3.3}$$

where ρ_p denotes the average specific gravity of the powder materials and can be obtained as a weighted average for blended powders. In a dispersed system of particles, each particle has on average a free cubic volume, l^3, available for expansion as shown in Figure 3.1. At a certain stage of hydration, the expanding outer clusters of each particle will start touching each other. This will be followed by further expansion and merging as the new hydration products start filling the empty spaces available in a cubic volume (cell). To make the analytical treatment of such a process simple, the

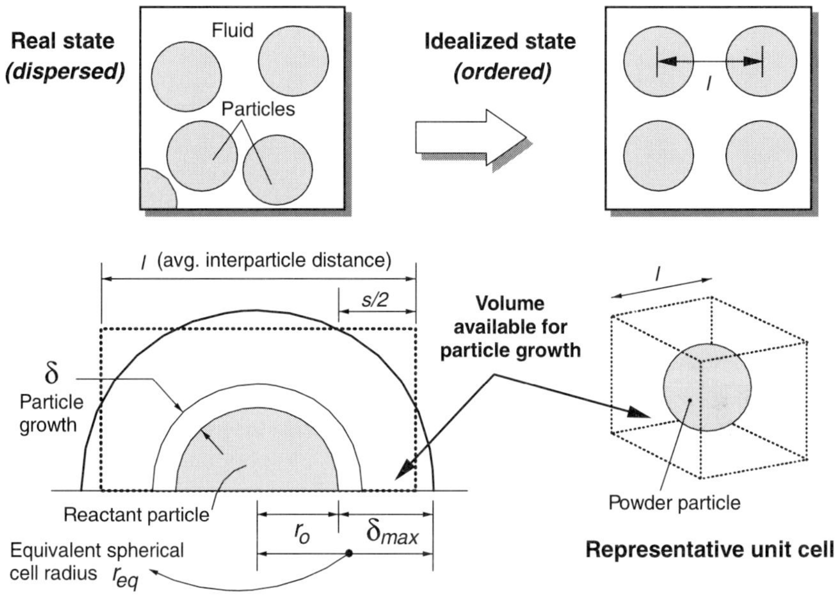

Figure 3.1 A representative unit cell of a powder particle.

initial volume of the cubic cell is transformed into a representative spherical cell of radius, r_{eq}, which can be simply obtained as

$$r_{eq} = \left(\frac{3}{4\pi}\right)^{1/3} l = \chi l \tag{3.4}$$

where l is the length of the edge of the cubic cell and χ is the stereological factor. In such a representation, we can find that touching of adjacent particles would occur when the thickness of the expanding cluster δ equals the spacing s (Figure 3.1). Furthermore, a complete filling of the cubic cell, or the equivalent representative spherical cell, would occur when the thickness of the expanding cluster equals $\delta_{\max}$ given by

$$\delta_{\max} = r_{eq} - r_o = r_o[2\chi \{G_o(1 + \rho_p \omega_o)\}^{1/3} - 1] \tag{3.5}$$

The geometrical parameters, r_{eq} and $\delta_{\max}$, represent the information pertaining to the stereological description of the initial state of a monosized particle dispersion.

3.2.2 Macroscopic volumetric balance of hydration products

Fresh cement paste is plastic in nature, consisting of a network of powder particles suspended in water. However, once the cement is set, very small changes in the overall volume take place. In this analysis, we ignore small changes in the gross volume that might take place due to bleeding or contraction while the paste is still plastic. Also, at any stage of hydration, the matrix contains the hydrates of various compounds, such as gel, unreacted powder particles, calcium hydroxide crystals, traces of other mineral compounds and large void spaces partially filled with water. As the hydration proceeds, new products are formed and deposited in the large water-filled spaces known as capillary pores. Moreover, the gel products also contain interstitial spaces, called gel pores, which are at least one order of magnitude smaller than the large voids of capillary pores. Due to the chemical reactions of hydration, water available in the capillary voids combines with the still unreacted powder minerals leading to the formation of new hydration products.

As a physical basis for pore structure development computation, let us subdivide the overall pore space into broadly three categories. These are the interlayer, gel and capillary porosity. Capillary porosity exists in the large interparticle spaces of powder particles, whereas gel porosity exists in the interstitial spaces of gel products or more specifically CSH grains. Interlayer porosity comprises the volume of water residing between the layer structures of CSH (Figure 3.2). For hydration reactions, only the water available in the capillary pores is relevant since it represents the space where new hydration products can be formed. The moisture existing in the interlayer

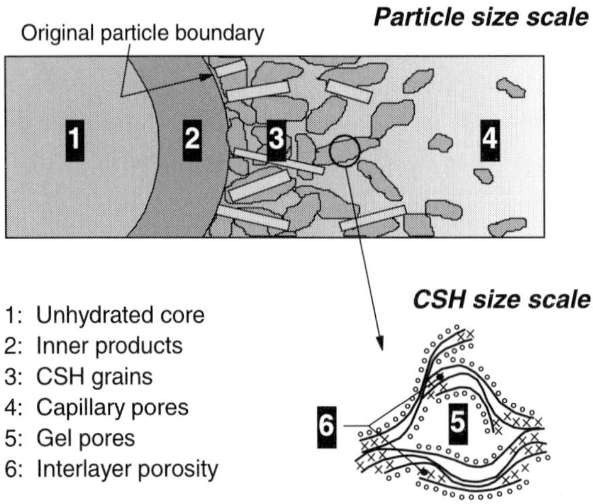

Figure 3.2 Schematic representation of porosity components of mortar.

porosity is a part of the physical structure of CSH and strongly bound to the surfaces. It is believed to be removable only under severe drying conditions, such as oven drying. The degree of hydration seems to be a useful parameter that can be used to compute the volume and mass of hydration products from a macroscopic point of view [2].

Bearing in mind these points, the weight, W_s, and volume, V_s, of gel solids per unit volume of the cement paste can be computed, provided the average degree of hydration, α, and the amount of chemically combined water, β, per unit weight of hydrated powder material, are known. The amount of chemically combined water is dependent on the chemical characteristics of the powder materials. In our model, the degree of hydration, α, is obtained as a ratio of the amount of heat liberated to the maximum possible heat of hydration for the mix. The amount of chemically combined water per unit weight of hydrated powder materials, denoted by β, is obtained based upon the stoichiometric balance of chemical reactions.

Computation of these parameters will be discussed in the multi-component heat of hydration model of cement (section 3.3). As described previously, we assume a layer structure for the CSH mass with an interlayer spacing of one water molecule [5, 6]. Moreover, it is assumed that the gel products have a characteristic porosity of ϕ_{ch} which is constant at all the stages of hydration. A value of 0.28 for ϕ_{ch} is usually reported for ordinary cement pastes [7]. The characteristic porosity contains the interstitial gel porosity as well as the interlayer porosity. Therefore, at any arbitrary stage of hydration, volume V_s of the gel products in a unit volume of the paste can be obtained as

$$V_s = \frac{\alpha W_p}{1 - \phi_{ch}} \left(\frac{1}{\rho_p} + \frac{\beta}{\rho_w} \right) \tag{3.6}$$

where W_p is the weight of powder materials per unit paste volume and ρ_w is the density of chemically combined water [7] ($\sim$1.25 g/cm^3). The interlayer porosity (ϕ_l) and gel porosity (ϕ_g) existing in the gel products are computed from the following expressions.

$$\phi_l = (t_w s_l \rho_g)/2 \qquad \phi_g = V_s \phi_{ch} - \phi_l \tag{3.7}$$

where ρ_g is the dry density of gel products $= \rho_p \rho_w (1 + \beta)(1 - \phi_{ch})/(\rho_w + \beta\rho_p)$, t_w is the interlayer thickness (2.8 Å), and s_l is the specific surface area of the interlayer.

At this stage, the dependence of the specific surface area of the interlayer on the mineralogical characteristics of the powder materials is not clear. While studies have reported values of the order of 500–600 m^2/g for Portland cements, only unreliable data exist for blended powder materials. In this book, tentatively, we propose the following relationship for s_l (m^2/g) as,

$$s_l = 510f_{pc} + 1500f_{sg} + 3100f_{fa} \tag{3.8}$$

where f_{pc}, f_{sg} and f_{fa} denote the weight fractions of the Portland cement, blast furnace slag and fly ash in the mix, respectively. Lastly, from the overall volume balance of the paste, capillary porosity (ϕ_c) or the volumetric ratio of larger voids can be obtained as

$$\phi_c = 1 - V_s - (1 - \alpha)(W_p/\rho_p) \tag{3.9}$$

Equation (3.7) and equation (3.9) macroscopically describe the microstructure of the hydrating paste in terms of the porosity of different phases, once the average degree of hydration and the amount of chemically combined water due to hydration are known.

3.2.3 *Expanding cluster model based on degree of hydration*

After initial contact with water, the powder particles start to dissolve and the reaction products are precipitated on the outer surfaces of particles and in the free pore solution phase. Figure 3.2 shows a schematic representation of various phases at any arbitrary stage of hydration. Precipitation of the pore solution phase on the outer surfaces of particles leads to the formation of outer products, whereas so called inner products are formed inside the original particle boundary where the hydrate characteristics are assumed to be more or less uniform. Diverse physical characteristic of the hydration products have been reported in the literature. The hydration products have been reported to appear to be like needles, fibres, tubes, rods, foilitic, tree-shaped, prismatic crystals, tabulated, acicular, scalular and so on. Moreover, development of the cementitious microstructure involves the formation of various hydration products, such as ettringite, CSH and CH at different degrees of hydration and time (Figure 3.3). The physical shape and size of these products might depend on various parameters, such as cement composition, temperature, degree of hydration or presence of additives. However, it seems a daunting task to take into account the effect of all these factors on the development of physico-chemical characteristics of the microstructure in an analytical method that could also be put into practical computational use.

For these reasons, we assume a simplified picture of the development of porosity and associated microstructure. For example, the microstructural properties of the inner product are assumed to be constant throughout the process of pore structure formation. Moreover, it is assumed that hydrates of similar characteristics are formed at all the stages of hydration. While this might not be strictly accurate in the very early stages of hydration, there is evidence that points to the fact that the hydration product formed in the later stages of hydration are almost identical [8].

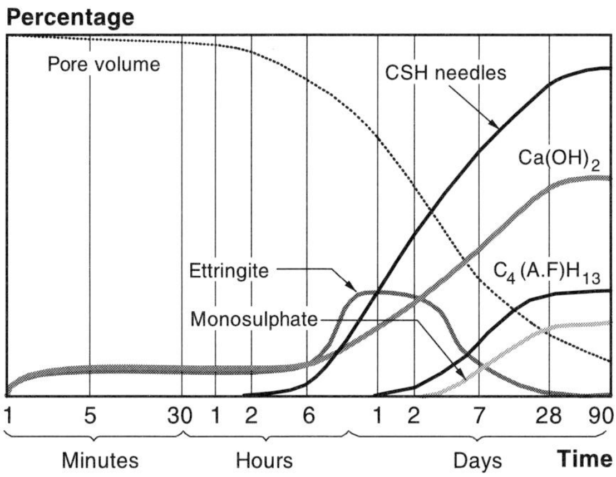

Figure 3.3 Formation of several hydration products with time.

The hydration products deposited outside the original particle boundary are assumed to have similar characteristics to the inner products. It is assumed that the conventional capillary porosity exists between the grains of hydrates of the outer product as well as in the spaces between growing clusters. The grains of CSH hydrates in the outer products as well as the inner products account for all of the finer gel and interlayer porosity. The characteristic porosity of the CSH mass, ϕ_{ch}, as well as the granular characteristics of the gel, are assumed to be constant throughout the hydration process. Here, a value of 0.28 is assumed for ϕ_{ch}. It has to be noted that this porosity includes the interlayer as well as the interstitial gel porosity. For this model of cluster expansion, the porosity distribution in the cluster around a partially hydrated particle would appear as shown in Figure 3.4. Moreover, the density at any point in this cluster, ρ_δ, would depend on the mass of the gel and the bulk porosity at that point as

$$\rho_\delta = \frac{(1+\beta)(1-\phi_\delta)}{(1/\rho_p + \beta/\rho_u)} \tag{3.10}$$

where ϕ_δ is the bulk porosity at any point in the outer product and ρ_u is the specific gravity of the unevaporable or chemically combined water. The bulk porosity distribution in the outer products is modelled by a power function as (Figure 3.4)

$$\phi_\delta = (\phi_{ou} - \phi_{in})\left(\frac{\delta}{\delta_m}\right)^n + \phi_{in} \tag{3.11}$$

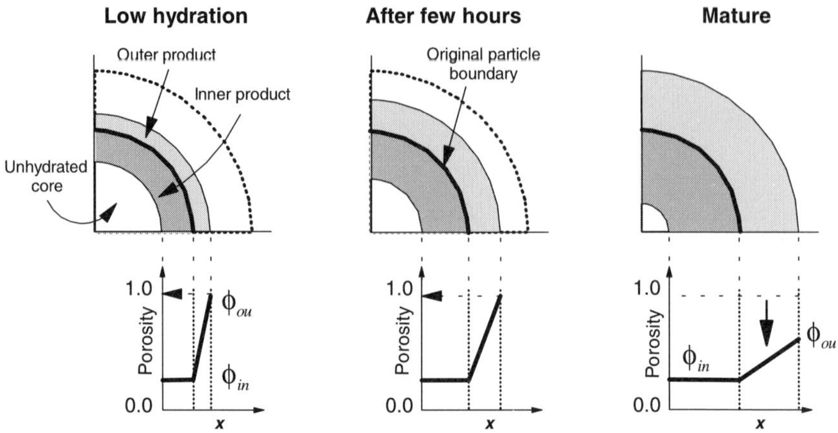

Figure 3.4 Bulk porosity function of outer product and cluster expansion.

where δ denotes the distance of any point in the cluster from the outer boundary of the particle, δ_m is the cluster thickness at the instant of hydration and ϕ_{in} is the porosity of the inner products. In our model, this would be the characteristic porosity of the gel, ϕ_{ch}, and ϕ_{ou} is the porosity at the outermost boundary of the expanding cluster and equals unity before the expanding cluster reaches the outermost end of the representative spherical cell of the particle. The parameter n is the power-function parameter that takes into account a generic pattern of deposition of products around the particle and might be used to take temperature effects into account. However, in our computations, we have adopted n as unity.

3.2.4 *Surface area of capillary and gel components*

The macroscopic balance of the hydration products provides us with an estimate of the macroscopic quantities, such as porosity, of various phases at different stages of hydration. However, several phenomena linked to the durability of concrete, such as transport processes, also require additional microstructure information, e.g. the specific surface areas or porosity distributions of each phase. It is not possible to obtain these parameters from the macroscopic discussions alone.

The specific surface area of gel can be estimated from adsorption experiments. Furthermore, if it is assumed that the hydration products formed at different stages of hydration are more or less identical, the porosity distribution parameter of gel products will be constant at all the stages of hydration. An important implication of this fact is that gel microstructural properties can be estimated without requiring knowledge of the progress of hydration. From the point of view of the transport processes, however, it is the larger capillary pores that are important. Unlike gel porosity, no direct estimation of the capillary porosity distribution is possible, since the capillary pores undergo dynamic changes, such as subdivisions, branching and filling up with new hydrates, as the hydration proceeds. The specific surface area of capillary porosity must be computed at each stage of the hydration. This is accomplished by considering the degree of hydration based expanding cluster model described in the previous section.

Let us consider Figure 3.5 which shows a reactant particle at an arbitrary stage of hydration. In this case, the weight of the hydrates existing in the inner and outer products should be equal to the total amount of hydrates formed up to that stage of hydration. The total amount of hydrates, W_T, as well as the amount of hydrates residing as inner products, W_{in}, can be obtained as

$$W_T = \frac{4}{3}\pi r_o^3 \alpha \rho_p (1+\beta) \qquad W_{in} = \frac{4}{3}\pi r_o^3 \alpha \left[\frac{(1+\beta)(1-\phi_{in})}{(1/\rho_p + \beta/\rho_u)}\right] \tag{3.12a}$$

The amount of hydrate material existing in the outer products, W_{ou}, can be obtained by integrating the mass present at any arbitrary location in the cluster over the entire cluster thickness as (Figure 3.5)

$$W_{ou} = \int_{r_o}^{r_o+\delta_m} 4\pi r^2 \rho_r \, dr = \frac{4\pi(1+\beta)}{1/\rho_p + \beta/\rho_u} \int_o^{\delta_m} [1-\phi_\delta](\delta + r_o)^2 \, d\delta \tag{3.12b}$$

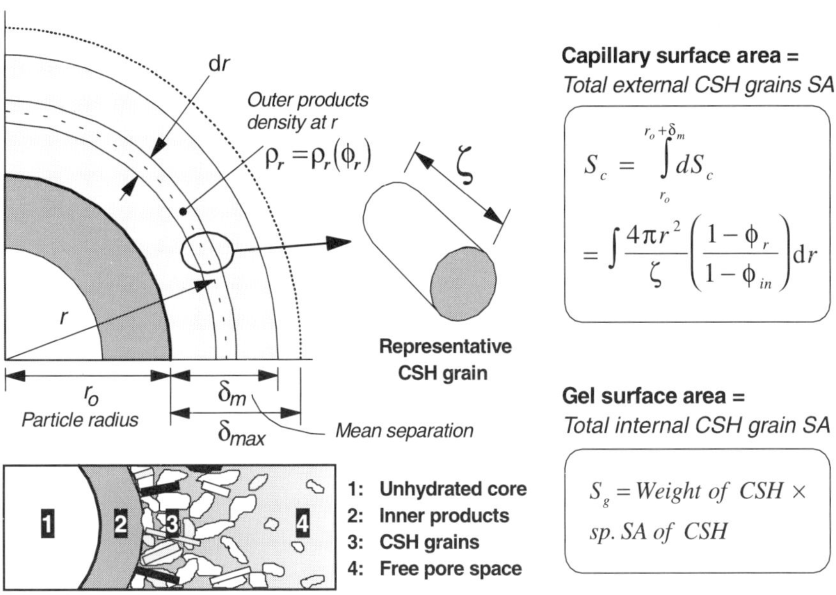

Figure 3.5 A reactant particle at arbitrary stage of hydration.

By applying the conservation of mass condition to equations (3.12a,b), such that $W_T = W_{in} + W_{ou}$ and simplifying the resulting expressions, we obtain the mass compatibility condition as

$$\begin{aligned}
&A\delta_m^3 + B\delta_m^2 + C\delta_m + D = 0\\
&A = \{n(1-\phi_{in}) + 3(1-\phi_{ou})\}/\{3(n+3)\}\\
&B = \{n(1-\phi_{in}) + 2(1-\phi_{ou})\}r_o/(n+2)\\
&C = \{n(1-\phi_{in}) + (1-\phi_{ou})\}r_o^2/(n+1)^2\\
&D = -(\alpha r_o^3/3)[\phi_{in} + \beta\rho_p/\rho_u]
\end{aligned} \tag{3.12}$$

In the cluster expansion model described above, two different conditions of interest would arise. The first case is when an unhindered free expansion of the cluster takes place. The second condition arises when the expanding front of cluster reaches the limit of the representative spherical cell, i.e. when $\delta_m = r_{eq}$. These two cases are discussed separately as below.

Case 1 Free expansion of the cluster

In such a case, the outer product cluster thickness, δ_m, can be simply computed from equation (3.12) by substituting the outermost cluster porosity, ϕ_{ou}, as unity and solving the resulting cubic equation.

Case 2 Cluster thickness equals free space available for expansion

In this case, the bulk porosity at the outermost boundary of the outer product, ϕ_{ou}, would gradually reduce so as to accommodate the new hydration products being formed. The porosity at the outermost end of the cluster, ϕ_{ou}, can be obtained by substituting $r_{eq} - s$ as the cluster thickness, δ_{max}, in equation (3.12). Such an exercise leads to the following expression for ϕ_{ou}.

$$\phi_{ou} = 1 - \frac{X+Y}{Z} \qquad \delta_{max} = kr_o$$

$$X = -n(1-\phi_{in})\left[\frac{k^3}{3(n+3)} + \frac{k^2}{n+2} + \frac{k}{n+1}\right]$$

$$Y = \frac{\alpha}{3}\left[\phi_{in} + \beta\frac{\rho_p}{\rho_u}\right] \qquad Z = \frac{k^3}{n+3} + \frac{2k^2}{n+2} + \frac{k}{n+1} \tag{3.13}$$

Once again, we emphasize that, in these models, it is implicitly assumed that the inner mass develops inside the original particle boundary with constant properties throughout the hydration process, whereas representative hydrate crystals of constant and uniform properties are deposited in the outer pore solution phase. With the maturity of the hydrating matrix, microstructural properties tend to become uniform in the outer as well as inner products. Consequently, differences of microstructural properties in the outer and inner products diminish with hydration. Once the variation of bulk porosity in the outer products is known, we can proceed to determine the capillary surface area contained in the outer products.

Let us consider a region of thickness dr located at a distance r from the particle centre (Figure 3.5). If dV_g and ρ_g represent the real volume and dry density of the hydrates in this region, and ϕ_r is the bulk porosity of the expanding cluster at r, then by equating the mass of the hydrates contained in this region, we have,

$$\rho_g \, \mathrm{d}V_g = 4\pi r^2 \, \mathrm{d}r \, \frac{(1+\beta)/(1-\phi_r)}{1/\rho_p + \beta/\rho_u} \tag{3.14a}$$

Furthermore, we assume that the volume to surface area ratio of the solid grains of hydrates contained in this region is ζ. This ratio may be in fact proportional to the average size (radius) of the hydration products. Since the CSHs constitutes the major proportion of the gel, it can be assumed that the overall average hydrate size would correspond to the CSH grain size. The average size of the particles of cementitious materials has been reported to be of the order of 0.01 μm. While the CSH size may range from 0.002 μm to 0.1 μm over the complete history of hydration, the later stages of hydration are dominated by finer grain sizes. Powers and Brownyard [9] mentioned the average gel particle size of 0.01 μm to 0.015 μm. Considering these discussions, we assume that ζ, which represents the ratio of volume to the external surface area of a typical grain, is constant during the course of hydration. Note that an assumption on the actual size of these grains is not made. In this book, the following empirical expression is used to obtain ζ (nm) for blended powders.

$$\zeta = 19.0 f_{pc} + 1.5 f_{sg} + 1.0 f_{fa} \tag{3.14b}$$

Using the volume-to-surface area concept, the capillary surface area dS_c contained in the region dr is obtained by applying the chain rule of differentiation to dV_g in equation (3.14a) and further simplifying as

$$\mathrm{d}S_c = \frac{\mathrm{d}S_c}{\mathrm{d}V_g}\,\mathrm{d}V_g = \frac{4\pi r^2}{\zeta}\left(\frac{1-\phi_r}{1-\phi_{in}}\right)\mathrm{d}r \tag{3.14c}$$

The specific surface area of capillary porosity, S_c, per unit volume of the matrix can be obtained by integrating the above expression over the entire cluster thickness and then normalizing it to the volume of the representative spherical cell. Similarly, the surface area of the gel, S_g, is simply obtained as a product of the mass of gel products to the specific surface area of gel mass. Then, it yields

$$S_c = \frac{3\delta_m}{\zeta r_{eq}^3(1-\phi_{in})}(A\delta_m^2 + B\delta_m + C) \qquad S_g = W_s s_g \tag{3.14}$$

where s_g is the specific surface area of the hydrates and A, B, C are similar to equation (3.12). The specific surface area of hydrates is assumed to be 40 m^2/g in this study [7].

3.2.5 Computational model of the total paste microstructure

Typically, the size and range of pores present in the cement paste are distributed over several orders of magnitude. Also, the distribution of moisture in such a porous system is different for each different range of pore sizes considered. As the hydration proceeds, large spaces initially taken up by water are replaced by the increasing network of hardening mass which is colloidal in nature [7]. This hardening mass suspended in the liquid constitutes what is called a cement gel. Hydration progress leads to more and more liquid space being taken up by the gel and eventually a discontinuity in the network of large voids filled up with water occurs. The pore structure thus formed in the hydrated mass is highly complex in nature due to its complicated shape, interconnectivity and spatial distribution. These pores have been traditionally classified according to their sizes. For example, the complete pore classification for hardened cement paste is shown in Figure 3.6, which applies the International Union of Pure and Applied Chemistry (IUPAC) classification [10]. The capillary pores can be considered to be the remnants of the water-filled space existing between the original cement grains, while micropores are part of the CSH component. The usual measurement of these pores using a mercury intrusion or helium pycnometry technique assumes a uniform pore shape, whereas the pores form a network of highly irregular geometry.

In the past, researchers have argued over those components of the overall pore structure that actually contribute to the moisture transport. In this study, from the point of view of moisture transport, the authors have subdivided the overall cementitious micropore structures into three basic components. The first of these components are the capillary pores that are actually empty spaces left between partially hydrated cement grains. Second are the gel pores that are formed as an internal geometrical structure of the CSH grains which deposit around a hydrating cement grain. Third, and important from the moisture transport point of view, is the physically fixed interlayer moisture that is actually a structural part of the CSH gel.

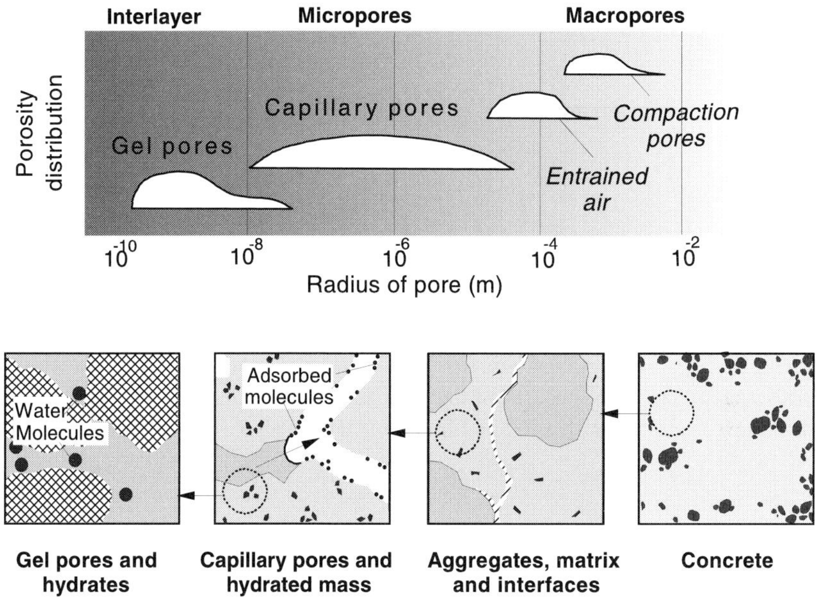

Figure 3.6 Schematic representation of porosity classification in concrete.

The authors believe that the actual transport of moisture takes place only into the gel and capillary components of the cementitious microstructure. This is because the moisture physically bound as interlayer water is unable to move under the usual pressure gradients. However, a local rearrangement of moisture from the interlayer structure to gel pores might take place depending on the severity of drying or wetting conditions. Also, the interlayer porosity might not influence the transport behaviour of fluids that do not interact with the layer structure of the CSH gel. Thus, for analytical purposes, the total porosity, ϕ, of hardened cement paste can be mathematically expressed as a summation of the capillary, gel and interlayer porosity. Moreover, if the distribution functions of porosity for these porosity systems are known, we can obtain the total porosity distribution of the paste as a function of the pore radius.

However, at present, the exact characteristics of the interlayer component of total porosity are not well understood. Therefore, the interlayer porosity is simply lumped with the porosity distributions of gel and capillary porosity, called *external* or *capillary-gel porosity*, to obtain the total porosity distribution of cement paste as

$$\phi(r) = \phi_{lr} + \phi_{cg}(r) \tag{3.15}$$

where, r is the pore radius, ϕ_{lr} is the interlayer porosity and $\phi_{cg}(r)$ is the combined porosity distribution of gel-capillary pores. In reality, $\phi_{cg}(r)$ is a complex and continuous distribution function. It can be imagined to be similar to the distributions as obtained by MIP (mercury intrusion porosimetry) tests on hardened cement paste. In our analytical methods, however, we have categorized this system broadly into capillary and gel pores. With such a classification, the noninterlayer porosity distribution $\phi_{cg}(r)$ of cement paste can be obtained as

$$\phi_{cg}(r) = \frac{1}{(\phi_{cp} + \phi_{gl})}[\phi_{cp}(r) + \phi_{gl}(r)] \tag{3.16}$$

where ϕ_{cp} is the total capillary porosity and ϕ_{gl} is the total gel porosity. Each of the capillary and gel porosity distributions, $\phi_{cp}(r)$ and $\phi_{gl}(r)$, are represented by a simplistic Rayleigh–Ritz (R–R) distribution function as

$$V = 1 - \exp(-Br) \qquad \mathrm{d}V = Br\exp(-Br)\,\mathrm{d}\ln r \tag{3.17}$$

where V represents the fractional pore volume of the distribution up to pore radius, r, and B is the sole porosity distribution parameter, which in fact represents the peak of porosity distribution on a logarithmic scale. A simplistic porosity distribution has been chosen to ease numerical computations and parametric evaluations. A bimodal analytical porosity distribution of the cement paste can be therefore obtained as

$$\phi(r) = \phi_{lr} + \phi_{gl}[1 - \exp(-B_{gl}r)] + \phi_{cp}[1 - \exp(-B_{cp}r)] \tag{3.18}$$

where the distribution parameters, B_{cp} and B_{gl}, correspond to the capillary and gel porosity components respectively. Interlayer porosity, ϕ_{lr}, is believed to exist in much smaller pores than gel porosity. Apart from the cement paste pores, there may be a distinct class of pore present in the aggregates. However, it is generally believed that these pores do not directly take part in the transport process in liquid and gases. Moreover, the aggregate cement paste interface shows distinct zones of high porosity and coarse microstructure. Characteristics of such pores will be discussed in Chapter 5. These pores may play an important part with regard to the transfer of liquid between cement paste and aggregates. Our immediate target is to obtain the pore structure parameters that will enable us to compute the dynamic porosity distributions with time, based upon the degree of hydration.

The use of a simplistic computational pore structure enables easy evaluation of the surface area parameters. Let us consider a unimodal R–R distribution function given by equation (3.17). If we assume a cylindrical pore shape in such a distribution, then the pore distribution parameter B can be obtained from the following relationship, if S the surface area per unit volume of the matrix is known (Figure 3.7), i.e.

$$S = 2\phi \int r^{-1}\,\mathrm{d}V = 2\phi \int_{r_{\min}}^{\infty} B\exp(-Br)\,\mathrm{d}\ln r \tag{3.19}$$

where $r_{\min}$ is the minimum pore radius. Unfortunately, this expression cannot be evaluated analytically as a closed-form solution. Therefore, the authors use an explicit relationship which has been obtained by fitting the accurate numerical evaluations of the above integral for a large number of data-sets which relate B as a function of S/ϕ.

From earlier discussions, the overall cementitious microstructure can be represented as a bimodal porosity distribution by combining the inner and outer products contributions to the total porosity function $\phi(r)$ as

$$\phi(r) = \phi_l + \phi_g\{1 - \exp(-B_g r)\} + \phi_c\{1 - \exp(-B_c r)\} \tag{3.20}$$

where r is the pore radius. The distribution parameters B_c and B_g can be easily obtained from equations (3.14) and (3.19) and Figure 3.7, once the surface areas and relevant porosity of the gel and capillary components are computed. The overall scheme of pore structure development is outlined in Figure 3.8 . This numerical model of cementitious microstructure is dynamic in nature so that the effects of hydration can be dynamically traced with time.

Figure 3.9 shows a comparison among the computed micropore structures of mortar of water-to-cement ratios of 25%, 45% and 65%. The pore structure has been computed at a degree of hydration of 0.15 and at the maximum achievable for each mortar in sealed curing conditions (60 days).

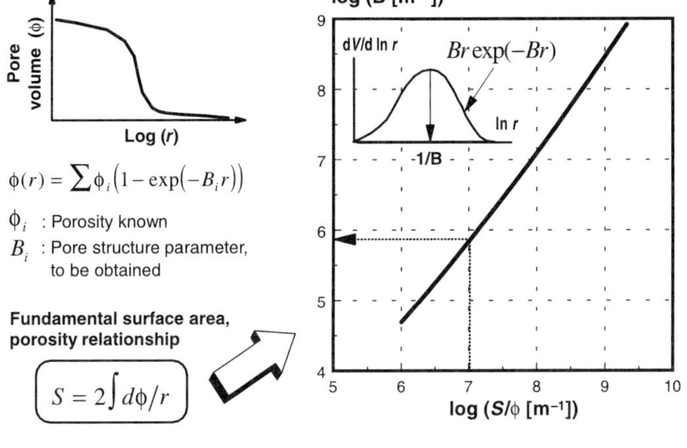

Figure 3.7 Relationships of pore structure parameter B and S/ϕ.

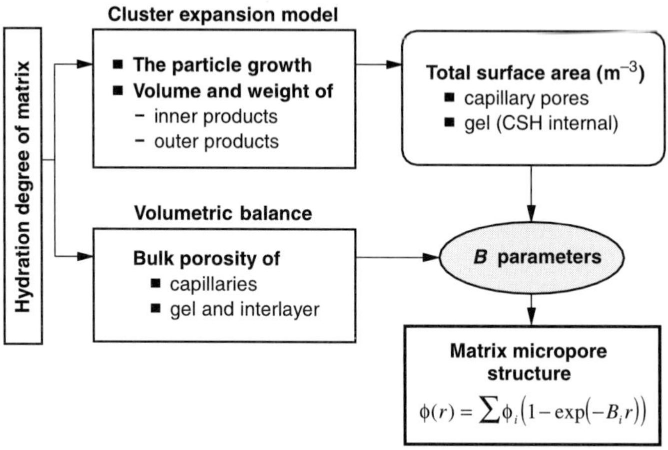

Figure 3.8 Schematic representation of the pore structure development computation.

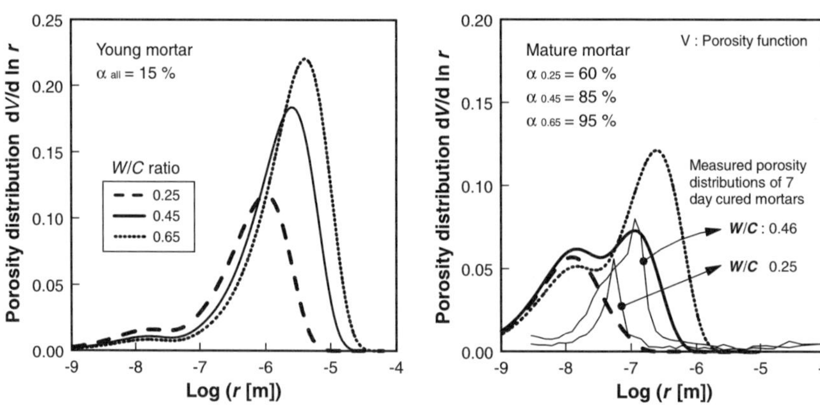

Figure 3.9 Comparison of computed microstructures for different mortars.

It can be observed that an increase in the gel pore volume fractions occurs gradually; also the ratio of finer pore porosity to coarser porosity increases for low W/P mortars. It appears that generally the peak of porosity distribution lies between 20 nm and 120 nm for most of the mature concrete. The microstructure development model described in this chapter will enable us to consider the coupling of moisture transport and hydration phenomena in a rational way, where the degree of hydration dependent effects are automatically considered in the overall computational scheme.

3.3 Multicomponent heat hydration model

This section describes an outline of the heat of hydration model of cement chiefly developed by Kishi *et al.* [4, 11, 12]. A detailed discussion of the model, its background, verifications and application to thermal crack control design is given in Chapter 7. This section presents the main points of interest. This model is used as a basis to explain the heat generation for arbitrary powder types with a wide variety of constituent minerals. It has been incorporated in the general scheme of microstructure development, hydration and moisture transport phenomena. As a factor responsible for the creation of the hardened cement matrix of concrete and the sole source of temperature rise during hardening, the heat of hydration of cement in concrete needs to be modelled. This factor is an essential component in concrete performance evaluation used in a rational durability design system which incorporates thermal crack control as well as other effects, such as drying shrinkage, development of strength and microstructure, etc.

Moreover, recently, the usage of concrete having a low W/P ratio and a large amount of powder, such as blast furnace slag and fly ash, in real construction systems has increased. This has necessitated the development of a versatile model that can be used for durability design, incorporating the effect of the constituent materials achieved performances on the vulnerability of a structure in an accurate and integrated way. On these lines, the hydration model presented here proposes a multicomponent model for the heat of hydration generation based on Arrhenius' law of chemical reaction.

3.3.1 Multicomponent concept for heat of hydration of cement

Cement is a polysize, polymineral material, where the major phases hydrate at different rates. The specific heat rate in such a system will clearly consist of the summation of heat rates of the different clinker minerals. The heat of hydration model proposed here is based on the multicomponent concept. Hence, the effect of arbitrary types of cement or powder materials can be rationally taken into account to predict the overall heat generation rate. The chemical compounds of cement considered in this model are aluminate (C_3A), alite (C_3S), belite (C_2S), ferrite (C_4AF) and gypsum. Other blending materials, such as slag or fly ash, are incorporated in the model as pseudoclinker mineral components. The specific heat generation rate of such a blended cement, denoted by H_C, is expressed as

$$\begin{aligned} H_C &= p_{C_3S}H_{C_3S} + p_{C_2S}H_{C_2S} + p_{SG}H_{SG} + p_{FA}H_{FA} \\ &\quad + p_{C_3A}(H_{C_3AET} + H_{C_3A}) + p_{C_4AF}(H_{C_4AFET} + H_{C_4AF}) \\ &= \sum p_i H_i \end{aligned} \tag{3.21}$$

where H_i is the specific heat generation rate of individual clinker component and p_i is the corresponding mass ratio in the cement such that $\sum p_i = 1$. Also, H_{C_3AFT} and H_{C_4AFET} represent the contribution due to the formation of ettringite from C_3A and C_4AF reactions with gypsum.

The temperature dependent heat generation rate, H_i, of each clinker component is based on a modified Arrhenius' law, where various coefficients represent the interaction among constituents as

$$H_i = \beta_i \mu \xi \gamma s_i H_{i,\,T_0}(Q_i) \exp\left[-\frac{E_i(Q_i)}{R}\left(\frac{1}{T} - \frac{1}{T_0}\right)\right]$$
$$Q_i = \int H_i \, \mathrm{d}t \tag{3.22}$$

where E_i is the activation energy of ith component reaction, R is gas constant and $H_{i,\,T_0}$ is the reference heat rate of ith component when temperature is T_0 (293 K) and mineral composition corresponds to that of ordinary Portland cement. The ratio $-E_i/R$ is usually called the *thermal activity*, β_i represents the reduction of probability of contact between unhydrated compounds and free pore water, μ represents the effect of mineral composition of Portland cement, ξ represents the reduction of the pozzolan's reaction due to the shortage of calcium hydroxide, s_i represents the relative fineness expressed as a ratio of Blaine index, γ represents the retardation in overall rate of cement and slag reaction caused by the presence of fly ash and organic admixtures. Each of these parameters are briefly discussed below. The heat of hydration model discussed in Chapter 7 considers the fineness of powder materials given by parameter s_i in greater detail. In this chapter, it will be ignored for the sake of brevity.

The reference heat rates of each phase as shown in Figure 3.10 are the material parameters of this model. These models have been obtained after careful comparisons of adiabatic temperature rise experiments with the proposed theory in a reverse data analysis method. The maximum theoretical specific heat generation of each phase is given by the term $Q_{i,\,\infty}$. Broadly, there are three distinct stages of hydration. Stage 1 represents the dormant period and Stage 2 is the period of hydration with a large evolution of heat and rapid microstructure formation. Stage 3 is the period during which the hydration is determined by the ion diffusion through the cluster consisting of hydration products around the reactant cement particle and thereby the rate of hydration is reduced.

3.3.2 Interaction among cement clinkers and pozzolans

The rate-controlling parameters of equation (3.22) are discussed briefly in

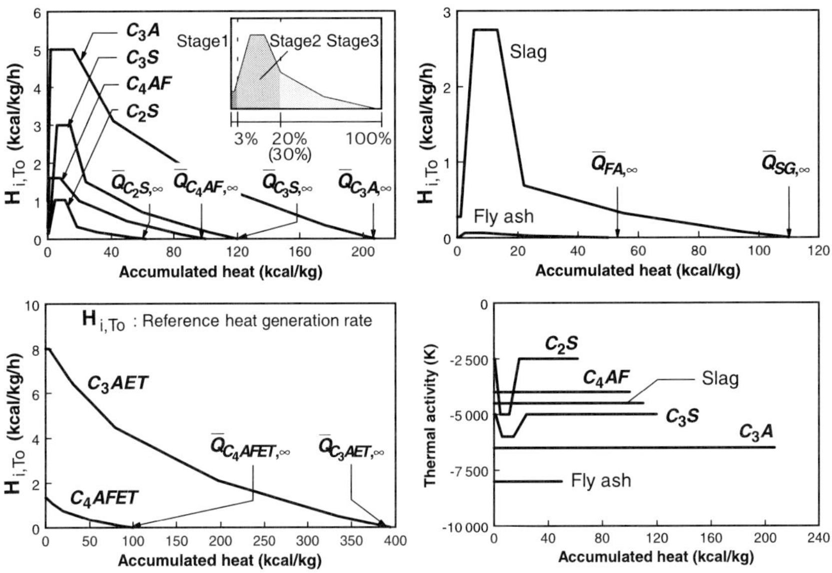

Figure 3.10 Reference heat generation rates of cement and pozzolans concretes.

this section together with some background on the chemical reactions defining the hydration process.

Modelling of ettringite formation

To model the initial stages of hydration accurately, it is necessary to understand the phase conversion processes that take place. Owing to the presence of gypsum in cement, which is usually added to retard the rapid hardening due to C_3A hydration, ettringite is produced from the reactions of C_3A and C_4AF with gypsum. This is accompanied by the evolution of considerable heat and conversion of ettringite to monosulphate due to the shortage of gypsum in the solution. The heat evolution from ettringite formation is over by the time gypsum disappears from the solution. The amount of gypsum consumed in the formation of ettringite is computed from the ratio $Q_{i,\,ET}/Q_{iET,\,\infty}$ and the chemical stoichiometric balance which is obtained from

$$\begin{aligned} &C_3A + 3C\overline{S}H_2 + 26H \rightarrow C_3A \cdot 3C\overline{S} \cdot H_{32} \\ &C_4AF + 3C\overline{S}H_2 + 27H \rightarrow C_3(AF) \cdot 3C\overline{S} \cdot H_{32} + CH \end{aligned} \tag{3.23}$$

where C is CaO, A is Al_2O_3, F is Fe_2O_3, H is H_2O, CH is $Ca(OH)_2$, and $\overline{S}$ is SO_3.

The accumulated heat Q_{C_3AET} and Q_{C_4AFET} are obtained from the reference heat generation rates as shown in Figure 3.10. The thermal activities assumed in this case are the same as that of C_3A and C_4AF. After the ettringite formation, the conversion to monosulphate and the subsequent initiation of C_3A and C_4AF hydration takes place. The conversion to monosulphate is included in the hydration by taking into consideration the material models as shown in Figure 3.10.

The effect of mineral composition

The mineral composition of cement vary widely, depending on its type. Cements range from early hardening type (C_3S rich) to super low heat type (C_2S rich). Although the thermal activity and reference heat generation rate of each mineral clinker can deal with these variations to a large extent, it is found that mutual interactions among these components must be taken into account to build a versatile model for any type of cement. Thus, a parameter μ is introduced which indicates the change of the reference heat rate in accordance with the cement type at Stage 3 of the hydration progress. It is defined as

$$\mu = 1.4[1 - \exp\{-0.48(p_{C_3S}/p_{C_2S})^{1.4}\}] + 0.1 \tag{3.24}$$

The ratio C_3S/C_2S is used an indicator representing the cement type. Moreover, this factor is used only for computing the heat generation rate of cement clinker minerals.

Calcium hydroxide as activator for pozzolans' reaction

The reactions of pozzolans, such as blast furnace slag and fly ash, are primarily governed by the calcium hydroxide released due to cement hydration which also acts as an activator. The reduction in the pozzolans' rate of reaction caused by the shortage of CH is formulated as

$$\xi = 1 - \exp\left\{-2\left(\frac{F_{CH}}{R_{CH}}\right)^5\right\} \tag{3.25}$$

where F_{CH} is the available CH in the solution, R_{CH} is the CH necessary for the reactions of pozzolans at any stage of hydration. The amount of CH released from C_3S and C_2S is computed from equation (3.32) (discussed later). The amount of CH consumed is obtained by considering the degree of the pozzolans' reaction and an assumed consumption ratio of CH per unit weight of pozzolans.

Retarding effects of fly ash and organic admixture

The addition of fly ash retards the hydration of Portland cement especially at early stages of hydration and makes the dormant period longer. Mixing of fly ash with cement leads to a depression of Ca^{2+} ions in the pore solution, due to its removal by aluminium ions existing on the fly ash particle's surface. This depression of concentration retards the formation of a calcium-rich surface layer on clinker minerals which is a precursor of reactivity. The organic admixtures also contribute to a retardation effect, where Ca^{2+} ions are restrained by the organic admixtures.

In the proposed model, the retardation effect to the rate of reaction of clinker minerals is considered only in the Stage 1 of the hydration. The retardation effect is modelled by a factor γ as

$$\gamma = \exp\left(-2\frac{1000\vartheta_{ef}}{20p_{C_3S} + 10p_{C_2S} + 5p_{SG}}\right) \tag{3.26}$$

where ϑ_{ef} represents the effective ability for retardation, and we have

$$\vartheta_{ef} = 0.02p_{FA} + \vartheta_{total}\chi_{sp} - \vartheta_{waste} \tag{3.27}$$

where ϑ_{total} is the amount of admixture, χ_{sp} is the degree of retardation effects by admixture, ϑ_{waste} indicates a reduction in the ability of retardation effects of admixtures due to the physico-chemical bindings offered by C_3A, C_4AF, slag and fly ash as

$$\vartheta_{waste} = \frac{1}{200}(16p_{C_3A} + 4p_{C_4AF} + p_{SG} + 5p_{FA}) \tag{3.28}$$

3.3.3 The effect of reduced free water and its relationship to moisture transport

The reference heat rate of each reaction embodies the probability of molecular collision with which hydration proceeds. Here, it is assumed that the collisions will be dependent on the amounts of free water, ω_{free}, and the nondimensional thickness of the cluster, η_i, made by already hydrated products and unhydrated chemical compounds. The parameter β_i indicates the reduction of hydration rate with respect to the increasing thickness of the cluster made of already hydrated product and the decreasing free water during hydration as

$$\beta_i = 1 - \exp\left\{-r\left(\frac{\omega_{free}}{100\eta_i}\right)^S\right\} \tag{3.29}$$

where the constants r, s are material parameters ($r = 5.0$; $s = 2.4$), also η_i is obtained from

$$\eta_i = 1 - (1 - Q_i/Q_{i,\,\infty})^{1/3} \tag{3.30}$$

where $Q_i/Q_{i,\,\infty}$ defines the degree of hydration of each component. In the past, researchers have debated whether the hydration reaction proceeds topochemically or according to a thorough solution mechanism. While no clear-cut consensus exists, it is believed that in the initial stages of hydration it is the topochemical reaction that dominates, whereas for the later stages when a significant cluster thickness is formed around particles, thorough solution mechanisms of hydration might predominate. For the same reason, small particles are supposed to follow a topochemical behaviour of reaction leading to quick dissolution, while larger particles might continue to hydrate by a thorough solution mechanism. The hydration model adopted in this book does not takes these factors directly into account.

However, the rate reduction factor β_i would be important in the later stages of hydration when a significant amount of cluster has built up around the particles. The amount of free water per unit weight of cement, ω_{free}, in this model is taken as the total condensed water in the developing microstructure. Since the water existing in the hydrate crystals or the gel structure cannot be taken up for hydration, the free water for hydration is taken as the water existing in the capillary pores in a condensed state. In the computational model, it is obtained as

$$\omega_{free} = \frac{\rho \phi_{cp} S_{cp,\,cond} V_m}{C} \tag{3.31}$$

where, ϕ_{cp} is the capillary porosity at a given point and time, $S_{cp,\,cond}$ is the degree of saturation of capillary or larger pore systems considering only condensed water (i.e. completely saturated pores), V_m is the fractional volume of mortar in the mix, and C is the cement content (kg/m^3).

It can be noticed, that ω_{free} is not equivalent to the classical definition of evaporable water. A significant amount of evaporable water exists beside ω_{free} in the hydrating structure, but is not available for hydration. Some of this unavailable water exists in the adsorbed state whereas other exists as a physically bound water in the CSH structure or the water existing in gel pores. The free water is the primary source of the nonlinear coupling between moisture transport and heat of hydration models. The degree of saturation of individual porosity components is obtained as a part of the moisture transport formulation discussed in the next chapter. The total amount of water consumed per unit volume of concrete, $(\alpha W_p \beta)$ due to the chemical reactions with clinker components is incrementally computed at any point of hydration from the following set of stoichiometric equations of hydration.

$$\begin{aligned}
&C_3A + 6H \rightarrow C_3AH_6\\
&C_4AF + 2CH + 10H \rightarrow C_3AH_6 - C_3FH_6\\
&2C_3S + 6H \rightarrow C_3S_2H_3 + 3CH\\
&2C_2S + 4H \rightarrow C_3S_2H_3 + CH
\end{aligned} \tag{3.32}$$

where S means SiO_2.

The heat of hydration generation model described above not only provides us with the heat generated in hydration but also the average degree of hydration of each clinker mineral as well as the amount of water fixed as chemically bound water in hydration. Finally, the temperature distribution and the degree of hydration of concrete can be obtained by applying thermodynamic energy conservation to the space and time domain of interest and using the heat of hydration generation model as

$$(\rho c)\frac{\partial T}{\partial t} = \mathrm{div}(k\nabla T) + H \tag{3.33}$$

where k is the mean thermal conductivity of concrete, ρc is the heat capacity of concrete and H is the concrete heat generation rate computed from various considerations described above (equations (3.21)–(3.32)) as

$$H = CH_C \qquad H_C = \sum p_i H_i \tag{3.34}$$

where C is the powder content per unit volume of concrete, and H_C is the specific heat generation rate of cement as obtained from equation (3.21).

3.3.4 Strength development in high-performance concrete

In engineering analysis, several approaches are used to describe mechanical properties, such as the strength development process. These are: the gel-space ratio concept, the total porosity concept, degree of hydration concept, maturity laws and others. A very good correlation has been recognized between the degree of hydration and development of strength parameters since the beginning of this century. Furthermore, for a fully compacted concrete, strength is found to be inversely proportional to the water-to-cement ratio and expressed by well-known Abram's law. This empirical law states that the compressive strength of concrete f_c equals $K_1/K_2^{W/C}$ where K_1 and K_2 are empirical constants and W/C is the water-to-cement ratio. As with several other brittle materials, it appears that the amount of porosity has a close relationship with the strength. To a certain degree, it is also possible to express the relative strength and porosity of several materials with the same relationship, where relative strength is the fraction of strength of the material at near zero porosity [7]. The water-to-cement ratio of a well-compacted concrete is the main factor, besides the degree of hydration, which governs the porosity of the cementitious matrix, thereby influencing its strength.

These concepts have been furthered here by taking into account the total heat liberated by the major mineral components of the individual powder materials as the parameter that correlates directly with the measured compressive strengths. Therefore, the compressive strength can be probably expressed as a linear weighted summation of the degrees of hydration of the major components of blended cements where the weighting coefficients can be obtained by comparisons of computed results with measured data.

Kato [12] has proposed such a model by reporting that a good fit could be obtained by expressing the differential increase in the strength as a weighted linear summation of the differential amount of heat liberated by the major chemical compounds of blended cement. Despite its simplicity, the proposed model can predict the development of compressive strength of powder materials, rich mortars and concrete successfully [12]. The model defines the incremental average compressive strength df_c' (MPa) as a function of the average degree of hydration α_i of major clinker mineral components of the binder materials in a linear relationship as

$$\begin{aligned} df_c' &= 25\ dQ_{C_3S} + 40\ dQ_{C_2S} + 27\ dQ_{SG} + 40\ dQ_{FL} \\ dQ_i &= w_i\ d\alpha_i \end{aligned} \tag{3.35}$$

where w_i is the weight ratio of the ith clinker mineral in the powder to mix water, $d\alpha_i$ is the incremental increase in the degree of hydration of the ith clinker mineral component. The details and the experimental background regarding the above strength development equation are discussed in Chapter 7.

This strength model can be described as an extended water-to-cement ratio law for concrete composites. It will be used in our combined systems of concrete performance evaluation. The validity of the model has been verified for powder-rich, low-water and aggregate content mixes. Currently developed high-performance concrete [13] with the selfcompacting feature and/or enhanced strengths remains within the range of validity. This implies that cement hydration based models for strength may be valid provided a high cement paste matrix constitutes the concrete composite and macroscopic defect free structures are formed with great uniformity. As a matter of fact, the strength of the concrete composite depends not only on the mechanical properties of the cement-paste matrix but also on the properties of dispersed aggregates and defects inherently present. Then, a generic extension of this strength model should probably take into account the effect of aggregate and initial flaws, and associated fracture mechanics of progressive cracking.

As discussed earlier, it is also possible to express the strength of hardened media as a direct function of porosity that implicitly takes into account the water-to-powder ratio as well as the degree of hydration. An attempt was made to fit experimental data of several mortar and concrete compressive strengths at various stages of hydration with the computed capillary porosity. It was found that the empirical fit gives quite good predictions for concrete under diverse curing conditions, as long as the total porosity is known. An alternative model of mechanical properties based upon the computed capillary porosity of the mortar is discussed in Appendix A. The porosity based model seems to give much better correlation than equation (3.35) [14] though this needs to be investigated further.

Early age development involves a simultaneous occurrence of hydration, structure development and moisture transport processes. In this chapter, two of these phenomena have been discussed. In the next chapter, the formulations of moisture transport will be highlighted as the essential phenomena needed to model concrete performance in structures surrounded by natural and artificial environments.

3.4 Summary

The phenomena of microstructure development and hydration have been analytically formulated. These formulations attempt to describe the salient features of the phenomena arising during the first few days of hardening in terms of the morphological changes occurring in the microstructure and the accompanying heat generation. The highlights of the analytical formulations for microstructure development and cement hydration are:

- The microstructure development is primarily based on the concept of cement particle growth which is computed from the knowledge of the state of maturity of the hydrating matrix. The overall microstructure can be computed based upon a few microscale characteristics, such as the average grain volume-to-surface ratio and the specific surface area of the gel products. The microstructural porosity is primarily subdivided into interlayer, gel and capillary pores.
- Regarding hydration, the cement clinkers are classified into four minerals within which five patterns of hydration are embodied. The hydration rate is directly coupled to the free water and temperature, which represent the thermodynamic environment of cement in concrete. Also, mutual interactions between cement powder components are considered in terms of the effect of alkalinity and ionic concentrations on the rate of hydration reactions.

References

[1] Dhir, R.K., Hewlett, P.C. and Dyer, T.D., Influence of microstructure on the physical properties of self-curing concrete, *ACI Journal*, 1996, **93**(5), 465–471.

[2] van Breugel, K. Simulation of hydration and formation of structure in hardening cement-based materials, Ph.D thesis submitted to Delft Technological Institute, Netherlands, 1991.
[3] Ulm, F.J. and Coussy, O., Strength growth as chemo-plastic hardening in early age concrete, *Journal of Engineering Mechanics*, 1996, **122**(12), 1123–1132.
[4] Kishi, T. and Maekawa, K., Multi-component model for heat of hydrationing of blended cement with blast furnace slag and fly ash, *Concrete Library of JSCE*, 1997, No. 30, 125–139.
[5] Powers, T.C., Mechanism of shrinkage and reversible creep of hardened concrete, Proceedings of the International Symposium Structural Concrete, London, Cement and Concrete Association, London, 1965, pp. 319–344.
[6] Feldman, R.F. and Sereda, P.J., A model for hydrated Portland cement paste as deduced from sorption-length change and mechanical properties, *Mater. Constr.*, 1968, **1**, 509–519.
[7] Neville, A.M., *Properties of Concrete*, Elsevier Science, Amsterdam, 1991.
[8] Copeland, L.E. and Hayes, J.C., Porosity of hardened Portland cement paste, *ACI Journal*, 1956, No. 52–39, 633–640.
[9] Powers, T.C. and Brownyard, T.L., Studies of physical properties of hardened Portland cement paste, *ACI Journal*, 1946–47, No. 1–9.
[10] CEB, *Durable Concrete Structures, CEB Design Guide*, Thomas Telford, UK, pp. 3–7, 1992.
[11] Kishi, T., Shimomura, T. and Maekawa, K., Thermal crack control design of high performance concrete, in Proceedings of International conference on CONCRETE 2000, Dundee, Scotland, E&FN Spon, 1993, Vol. 1.
[12] Kato, Y. and Kishi, T., Strength development model for concrete in early ages based on hydration of constituent minerals, *Proceedings of the JCI*, 1994, **16**(1), 503–508.
[13] Okamura, H., Maekawa, K. and Ozawa, K., *High Performance Concrete*, Giho-do Press, Tokyo, 1993.
[14] Ishida, T., Chaube, R.P., Kishi, T. and Maekawa, K., Analytical prediction of autogenous and drying shrinkage of concrete based on microscopic mechanisms, Proceedings of International Conference on concrete under severe conditions, Tromso, Norway, E&FN Spon, London, 1998.

Chapter 4

Moisture transport in cementitious materials

- Coupled liquid and vapour transport formulations
- A unified isotherm model for arbitrary drying–wetting path
- Nonconventional permeability models for cementitious materials

4.1 Introduction

In this chapter, a moisture transport model for cementitious materials is formulated which considers the multiphase dynamics of liquid and gas phases. The contributions of the diffusive movement of moisture as well as bulk movements of moisture have been combined in the model. The total moisture porosity of cementitious materials is divided into interlayer, gel and capillary porosity components. The transport model considers the contributions from each of these components from a thermodynamic viewpoint. The macroscopic moisture transport characteristics of flow, such as conductivity and isotherms, are obtained directly from the microstructure of the porous media. This is achieved by considering the thermodynamic equilibrium conditions which are typical for extremely slow flow and the random geometrical nature of the pores spread over a large order of magnitude in cementitious materials. Special attention is directed to wetting and drying cycles for practical use of the model for concrete performance under natural environments.

In particular, attention is drawn to phenomena usually not well investigated. These include the hysteresis phenomena observed in the moisture isotherms of porous media and the anomaly in observations of the intrinsic permeability of hardened cement paste. The hysteresis is accounted for by analytical models that consider the effect of entrapment of water in pores. The same models can be used to predict the moisture content of concrete under an arbitrary wetting or drying history. The discrepancy in the intrinsic permeability and nonconformance to the square-root law of moisture sorption is explained by a history dependent viscosity of pore water. Various models have been simplified for computational use by assuming a simplified mathematical porosity distribution of hardened cement paste.

4.2 Coupled liquid and vapour transport formulation

The mechanisms that are considered to be the driving force for species transport in an isothermal porous media are viscous forces, gravity, hydrodynamic dispersion and pressure forces. Viscous forces may arise by virtue of the motion of the species and resulting shear stress acting along the fluid surface due to viscosity. Hydrodynamic dispersion involves mechanical dispersion, and molecular diffusion, which may occur due to concentration and temperature gradients. Pressure forces also arise in an unsaturated system due to capillary forces. The capillary forces depend upon the interfacial tension between different phases, on the angle of contact of the interface to the solids and on the radius of curvature of the interface. In general, it is the difference of energy or potential that causes the mass transport; movement always occurring from higher to lower energy states. In this chapter, we will be primarily concerned with the viscous and capillary driving forces; the capillary force plays the role of the primary driving force for the liquid phase and viscous drag acts as the retarding force. The origin of capillary traction forces in the porous microstructure can be understood by considering the following explanation of the progress of absorption in an unsaturated porous medium.

The process of moisture transport in a porous medium can be divided into five major steps. For the sake of simplicity, let us idealize the porous medium as a single pore with a neck at each end (Figure 4.1). The very first step is adsorption of water molecules to the solid pore surface. Until a complete monolayer of water molecules is formed, vapour flux cannot be transmitted. Under quasi-equilibrium conditions, the thickness of the adsorbed layer depends upon the vapour pressure in the pore. Subsequent to this step, an unimpeded flux of vapour takes place where the vapour behaves like an ideal gas. As the vapour pressure gradually increases due to the vapour flux in pore, condensation occurs at the neck portions of the pore. In this situation, the system is impervious to inert gases and pervious to vapour through a process of distillation, in which necks act as short-circuited paths for vapour movement. This step can also be termed *liquid assisted vapour movement*. In Stage 4 with substantial thickness of the surface adsorbed water formed, flow in thin liquid films may occur. Eventually, there is a transition to Stage 5 where the vapour pressure increases to a point such that condensation occurs within the pores. Subsequent to this step, saturated flow begins in the porous system. The actual porous medium contains pore sizes spread over several orders of magnitudes therefore, at any given humidity, different steps of moisture transport would be in progress in a small representative elementary

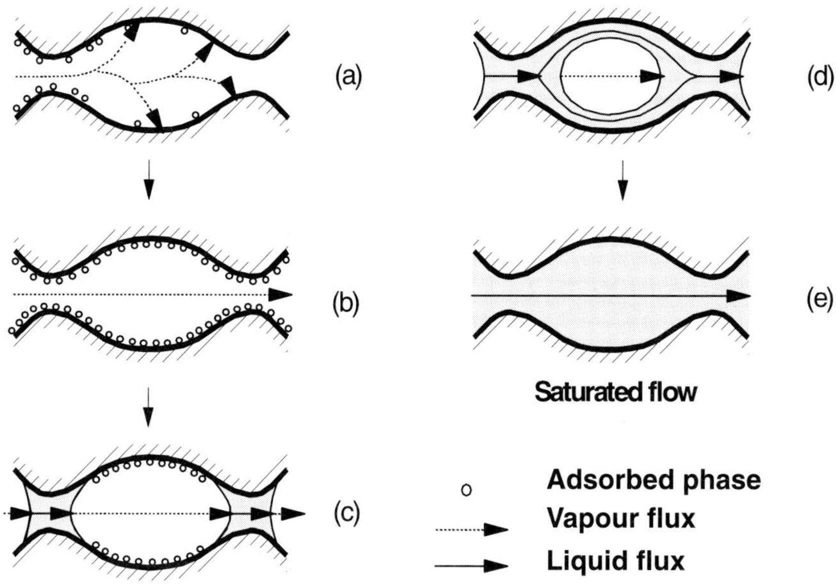

Figure 4.1 Progress of moisture sorption in an idealized pore.

volume (REV). Flow in the saturated pores would be generally in the laminar regime and can be explained by Hagen–Poiseuille flow behaviour. Integrating the moisture transport over all the saturated and unsaturated pores present in the REV of concern, the total moisture transport behaviour in the porous medium can be obtained.

4.2.1 *Mass and momentum conservation*

In this section, general equations of mass and momentum conservation for the moisture transport in an isotropic porous medium are developed under an assumption of an isothermal and nondeformable porous medium. A more general case for moisture transport in a developing microstructure will be discussed later in this chapter. All the equations discussed in subsequent sections are a representation of macroscopically averaged properties, such as velocity, saturation or pressure. In other words, the porous media is viewed as a continuum and balance equations are developed over a representative elementary volume which represents an average of the properties existing at microscopic level. All of the discussion is limited to the one-dimensional case only. However, the extension of these equations to three dimensions is straight forward.

A general notation is used in the formulation for easy identification of various terms. A single subscript denotes a particular component or a phase

Table 4.1

Symbol	Property	Units [dimension]
S_i	Saturation of the *i*th phase	$m^3/m^3[L^3/L^3]$
P_i	Pressure of the *i*th phase	$Pa[L^{-1}MT^{-2}]$
u_i	Velocity of the *i*th phase or component	$m/s[LT^{-1}]$

of concern, e.g. l represents a liquid water or liquid phase, g represents the gaseous phase, a represents the dry air component of the gas phase, and v represents the vapour (steam) component of the air phase. Italic script is used to denote the major properties of interest in the porous media. The three major properties identified here are given in Table 4.1

Mass conservation equations

In the REV as shown in Figure 4.2, only liquid and gas phases are present. They are assumed to fill up the network of pores. Thus, the volumetric compatibility equation can be stated as

$$S_l + S_g = 1 \tag{4.1}$$

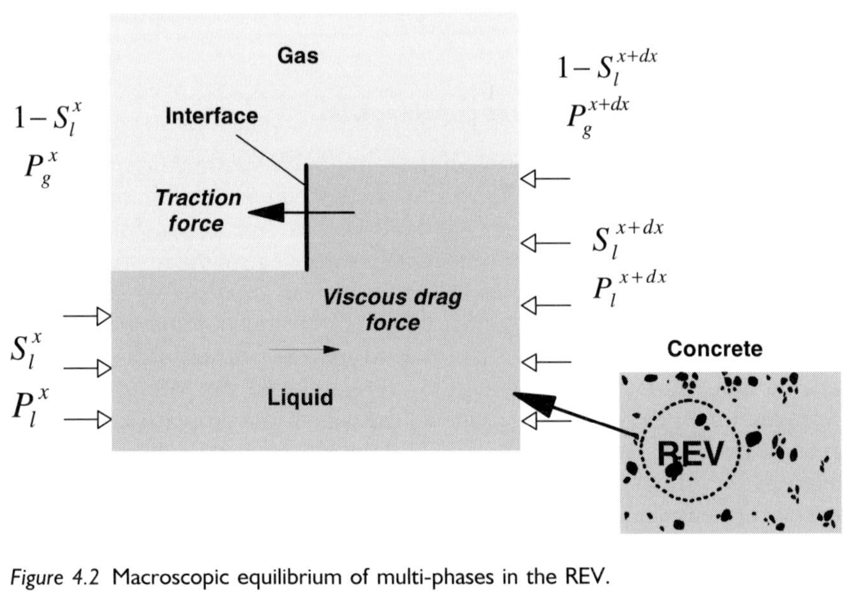

Figure 4.2 Macroscopic equilibrium of multi-phases in the REV.

where S_g represents the fraction of pores filled up by gaseous phase and S_l represents the volumetric saturation level of pores due to liquid water. Using this relationship, mass conservation for liquid and gaseous phase components can be obtained in a differential form as expressed below. Due to the nonreacting assumption made of the porous media, no account is made here of the water that might be consumed in hydration or other biochemical processes. Such general cases will be discussed in Chapter 6 when we review those processes exclusively.

For liquid water

$$\frac{\partial}{\partial t}(\rho_l \phi S_l) + \frac{\partial}{\partial x}(\rho_l \phi S_l u_l) + \upsilon = 0 \tag{4.2}$$

where ϕ is the bulk porosity of concrete, S_l is the volumetric saturation of pores due to liquid water expressed as a fraction, ρ_l is the the density of liquid water, u_l is the averaged velocity of pore water in the REV, and υ is the rate of phase transition (mass transformation) of moisture from liquid to vapour phase per unit porous volume. A moisture loss term due to hydration and changes in bulk porosity will be discussed later.

The gas phase is actually composed of vapour and air components. Therefore, mass conservation of these two components can be added together to give total mass conservation of the gaseous phase. Here, it is assumed that these components, that include H_2O, O_2, N_2 or other possible oxides, are not reacting with the solid porous matrix. Then, we have

$$\frac{\partial}{\partial t}\left(\rho_g \phi (1 - S_l)\right) + \frac{\partial}{\partial x}\left(\rho_g \phi (1 - S_l) u_g\right) - v = 0 \tag{4.3}$$

where ρ_g is the bulk density of gas phase, and u_g is the averaged gaseous phase velocity in the REV. The gas phase velocity can be determined from the molar concentrations of vapour and dry air components present in the gas phase. Next, the momentum conservation for gaseous and liquid phases which describes the equation of motion of multiphases in a porous media is discussed.

Momentum conservation of phases

Primary mechanisms considered here for liquid and vapour transport are pressure forces that may arise in an unsaturated system due to capillary tension or bulk hydraulic heads for a completely saturated system. Also, viscous forces will act on the phase of interest by virtue of the motion of species and resulting shear stresses. If P_l and P_g denote the bulk liquid and gas pressure, we can obtain the momentum conservation of liquid and gas phase as described below (see Figure 4.2).

In the liquid phase:

$$\frac{\partial}{\partial t}(\rho_l S_l u_l) + \left(\rho_l S_l \frac{\partial u_l}{\partial x}\right) u_l + u_l \frac{\partial}{\partial x}(\rho_l S_l u_l) + \nu u_l = -\frac{\partial}{\partial x}(P_l S_l) + P_g \frac{\partial S_l}{\partial x} - u_l K_{Dl} S_l + T_{gl} \quad (4.4)$$

where K_{Dl} represents the average viscous drag force acting over a unit volume of the liquid phase per unit liquid velocity. Here, a linear relationship of viscous drag with velocity is assumed, since the rate of moisture progress is very slow and the flow can be assumed to be well within the laminar regime. Also, T_{gl} denotes the averaged traction force acting over the liquid phase arising due to surface tension forces. This force is assumed to be the primary driving force for the liquid moisture movement in an unsaturated porous medium. Equation (4.4) can be further simplified using the mass balance equation (4.2) as

$$\rho_l S_l \frac{\mathrm{D}u_l}{\mathrm{D}t} = -\frac{\partial}{\partial x}(P_l S_l) + P_g \frac{\partial S_l}{\partial x} - u_l K_{Dl} S_l + T_{gl} \quad (4.5)$$

where $\mathrm{D}/\mathrm{D}t$ is the substantial derivative term and equals $\partial/\partial t + u_l \partial/\partial x$. The viscous drag coefficient K_{Dl} would generally depend on the porous medium microstructural properties and the viscosity characteristics of the liquid. It has to be noted here that this parameter may depend on the interaction of fluid with the surrounding porous media. The resulting implications of this phenomenon on the fluid conductivity as well as overall transport behaviour will be taken up in section 4.4.

As in the liquid phase, momentum conservation for the gas phase can be obtained as

$$\rho_g(1 - S_l)\frac{\mathrm{D}u_g}{\mathrm{D}t} = -\frac{\partial}{\partial x}(P_g(1 - S_l)) - u_g K_{Dg}(1 - S_l) \quad (4.6)$$

where K_{Dg} is the average viscous drag coefficient acting over the unit volume of gas phase and u_g is the average convective bulk gas phase velocity. Bulk pressure gradients are assumed to be the only driving force for the above derivation. The average velocity of vapour and air components of the gas phase can be obtained as

$$u_v = u_g - \frac{D_v}{\rho_v}\frac{\partial \rho_v}{\partial x} \qquad u_a = \frac{1}{\rho_a}(\rho_g u_g - \rho_v u_v) \quad (4.7)$$

where, u_v, u_a are the bulk vapour and air velocity respectively, ρ_v, ρ_a are the bulk density of vapour and air components, and D_v is the bulk molecular diffusivity of the vapour component in the porous media.

4.2.2 Reduction to classical formulation

The system of equations presented in previous sections contains many variables as well as numerous unknown constitutive laws involving various terms. To be able to solve the mass transport problem in concrete without compromising significantly on the accuracy, we must make some approximations concerning the nature of mass transport.

Our first approximation is that the total gas pressure in the porous medium is assumed to be constant and equal to the atmospheric pressure. Since concrete is a porous medium consisting of a very fine porous network, the change in bulk liquid pressure is usually a few orders of magnitudes greater than the change in gas pressure. Therefore, the constant gas pressure assumption appears to be accurate enough for most of the practical cases of interest. Thus, we assume that the only movement in the gas phase occurs due to the molecular diffusion and convective movement is insignificant. Therefore, the flux of moisture due to vapour phase movement can be obtained from equation (4.8) as

$$q_v = -D_v \frac{\partial \rho_v}{\partial x} \tag{4.8}$$

Secondly, we assume that the total mass of vapour present in the REV of concern can be neglected compared to the liquid mass. In other words, to obtain the total mass conservation of moisture (liquid and vapour), the conservation equations for liquid and vapour phases (equations (4.2) and (4.3)) are added up such that the phase transition term is cancelled out and the mass of the vapour component is neglected. Therefore, total mass conservation for the moisture in concrete can be obtained as

$$\frac{\partial \theta_w}{\partial t} + \operatorname{div} q_w = 0 \tag{4.9}$$

where $\theta_w \approx \phi \rho S$ is the total mass of water present in a unit volume of concrete expressed in kg/m^3 (as the mass of moisture in vapour form is neglected). Here q_w is the total flux of moisture expressed in kilogrammes per unit area per second.

Thirdly, it is assumed that the velocity of liquid phase movement is very small and no acceleration of the liquid phase takes place as it moves in the porous medium. Also, due to the quasi-equilibrium conditions of movement, the D/Dt terms of equation (4.5) are neglected. In other words, convective terms of the liquid phase are ignored. With these assumptions, equation (4.5) reduces to

$$-\frac{\partial}{\partial x}(P_l S_l) + P_g \frac{\partial S_l}{\partial x} - u_l K_{Dl} S_l + T_{gl} = 0 \tag{4.10}$$

Furthermore, the equilibrium condition of the nonaccelerating liquid–gas interface enables us to obtain an expression for the traction force, T_{gl}, for a unit volume of porous body as

$$(P_g - P_l)\frac{\partial S}{\partial x} = -T_{gl} \tag{4.11}$$

Thus, there is no driving force when either the saturation gradient is zero or when capillary pressure difference equals zero. Substituting the above expression for the traction force into the approximate momentum balance of equation (4.10), we obtain the following expression for the averaged flux, q_l, of moisture in the liquid phase.

$$q_l = \rho\phi S u_l = -\frac{\rho\phi S}{K_{Dl}}\frac{\partial P_l}{\partial x} = -K_l \frac{\partial P_l}{\partial x} \tag{4.12}$$

where K_l is the saturation content dependent permeability coefficient and is a parameter that represents the ease with which a liquid can flow through the complex porous network. Using equations (4.8) and (4.12), we have total moisture conservation for concrete in one dimension as

$$\frac{\partial \theta_w}{\partial t} = \frac{\partial}{\partial x}\left[D_v \frac{\partial \rho_v}{\partial x} + K_l \frac{\partial P_l}{\partial x}\right] = \frac{\partial}{\partial x}\left[D(\theta_w)\frac{\partial \theta_w}{\partial x}\right] \tag{4.13}$$

where $D(\theta_w)$ is the bulk diffusivity of concrete. This parameter can be obtained by considering the Kelvin's thermodynamic equilibrium condition, which gives a relationship between the relative liquid pressure P_l and the relative humidity of the atmosphere with which it is in equilibrium as

$$\Delta P_c = P_l = \frac{\rho R T}{M} \ln h \tag{4.14}$$

where ΔP_c is the pressure difference due to capillarity, ρ is the density of the liquid, M is the molecular mass of the liquid, R is the universal gas constant, T is the absolute temperature in K, and h is the relative humidity, i.e. the ratio of vapour pressure to the fully saturated vapour pressure. Equation (4.14) is valid only under the assumptions made when defining the capillary pressure difference. The change in liquid pressure given by equation (4.14) is in fact equivalent to the change in free energy of the homogenous bulk phase, i.e. water. Using equation (4.14), we obtain the conventional diffusivity parameter $D(\theta_w)$ as

$$D(\theta_w) = \left(\rho_s D_v + K_l \frac{\partial P_l}{\partial h}\right) \frac{\partial h}{\partial \theta_w} \tag{4.15}$$

where ρ_s is the saturated vapour density such that $h = \rho_v/\rho_s$. The diffusivity parameter, $D(\theta_w)$, combines the contributions of moisture transfer in the liquid as well as vapour phase. Also, the underlying parameter contains all the details regarding the geometrical characteristics of pore structure and the moisture retention characteristics defined by the moisture isotherms. In the conservation equation (4.13), secondary effects, such as gravity terms, are neglected since the capillary pore pressure gradients are usually a few orders of magnitude higher than the gravity terms.

For analysing the generic saturated–unsaturated flow in a porous medium, a pore pressure based formulation is usually desired, since a diffusivity based model is only valid in the unsaturated flow regime. A pore pressure based formulation is also more robust as the transport coefficients are continuous with respect to the pore pressure. In the transformation of equation (4.13) to a pore pressure based formulation, special attention must be paid to the hydration and microstructure development phenomena. During the early age hardening of concrete, pore water content is dependent not only on the pore water pressures but also on the microstructural characteristics of the cementitious matrix. Based on our definition, the rate of change of unit water content $\mathrm{d}\theta_w/\mathrm{d}t$ in concrete can be expressed as a summation of the rate of change of bulk porosity $\mathrm{d}\phi/\mathrm{d}t$, density of water $\mathrm{d}\rho/\mathrm{d}t$ as well as the degree of saturation $\mathrm{d}S/\mathrm{d}t$.

Furthermore, a change in the microstructure will result in a corresponding change in the degree of saturation, S, even if the pore pressures is kept constant. If the cementitious microstructure at a particular stage of hydration could be represented by the microstructure related parameter B_m (Chapter 3). Then, the degree of saturation should be expressed as a function of the pore pressure P as well as of the microstructure, denoted by B_m. The generic form of unit water content of concrete would be therefore expressed as $\theta_w = \rho\phi S(P, B_m)$. If a sink term of moisture content changes in a control volume, e.g. Q_p, which represents the rate of loss of moisture in the porous media due to chemical fixation during hydration, changes in bulk porosity and microstructure or other such processes are also included. Then, equation (4.13) can be extended using the above-mentioned discussions and alternatively expressed in liquid pore pressure terms as

$$\alpha_p \frac{\partial P}{\partial t} - \mathrm{div}(D_P \nabla P) + Q_P = 0 \tag{4.16a}$$

$$\alpha_P = \phi\left(S\frac{\partial \rho}{\partial P} + \rho\frac{\partial S}{\partial P}\right)$$

$$D_P = K_l + D_v\frac{\partial \rho_v}{\partial P} = K_l + K_v$$

$$Q_P = Q_{pd} + Q_{hyd} + Q_{other} \qquad Q_{other} \approx 0 \tag{4.16b}$$

where α_P is the specific fluid mass capacity parameter that approximately represents the amount of water that the concrete can absorb or release for a unit change in the liquid pore pressure potential. It can be noted that the extended definition of unit water content of concrete has been used in deriving the equations (4.16). The terms denoting the change in the shape of the microstructure during hydration contribute significantly in equation (4.16b) and must be included for a microstructure based mass transport theory in a porous media. Here D_P represents the macroscopic moisture conductivity of concrete owing to the pore pressure potential gradients. The moisture conductivity due to temperature gradients is generally quite small compared with that due to pore pressure gradients under normal conditions. This is another reason for not including the temperature effects in the above formulation. Here α_P and D_P are obtained based on microstructure-based models of moisture isotherms and conductivity, respectively. A summarized representation of the approach adopted in this book to obtain various coefficients of equation (4.16) is shown in Figure 4.3. Various constitutive material models of the transport behaviour in concrete are explained later.

The sink term Q_P

The sink term Q_P represents the rate of internal moisture loss due to hydration and related effects. It is obtained by considering the changes in the bulk porosity distribution as well as the amount of water bound to the hardened cement paste structure by dynamically coupling the moisture transport model to a multicomponent heat of hydration model and microstructure formation model discussed in Chapter 3.

During the early age hydration, a gradual reduction in the bulk porosity of concrete takes place, accompanied by a shift of porosity distribution to finer pore radii. Also, a substantial amount of free moisture is consumed in the chemical reactions of hydration; this which gets fixed as chemically combined water of hydration products. The first term of Q_P in equation (4.16b), Q_{pd}, represents the contributions due to the change in the bulk porosity distribution of the hardened cement-paste matrix. It includes changes in the bulk porosity as well as distributions of the interlayer and

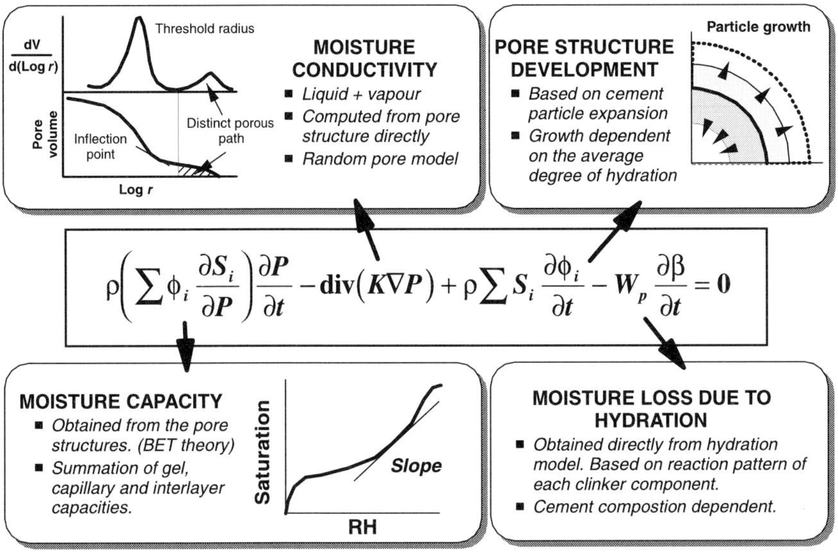

Figure 4.3 Schematic representation of moisture transport modelling in concrete.

gel-capillary components. This reduction in the change of individual porosity results from the continual formation of new hydration products around cement grains, depending on the degree of hydration of each mineral clinker component of cement. The updated porosity at any stage of hydration can be computed from the pore structure formation model. Therefore, the first term is explicitly obtained as

$$Q_{pd} = \rho\left(S\frac{\partial \phi}{\partial t} + \phi\frac{\partial S}{\partial B_m}\frac{dB_m}{dt}\right) \tag{4.17}$$

A rigorous treatment of the effect of the dynamic nature of porosity distribution on the moisture transport formulation can be found in section 4.6. Primary coupling of the hydration and moisture transport phenomenon occurs due to the second term of Q_P in equation (4.16b), i.e. Q_{hyd}, which represents the rate of fixation of pore water as a chemically combined part of the CSH due to hydration and requires an accurate computation especially during the early stages. The rate of moisture loss due to selfdesiccation, Q_{hyd} ($= \mathrm{d}(W_p\alpha\beta)/\mathrm{d}t$), is dependent on the degree of hydration, α, which is in turn dependent on the available free water (taken as capillary condensed water in this model [see section 3.3.3]). The incremental amount of combined water due to hydration is obtained from equation (3.32) (Chapter 3) using a multicomponent heat of hydration model of cement. This interdependency of moisture transport and hydration along with a dynamic change in the cementitious microstructure makes the early age hydration problem dynamically coupled.

Mass transport characteristics of concrete

In the past, numerous models have been proposed that describe the interrelationships of transport coefficients, such as permeability and fluid capacity, of the porous media to the underlying microstructure [1, 2, 3]. Due to these inherent relationships, a description and understanding of concrete microstructure that is relevant to the mass transport phenomena is necessary. In this book, we consider a mathematical representation of the microstructure discussed in Chapter 3. The constitutive models representing the moisture transport characteristics of concrete have been developed for a generic concrete microstructure. However, to convert these models to a computational procedure, a bimodal porosity distribution function is used, whereas key concepts are illustrated using a unimodal porosity distribution; both approaches are discussed in section 3.2.5. In the following sections, we will focus on the evaluation of various parameters in the diffusivity term of equation (4.15) as well as various parameters in equations (4.16) based on the physical nature and characteristics of the cementitious microstructure. A physical approach to obtain transport parameters would enable us to understand the relationships of history dependent conductivity and moisture isotherms with the random nature of the cementitious microstructure.

4.3 Isotherms of moisture retention

Due to the various deterioration processes associated with the presence of moisture in concrete, an accurate prediction of moisture content in concrete under arbitrary environmental conditions is essential for developing a rational and quantitative performance evaluation system for concrete structures. Moisture in the cementitious microstructure can be present in both the liquid and vapour phases. Usually, the volumetric content and thermodynamic behaviour of these two phases of moisture dispersed in capillary and gel pores are determined by the Kelvin's equation, which expresses the thermodynamic equilibrium between liquid water and the vapour phase. The problem of predicting the water content of concrete under arbitrary environmental conditions is addressed at this stage by considering the saturation states of different classes of micropores of concrete and integrating their saturation. A schematic description of the isotherm computation problem and the methodology required is illustrated in Figure 4.4.

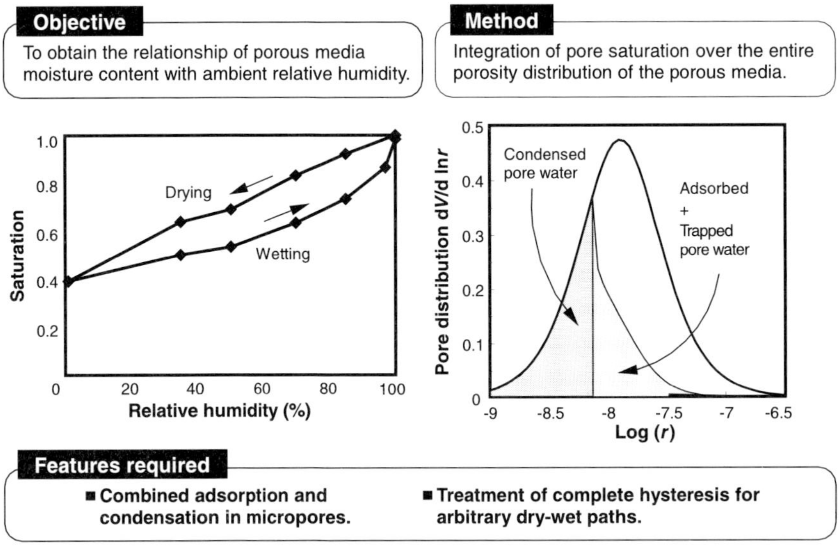

Figure 4.4 Definition of the problem of concrete water content prediction.

4.3.1 *Thermodynamic equilibrium of phases*

Under normal conditions, some of the pores of the cementitious microstructure will be completely filled with water whereas others might be empty or contain a thin adsorbed layer of water molecules on the internal solid surfaces. Under static and isothermal conditions, the liquid moisture dispersed in the cementitious microstructure will be generally in equilibrium with the vapour phase surrounding the completely filled pores. For such a system, a small change in the vapour pressure or relative humidity of the gas phase present in microstructure will cause a corresponding change in the liquid pressure phase, so as to satisfy the equilibrated energy state principle of thermodynamic equilibrium. If we neglect the surface energy effects of the adsorbed water, the change in liquid pressure can be treated as equivalent to the change in the free energy of the homogenous bulk phase, i.e. pore water. By arbitrarily assigning the saturated state of the porous media to a zero capillary pressure or the reference state, we can obtain the pressure difference due to capillarity across a liquid–vapour interface as

$$\Delta P_c = P_l = \frac{\rho RT}{M} \ln h \tag{4.18}$$

where ΔP_c is the pressure difference due to capillarity, and P_l is the liquid pressure (relative to the vapour pressure that is usually small and constant, and hence usually set as zero for computational ease). Equation (4.18) predicts a large drop in liquid pressure associated with a small decrease in the vapour pressure, especially at low relative humidity (RH). From a microscopic point of view, the pressure difference between the liquid–vapour phases would be balanced by the curved vapour–liquid interfaces formed in numerous pores of the microstructure. The stress continuity across such liquid–vapour phase interfaces, which shows a large difference in bulk pressures, can be obtained by considering the interface equilibrium. We will primarily confine our discussion of interface equilibrium to the system of hydrated solid matrix and moisture only. The continuity of stress field across such an interface is modified by the effect of surface tension. For a generic curved interface, the pressure on one side differs from the pressure on the other by (Figure 4.5)

$$p_2 - p_1 = P_g - P_l = \gamma \left(\frac{1}{R_a} + \frac{1}{R_b} \right) \tag{4.19}$$

where γ is the surface tension of the liquid, and R_a and R_b are radii of curvature of the interface in a pair of perpendicular directions. If the surface tension is uniform, the shear stress is continuous across an interface. However, the presence of impurities or surface active agents or temperature gradients can set up the gradients of surface tension. These effects can be neglected for an ideal isothermal system. Equation (4.19) may also be termed the equilibrium form of the surface momentum equation at the microscale. If the vapour phase pressure is set as constant and equal to zero, we obtain the famous Kelvin's equation by combining

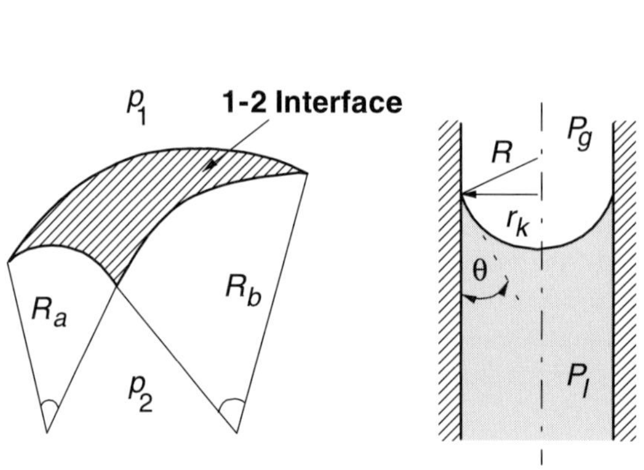

Figure 4.5 Interface equilibrium in an idealized pore.

equation (4.18) and equation (4.19) as

$$P_l = -\gamma \left(\frac{1}{R_a} + \frac{1}{R_b} \right) = \frac{\rho RT}{M} \ln h \tag{4.20}$$

where P_l is actually the relative liquid pressure. The curvature of the interface would depend on the assumptions that we make regarding the geometrical description of how a pore is filled. Clearly, equation (4.20) defines the potential or energy state of water in terms of relative liquid vapour pressure.

Next, we assume that the complex micropores are cylindrical in shape. For an interface formed at such a pore, radii R_a and R_b are equal to the radius of the pore r_s. Furthermore, under ideal conditions, the angle of contact of the liquid water to the solid surface is assumed to be zero. Therefore, from equation (4.20) we have

$$r_s = -\frac{2\gamma M}{\rho RT} \frac{1}{\ln h} \tag{4.21}$$

If the porosity distribution of the microstructure is known, equation (4.21) enables us to obtain the amount of water present in the microstructure at a given ambient RH. This is because, to satisfy the equilibrium conditions, all the pores of radii smaller than r_s would be completely filled whereas others would be empty. By integrating the pore volume that lies below pore radius r_s, we obtain an expression of water content in a porosity distribution V given by equation (3.17) as (Figure 4.6)

$$S = \int_0^{r_s} dV = 1 - \exp(-Br_s) \tag{4.22}$$

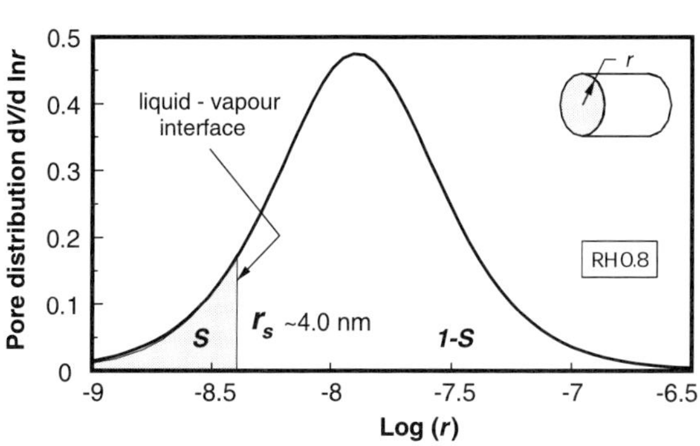

Figure 4.6 Moisture distribution in a porosity distribution.

It might be possible to obtain the water content of the concrete if the same principle as above is applied to a representative microstructure of concrete. Such a model completely neglects a significant amount of moisture in concrete present as adsorbed phase. Also, this model predicts an explicit one-to-one relationship of water content and the relative humidity. That is, the water content as obtained from this model would not be dependent on the history of drying and wetting. However, it is experimentally known that the water content in concrete is different under drying and wetting stages, even if exposed to the same relative humidity. This phenomenon is called *hysteresis* and can be explained by various mechanisms, such as the inkbottle effect, differences in the energy of adsorption and desorption of water molecules to micropore surfaces, and so on. A rational and quantitative method to express this irreversible hysteresis phenomenon for practical evaluations of moisture content in a porous body does not exist. Therefore, in the subsequent sections, we attempt to build a generic path-dependent absorption–desorption model of concrete, which considers the adsorption phenomenon as well as the path-dependence of water content under arbitrary drying–wetting conditions.

4.3.2 Ideal adsorption models and absorption isotherms

The moisture in a cementitious microstructure is generally dispersed into interlayer, gel and capillary porosity (section 3.2.5). The total moisture content of the hardened cement paste can be obtained by summing the amount of water present in each of these components. While moisture content in the capillary and gel components of porosity can probably be explained by physical thermodynamic models of condensation and adsorption, the same cannot be said about the interlayer component of the porosity. This is because the interlayer water probably exists in extremely small spaces between CSH sheets, which are of the order of only a few water molecule diameters thick [4]. The exact behaviour of such water in terms of its entry and exit into this layer structure cannot be explained by a physical theory. For this reason, the contribution of the interlayer component of the isotherm is obtained from direct empirical isotherms and is explained later.

The molecules at the surface of a solid porous matrix have higher energy due to the unattached molecules. In presence of molecules of a different substance, physical binding of these molecules takes place due to van der Waal's forces. This process of physical bond formation is called *adsorption*. Due to this interaction, a layer of finite thickness of molecules of different substances or adsorbates is attached over the adsorbent surfaces. For, hardened cement paste–moisture systems, various models have been proposed that compute the thickness of this adsorbed water film. The Bradley's equation whose constants for silicate materials have been obtained by Badmann [23] can be stated as

$$t_a = [3.85 - 1.89\ln(-\ln h)] \times 10^{-10} \tag{4.23}$$

where t_a is the statistical thickness of the adsorbed water layer in metres, and h is the actual pore relative humidity. Another theory that deals with the general case of adsorption of water molecules over a plane surface was developed by Brunauer, Emmet and Teller and is popularly known as the BET theory [6]. The main drawback of this theory relating to the porous media is that the pores cannot be assumed to be a planar surface of infinite radius. A modification in the original theory to take into account the shape effect of pores was proposed by Hillerborg [7]. With this model, adsorption and condensation phenomena, classically separated as different phenomena, become essentially a single continuous mechanism where no distinction is made between the adsorbed and condensed water. For an ideal case, neglecting osmotic effects, the thickness denoted by t_a of adsorbed layer is given by (see Appendix D for the details)

$$t_a = \frac{0.525 \times 10^{-8} h}{(1 - h/h_m)(1 - h/h_m + 15h)} \tag{4.24}$$

where h_m is the humidity required to fully saturate the pore. If we assume a cylindrical shape of pores then from Kelvin's equation and Figure 4.7, we have

$$h_m = \exp\left(\frac{-\gamma M}{\rho R T r_1}\right) \tag{4.25}$$

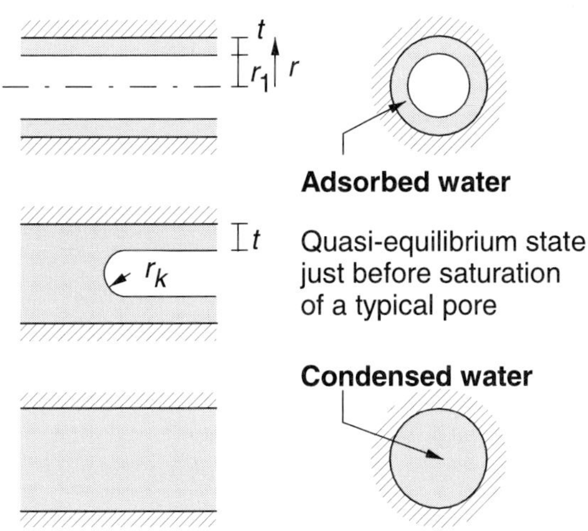

Figure 4.7 Adsorption phenomenon in a single pore.

where $r_1 = r - t_a$ is the actual interface radius and is smaller than the actual pore radius. As the relative humidity inside the pore increases, the thickness of the adsorbed layer increases. At the same time, the humidity, h_m, required to completely fill the pore decreases. Therefore, at some stage the situation becomes unstable and the pore becomes fully saturated. The degree of saturation, S_r, of a cylindrical pore in an unsaturated state can be obtained as

$$S_r = 1 - \left(\frac{r - t_a}{r} \right)^2 \tag{4.26}$$

The degree of saturation of individual pores can be integrated over a given porosity distribution to obtain the total saturation of the porous media. This integration can be significantly simplified if we consider the total saturation as contributions from fully filled and partially filled pores. Therefore, the degree of saturation of a model porosity distribution, V, given by equation (3.17) is obtained as

$$S = \int_0^{r_c} \mathrm{d}V + \int_{r_c}^{\infty} S_r \, \mathrm{d}V = S_c + \int_{r_c}^{\infty} BrS_r \exp(-Br) \mathrm{d} \ln r \tag{4.27}$$

where r_c denotes the pore radius with which the equilibrated interface of liquid and vapour is created. In other words, pores of radii smaller than r_c are completely saturated, whereas larger pores are only partially saturated. It should be noted that the pore radius, r_c, is larger than the pore radius, r_s, as determined by equation (4.21). An adsorbed film of thickness, t_a, exists in other unsaturated pores. Figure 4.8 shows moisture distribution in a pore structure considering adsorption and condensation. Comparing this figure with Figure 4.6, it can be noted that no consideration of adsorption would lead to an underprediction of moisture content, since a significant amount of moisture exists in the pores that are not completely filled but contain an adsorbed layer of water. This difference would generally be larger at higher humidity. Figure 4.9 shows the absorption saturation characteristics of individual pores as well as the moisture profiles in a model pore distribution at different RH.

Let us now consider some computational simplifications. The Hillerborg model [7] of adsorption given by equations (4.24)–(4.26) is implicit in t_a. Therefore, some iterative scheme, such as the Newton–Raphson (NR) method, must be used to generate the isotherms for each individual pore. Moreover, for a given relative humidity, h, evaluation of r_c is not a straightforward task and requires the evaluations of isotherms for each of the pores.

Clearly, this is a daunting task and poses a practical limitation to the use of this model in a dynamic computational scheme where a change in the

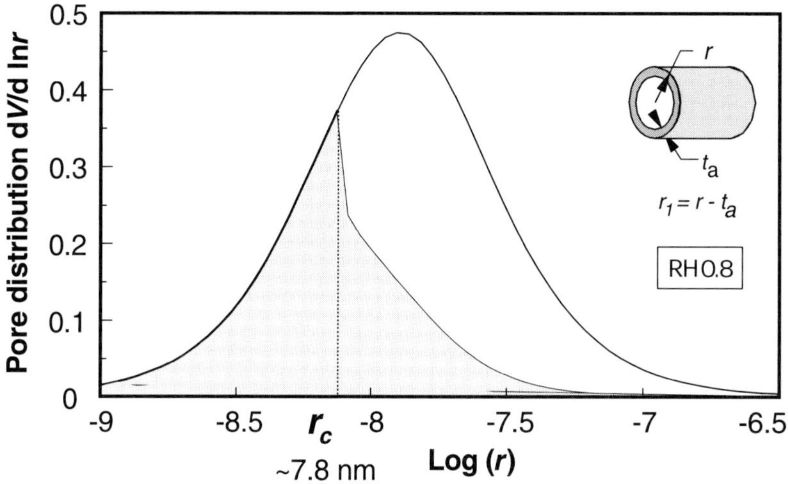

Figure 4.8 Moisture profile in a model pore distribution with unified adsorption and condensation.

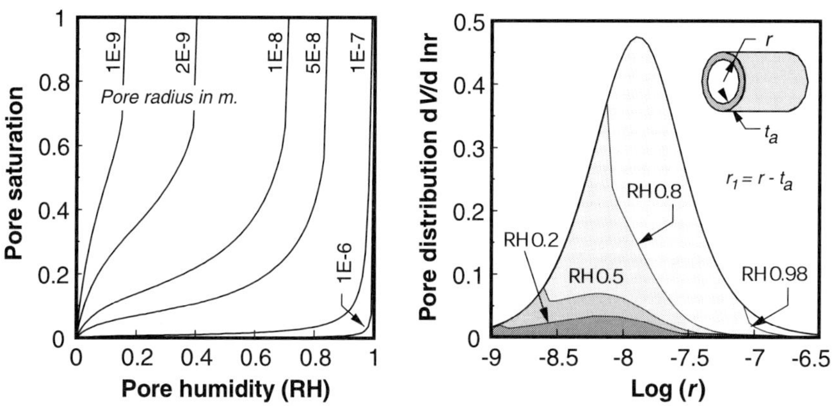

Figure 4.9 Moisture profiles in a model pore distribution at various humidities.

microstructure itself with time makes frequent evaluations of the total isotherms necessary. For this reason, the authors propose a simplification to the scheme for obtaining total saturation, where r_c is obtained from an explicit relationship given as

$$r_c = Cr_s = -\frac{2C\gamma M}{\rho RT}\frac{1}{\ln h} \qquad C = 2.15 \tag{4.28}$$

The constant C has been obtained after numerous comparisons of the analytical predictions of the equilibrated interface radii, r_c, as obtained from the modified BET theory and r_s as given by Kelvin's equation. The relationship holds good over the range of pore radii encountered in cementitious microstructure and does not appear to be very sensitive to the temperature. Also, the contribution from unsaturated pores is obtained not by evaluating the complete integral of equation (4.27) but from a modified expression as given below.

$$S_{ads} = \int_{r_c}^{\infty} S_r \, \mathrm{d}V \approx t_m \int_{r_c}^{r_{0.99}} 2r^{-1} \, \mathrm{d}V = t_m A_s \tag{4.29}$$

where t_m is the thickness of the adsorbed film of water given by equation (4.24) in a pore of radius r_m, r_m is the geometrical mean of r_c and $r_{0.99}$, A_s is the surface area of the pores lying between those of radius r_c and $r_{0.99}$, and $r_{0.99}$ corresponds to the pore radius below which 99% of the porosity distribution exists. Saturation as given by equation (4.29) requires only one iterative evaluation of t_a and gives reasonable estimates of moisture contents compared to the exact but highly computationally intensive equation (4.27).

4.3.3 Computational model of hysteresis behaviour of isotherms

Usually, the absorption and desorption curves of typical moisture isotherms are observed to follow different paths not only in concrete but also other porous media. Irreversibility of isothermal paths of water content in concrete under cyclic drying–wetting condition is shown in Figure 4.10. All desorption curves lie above the adsorption curves and hysteresis loops can be observed. In the past, the thermodynamic equilibrium condition given by equation (4.21) was used to describe the state of moisture in concrete. Such approaches are not adequate to describe the moisture state in concrete, since they fail to address the issue of hysteresis behaviour under generic drying/wetting conditions.

The hysteresis behaviour can be partially addressed if we consider a moisture isotherm model which takes into account the effect of entrapped liquid water in the microstructure during drying. So far, we have considered both the condensed and adsorbed phases of liquid water. For the sake of simplicity, the adsorbed component of moisture will be neglected in the following discussions as it has no effect on the hysteresis model explained later. Moreover, the moisture content contributions of the adsorbed water can be simply added to the contributions from fully saturated pores to give the total saturation. Also, to illustrate the hysteresis model discussions are limited to a model porous media that has a representative porosity distribution as given in equation (3.17).

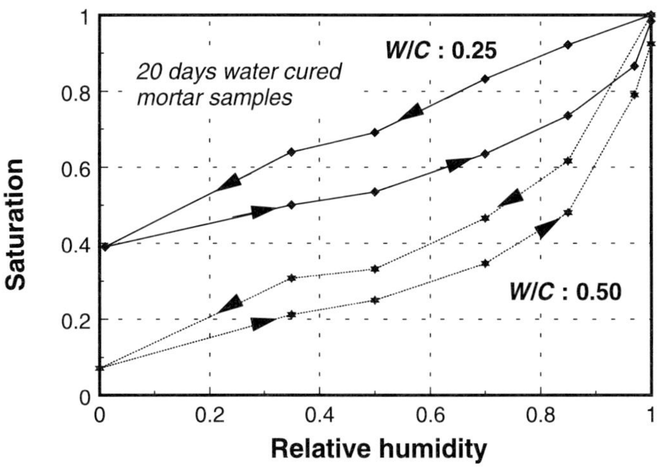

Figure 4.10 Irreversibility of saturation–humidity paths in typical moisture isotherms.

Entrapment of liquid water in the microstructure (inkbottle effect)

As already mentioned, the virgin drying curve of moisture isotherm is always found to be higher than the corresponding wetting curve. To describe this behaviour, we focus primarily on the entrapment of liquid water in larger pores, which might be one of the main reasons of hysteresis.

Owing to the complex geometrical characteristics of a random microstructure, many of the larger pores have interconnections only through pores that are much smaller in pore radii. For such pore systems, there exists a possibility of liquid becoming entrapped in the larger pores for arbitrary drying–wetting histories. The concept of entrapment of pore water in an idealized pore is shown in Figure 4.11, which shows that during drying we can expect some additional trapped water in the pores whose radii r are larger than r_c. Such pores have external openings only through the pores of smaller radii, and therefore cannot lose moisture as long as the connecting pores remain saturated. The additional water trapped in such pores gives rise to the hysteresis effect observed in isotherms. By adding the volume fraction of trapped water, S_{ink}, in such pores, to the volume fraction of water, S_c, as obtained in section 4.2.2, we can obtain the total saturation of the porous media under arbitrary history.

To consider the volume of entrapped water in such pores, we define a probability parameter, f_r, to take into account the hysteresis behaviour (Figure 4.12). The parameter f_r denotes the probability of water entrapment in a pore of radius r larger than the pores of radii r_c. In other words, f_r is the probability that a pore of radius r would be connected only to the pores whose radius is

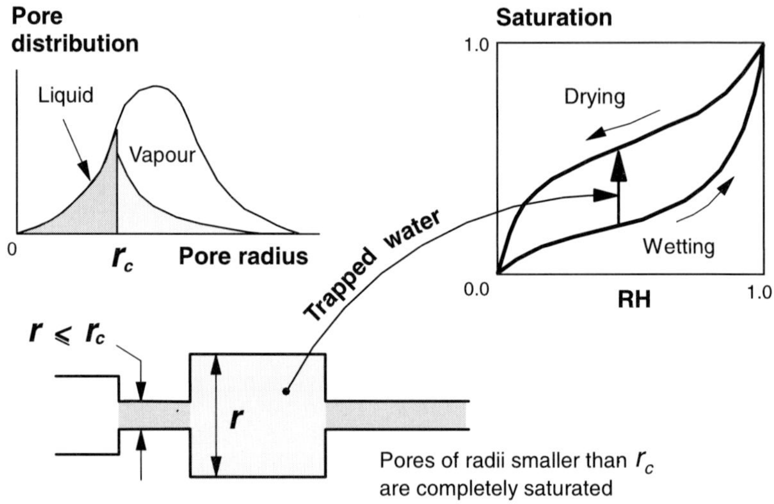

Figure 4.11 Entrapment of water in pores with restricted openings.

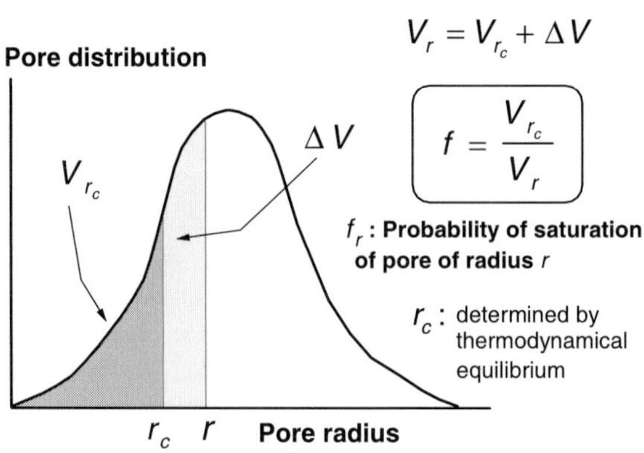

Figure 4.12 The definition of the probability of entrapment parameter, f_r.

smaller than r_c. Obviously, this probability will be dependent on the chance of intersection of the larger pores with the smaller completely filled pores. In its simplest form, we take this probability to be proportional to the ratio of the volume of completely saturated pores and the volume of pores with radii $\leqslant r$. Based on this assumption, a mathematical definition of f_r which statistically represents the structure of pore connection can be obtained as

$$f_r = \frac{V_{r_c}}{V_r} \tag{4.30}$$

where V_{rc} is the volume of the pores with radius $\leqslant r_c$, which is in fact the volume of fully saturated pores, and V_r is the volume of the pores with radius $\leqslant r$ in the porosity distribution. The moisture isotherms of a porous media can be completely described for any arbitrary drying–wetting histories, if we consider primarily four patterns of the drying–wetting isothermal paths. These are:

1. virgin wetting paths, from initially a completely dried state (primary wetting loop);
2. virgin drying paths, from initially a completely wet state (primary drying loop);
3. scanning curves, with a completely dried state as the starting point;
4. scanning curves, with a completely wet state as the starting point.

Primary drying–wetting loops of the moisture isotherm

WETTING STAGE

First of all, we consider the virgin wetting curve, where absorption starts from a completely dry state of the porous medium. Essentially, this condition is similar to the one discussed in section 4.2.2. Under equilibrium conditions, the pores of radii smaller than r_c would be completely saturated, whereas larger pores would only contain moisture in the adsorbed phase, which is neglected here for the sake of brevity. Therefore, the total saturation, S_{total}, of the porous media can be obtained by integrating the individual micropore saturation over the entire porosity distribution function as (Figure 4.13)

$$S_{total} = S_c = \int_0^{r_c} \Omega \, dr = \int_0^{r_c} dV = 1 - \exp(-Br_c) \tag{4.31}$$

where $\Omega = dV/dr$ is a representative porosity distribution function. In other words, the total saturation in this case is obtained by simply summing the condensed pore volumes and does not include any contribution due to hysteresis.

DRYING STAGE

In this case, the degree of saturation, S_{ink}, due to the additional water present in the trapped pores should be computed so that total saturation of the porous media can be obtained. This is done by summing the most probable degree of saturation of all pores having radius greater than r_c, over the porosity distribution ranging from pore radius r_c to infinity.

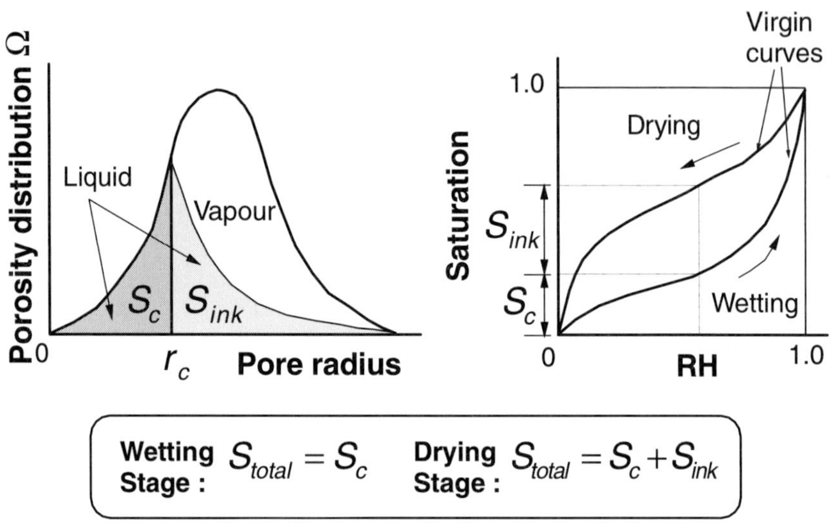

Figure 4.13 Hysteresis along virgin drying and wetting paths.

Mathematically, we have

$$S_{ink} = \sum_{r=r_c}^{\infty} f_r \Omega \Delta r = \int_{r_c}^{\infty} f_r \, \mathrm{d}V = \int_{r_c}^{\infty} \left(\frac{S_c}{V} \right) \mathrm{d}V = -\, S_c \ln(S_c) \tag{4.32}$$

Therefore, the total saturation under virgin drying conditions can be obtained by simply adding S_{ink} to S_c, which is the volume of completely saturated pores below radius r_c, as (see Figure 4.13)

$$S_{total} = S_c + S_{ink} = S_c(1 - \ln S_c) \tag{4.33}$$

It will be noticed, that the additional term $-S_c \ln S_c$ is always positive, which confirms the observation that drying curves are always higher than the absorption curves. It is important to note that this term, which represents the additional quantity of trapped pore water, does not depend on our assumptions regarding the mathematical description of the porosity distributions. In other words, equation (4.33) will give us an estimate of the trapped water, as long as we know S_c or the virgin wetting history of the porous media.

Scanning curves of the moisture isotherms

In the previous section, a computational model was presented for a quantitative representation of virgin wetting–drying behaviour of moisture in concrete. However, actual concrete structures are often exposed to variable environmental conditions, which are not virgin loops. Therefore, we need to extend the application of this concept from virgin wetting–drying paths to arbitrary environmental conditions, such as complicated cyclic wetting–drying conditions. As with the virgin wetting and drying stages, we consider the following two cases for discussion, which cover all possible scenarios of drying–wetting paths.

FROM VIRGIN WETTING PATH TO DRYING AND SUBSEQUENT LOOPS

During the monotonous wetting phase, the absorption path of the moisture isotherm will follow a similar curve to the virgin wetting curve (states a to c in Figure 4.14). However, when the relative humidity decreases, the desorption curve will never return to the virgin curve, but instead trace an inner scanning drying curve (states c to e in Figure 4.14). For scanning drying curves, the inkbottle model similar to the former drying case is applied and the saturation owing to the inkbottle effect, S_{ink}, can be obtained as

$$S_{ink} = \int_{r_c}^{r_{\max}} \frac{S_c}{V}\,dV = S_c[\ln S_{r_{max}} - \ln S_c] \tag{4.34}$$

where r_{max} is the pore radius of the largest pores that experienced a complete saturation in the wetting history of the porous media, and $S_{r_{max}}$ is the the highest saturation experienced by the porous media in its wetting history. In fact, r_{max} is the radius of pore that corresponds to the saturation state, $S_{r_{max}}$, of the porous media (state c in Figure 4.14). In equation (4.34), the summation to obtain the trapped water is carried out up to r_{max} since pores

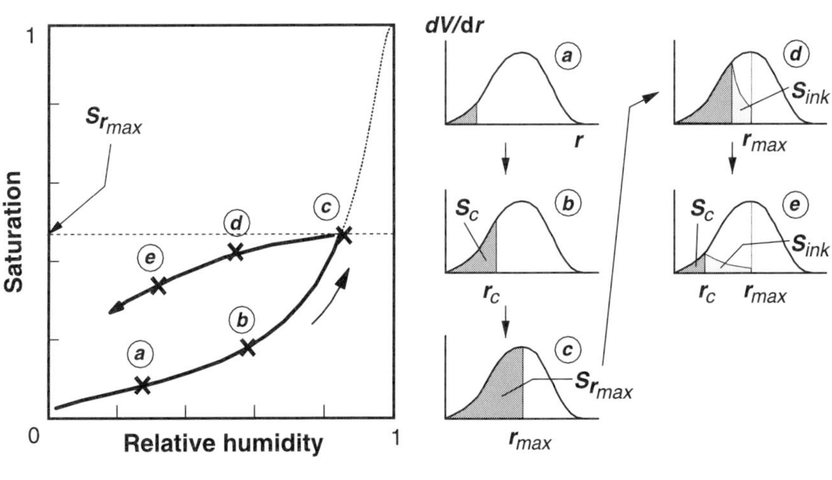

Figure 4.14 Scanning paths of isotherms along wetting paths.

whose radius is greater than r_{max} would have never experienced a complete saturation in their wetting history and therefore would not contain any trapped pore water. Thus, total saturation as a sum of the usual condensed and entrapped water is

$$S_{total} = S_c + S_{ink} = S_c(1 + \ln S_{r_{max}} - \ln S_c) \tag{4.35}$$

In the inner loops, absorption and desorption processes are assumed to be reversible so that the inner scanning curves follow a similar path. In other words, subsequent hysteresis in the scanning curves is neglected. We have assumed the reversibility of the inner loops primarily to obtain a closed form analytical solution to the hysteresis model. Of course, based on the inkbottle concept discussed earlier, it is possible to trace exact hysteresis behaviour in the inner loops. Here, this would require keeping trace of all the turning points in the drying–wetting history and therefore would limit the practical applicability of the model. In the history of the porous media, if wetting proceeds such that r_c exceeds r_{max}, the adsorption path will return to the virgin wetting loop.

FROM VIRGIN DRYING PATH TO WETTING AND SUBSEQUENT LOOPS

As in the previous case, during the first drying, saturation will decrease along the virgin drying loop (states a to c in Figure 4.15). When the ambient relative humidity increases, the scanning absorption loop will be formed, and a gradual moisture re-entry from the filling of smaller pores will take place (states c to e in Figure 4.15). In this case again, the total saturation yields

$$S_{total} = S_c + S_{ink} = S_c + \int_{r_c}^{\infty} \left(\frac{S_{r_{min}}}{V} \right) \mathrm{d}V = S_c - S_{r_{min}} \ln S_c \tag{4.36}$$

where r_{min} is the pore radius of the smallest pores that experienced an emptying in the drying history, and $S_{r_{min}}$ is the lowest saturation of the porous media in its wetting–drying history (state c in Figure 4.15). The integration in this case is carried out from pore radius, r_c, since all the pores below r_c would be completely saturated (Figure 4.15). Moreover, the probability parameter, f_r, in this equation takes $S_{r_{min}}$ as the volume of completely filled pores since the probability of entrapment of liquid in the pores of radii above r_c would correspond to a state of the porous media (state c in Figure 4.15) when it experienced its lowest saturation. The scanning curves of absorption and desorption paths are assumed to be similar owing to the reasons explained earlier. As drying proceeds such that when r_c becomes smaller than r_{min}, the desorption path will again return to the virgin drying loop.

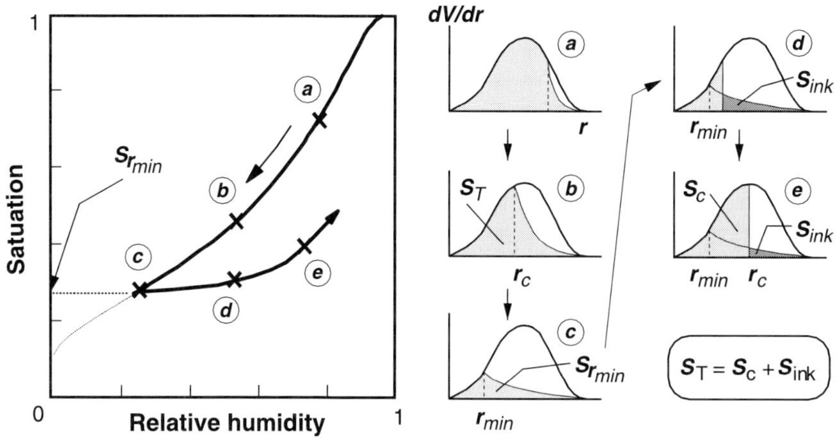

Figure 4.15 Scanning paths along virgin drying curves.

The results obtained above are fundamental in nature and do not depend on the assumptions of the porosity distribution functions. The probability of the water entrapment model for f_r is sufficient to build a complete description of the hysteresis phenomena. Consideration of porosity distribution functions are simply aids in the easy visualization of the phenomena. The final results in the previous derivations do not include the contributions from the adsorbed phase of pore water which we have kept out of the discussion for the sake of clarity. Therefore, the actual saturation of the porous media S is given by

$$S = S_{total} + S_{ads} \tag{4.37}$$

where S_{total} depends on the history of drying and wetting and can be obtained from equations (4.31)–(4.36). The adsorbed component of saturation, S_{ads}, can be obtained from equation (4.29).

4.3.4 *Interlayer moisture and its contribution*

While the discussion in the previous section can explain the hysteresis behaviour of isotherms at high relative humidity, it cannot fully explain the behaviour actually observed for cementitious materials at low relative humidity. Various reasons have been advanced to account for this difference at low relative humidity. One reason is thought to be rooted in the gel structure of hydrated Portland cement (Figure 4.16). The hysteresis is accounted for by the distinction of water present in the CSH structure as interlayer hydrate water. It has been reported that very large internal surface

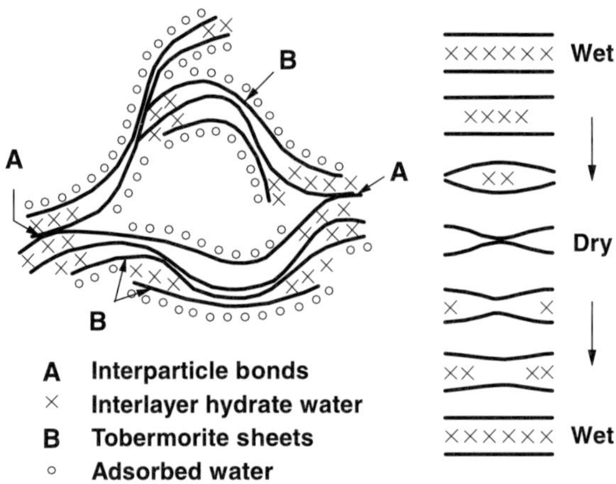

Figure 4.16 A schematic representation of CSH gel structure.

areas as predicted by the BET theory are not possible for pore distributions as obtained by mercury intrusions or other similar means. Moreover, the values of the internal surface areas of hardened cement paste as computed from water adsorption experiments are significantly higher than those computed from nitrogen adsorption experiments. This difference was accounted for by Feldman and Sereda [8] in the interlayer water concept. The interlayer water is a structural component of the gel structure and resides in the layer structures of CSH, accounting for a very high surface area as computed from the BET equation. The removal process of water in such layers is different from its re-entry during sorption and thus accounts for the hysteresis.

As shown in Figure 4.17, gradual removal of interlayer water from the edges occurs in the range of about 30–10% RH. Further removal which probably occurs in the range 10–2% RH is accompanied by a large change in the length of the specimen. The important point to note is that a significant amount of interlayer water resides in the CSH structure even at very low humidity during drying, and severe drying conditions are required to remove this water. During the subsequent wetting process, re-entry of interlayer water takes place gradually with most of it occurring at higher humidity.

Summarizing, during the drying phase most of the interlayer water can be removed only at very low RH whereas the re-entry is gradual over the entire range of RH. The exact thermodynamic behaviour associated with these processes is not clearly understood. Therefore, the authors propose an empirical set of equations to describe the generic drying–wetting behaviour of the interlayer component on the same lines as above by considering four scenarios of the drying–wetting histories. These relationships interrelate the interlayer saturation to the ambient humidity for different cases of drying–wetting paths and are described as (Figure 4.17)

$$
\begin{aligned}
&S_{lr} = h && \text{Virgin wetting loop}\\
&S_{lr} = h^{0.05} && \text{Virgin drying loop}\\
&S_{lr} = 1 + (h-1)\left(\frac{S_{min}-1}{h_{min}-1}\right) && \text{Scanning curve, Initially fully wet}\\
&S_{lr} = S_{max}h^{0.05} && \text{Scanning curve, Initially fully dried}
\end{aligned}
\tag{4.38}
$$

The constants in equations (4.38) have been chosen after comparison with the test results. The model is very simple in that it does not include the effect of temperature. The completely dried state in the above model corresponds to the state of hardened cement paste which is dried to constant weight at 105 °C. Clearly, heating to higher temperatures would lead to a higher loss of the structural component of CSH water. In this book, we have taken the definition of porosity of hardened cement paste as that corresponding to a maximum dried state of 105 °C. Moreover, due to the inclusion of interlayer phase as a part of the total hardened cement paste porosity, the porosity discussed throughout in this book is the suction moisture porosity only.

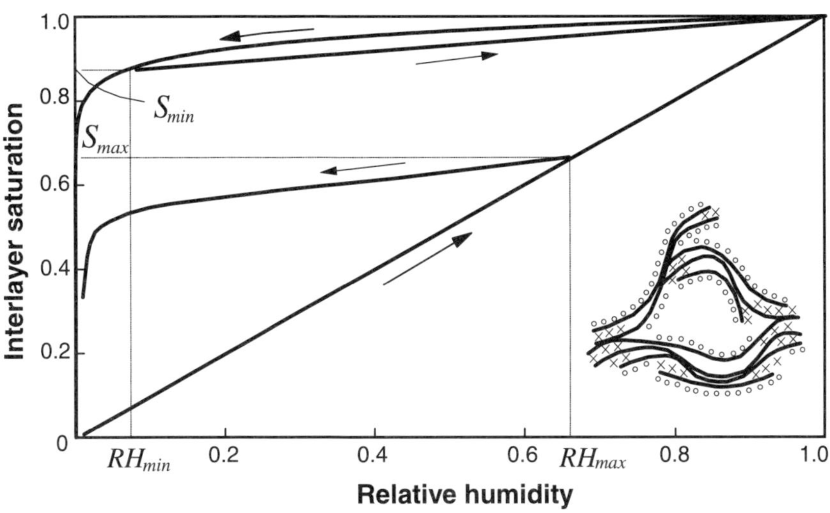

Figure 4.17 The computational isotherm model of the interlayer.

4.3.5 The total moisture isotherm of the hardened matrix

Various models discussed in sections 4.3.1–4.3.4 can be combined together to predict the actual moisture isotherms of the cementitious matrix. In our computational model, the microporosity of the capillary and gel pores are represented by simple mathematical distributions (equation (3.17)). These can be combined together to give the noninterlayer porosity distribution, V_{cg}, of hardened cement paste as

$$V_{cg} = \frac{1}{(\phi_{cp} + \phi_{gl})} \{\phi_{cp}\,[1 - \exp(-\,B_{cp}r)] + \phi_{gl}[1 - \exp(-\,B_{gl}r)]\} \qquad (4.39)$$

The degree of saturation, denoted by S_{cg}, of the porosity distribution given by equation (4.39) at any arbitrary drying–wetting history would be given by the isotherm model discussed in section 4.3.3 (equation (4.37)). Adding this to the interlayer component of moisture gives us the total moisture content of hardened cement paste denoted by θ as

$$\theta = \rho_l\,[\phi_{lr}S_{lr} + (\phi_{cp} + \phi_{gl})S_{cg}] \qquad (4.40)$$

The pore distribution parameters B_{gl} and B_{cp}, as well as the gel, capillary and interlayer porosity required in equation (4.40) are obtained from a combined pore-structure development, hydration and moisture transport model as described in Chapter 3. However, experimental data on porosity and an approximation of B_{gl} and B_{cp} based on MIP results can also be used to predict the moisture content of hardened cement paste for any arbitrary drying–wetting path.

4.4 Permeability of concrete

Darcy's law of saturated flow through porous media states that the flux through a porous medium is proportional to the pressure gradient applied to it, i.e.

$$Q = -\frac{k}{\eta}\frac{\mathrm{d}P}{\mathrm{d}x} \qquad (4.41)$$

where k is the intrinsic permeability of the porous medium, a material property, and η is the viscosity of the fluid. For extremely small rates of flows, typical for porous media, fluid flow characteristics are taken care of by the viscosity term. The parameter k is actually a transport coefficient dependent on the characteristics of the porous media only and is hence a constant. Due to the recognition of the fact that transport coefficients and pore structures of the porous media are interrelated, there are many approaches that build upon this correlation. However, abundant diversity in the geometrical as well as physico-chemical characteristics of porous structures has limited the universal applicability of these methods. The aim of this section is to give a brief overview of the existing microstructure-based permeability models and to propose a simple microstructure-based mathematical model of permeability of concrete, derived in part from work in soil science.

Many of the models put forward to interrelate permeability and microstructure include correction factors that apply only for a group of similar materials. Even for a specific material, such as concrete, different researchers have suggested different microstructure properties and empirical correction factors as parameters in their models to predict permeability. The empirical formulations have traditionally taken the porosity and critical pore diameter as some of the basic parameters. The theoretical approach varies from assuming simple Hagen–Poiseuille flow in the capillaries [2] to numerical models based on computer simulations. In this chapter, we will confine our discussion to those models that consider porous media microstructure as the basis of development. Mehta and Manmohan [9] suggested a rather empirical formula for the prediction of permeability as

$$K_l = \exp(3.8V_1 + 0.2V_2 + 0.56 \times 10^{-6} TD + 8.09 MTP - 2.53) \qquad (4.42)$$

where K_l is the permeability coefficient, V_l and V_2 are the pore volumes in the > 1320 Å and 290–1320 Å range, respectively, TD is the threshold diameter, and MTP is the total pore volume divided by the degree of hydration. It was suggested by Nyame and Illston [10] that the maximum continuous pore radius has a close relationship with permeability. Whereas, median or average pore radius of the porosity distribution have been suggested as the controlling factor by others [12]. Garboczi [11] reviewed many classical and current pore structure theories and suggested the Katz and Thompson (KT) theory [12] as a permeability model with universal appeal. This model as such has no adjustable parameters and all the quantities can be measured experimentally.

$$k = \frac{1}{226} \frac{d_c^2}{F} \qquad (4.43)$$

where k is the intrinsic permeability, d_c is the critical diameter obtained from the threshold pressure in a mercury injection experiment, and F is the formation factor, which is the ratio of brine conductivity in the pore space to porous media conductivity (measured experimentally). It appears that the KT model [12] gives reasonable estimates of intrinsic permeability. However, criticism of this theory has been directed towards the difficulty in measuring accurately the inflection point or threshold pressure in the mercury injection experiment, which due to crack-like pore characteristics of concrete may not even be present at all in some cases. Moreover, the theory is not so successful in predicting the permeability values of mortar with W/C ratios less than 0.4 [13]. A different approach is chosen here for permeability modelling because the KT theory itself does not povide information about the relative permeability or permeability of the partially saturated porous media. This means that the theory cannot be easily used in a computational scheme that combines partially saturated transport, microstructure formation and hydration in concrete [14]. Theoretically, the model described in the next section does not give very different results from those obtained by the KT theory but it is more conducive to a coupled computational scheme of hydration, mass transport and pore-structure formation.

4.4.1 Intrinsic permeability model of a porous medium

With regard to the transport of material within a complex system of randomly dispersed pores, two characteristics are of paramount importance. These are: (1) total open porosity, which constitutes a continuously connected cluster of pores and (2) pore size distribution, which influences the rate of the transport. Any of the pore structure theories (PST) should consider these parameters in one way or another. For intrinsic permeability, we consider both of these parameters, since the probability of permeation not only depends on the total open porosity but also upon the chances of interconnection of pores that contribute to the total porosity.

Let us consider a section of the porous media of small but finite thickness dx as shown in Figure 4.18. Our goal is to deduce the effective intrinsic permeability parameter of the porous medium based on the permeation probability through this section. Let Ω_A represent the average areal distribution function of pores exposed on any arbitrary face cut perpendicular to the flow, such that the total areal porosity ϕ_A is equal to

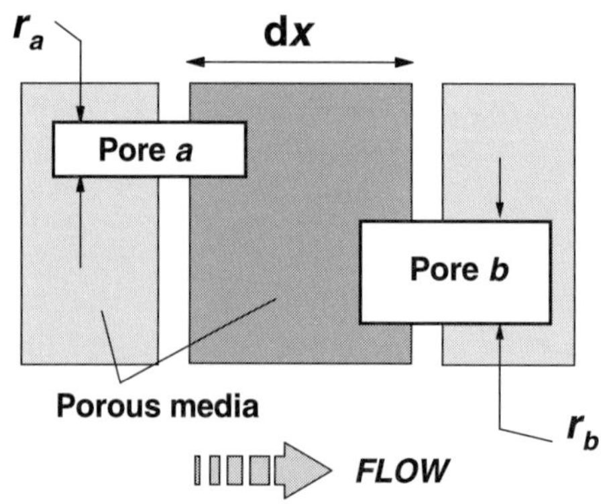

Figure 4.18 Flow across a finite thickness of porous media.

the term $\int \Omega_A \, \mathrm{d}r$. Areal and volumetric porosity are related to each other as

$$\phi = n\phi_A \tag{4.44}$$

where n is the tortuosity factor $= (\pi/2)^2$ for a uniformly random porous media, ϕ_A is the areal porosity, and ϕ is the volumetric porosity of the porous medium. The fractional area of pores of radius r_a and pores of radius r_b at either face of the section can be obtained as

$$\mathrm{d}A_a = \Omega_A \mathrm{d}r_a \qquad \mathrm{d}A_b = \Omega_A \mathrm{d}r_b \tag{4.45}$$

From a statistical viewpoint, the probability of permeation $\mathrm{d}p_{ab}$ through a and b pores of this section is the product of the normalized areas $\mathrm{d}A_a$ and $\mathrm{d}A_b$ (or areal porosity) of the pores of radius r_a and r_b

$$\mathrm{d}p_{ab} = \mathrm{d}A_a \mathrm{d}A_b \tag{4.46}$$

It has to be noted that we assume an independent arrangement of pores in this section. For such a case, integrating equation (4.46) yields the total penetration probability as ϕ^2. If the arrangement of pores in the porous medium were to be constant, the penetration probability $\mathrm{d}p_{ab}$ as given by equation (4.46) would be unity for a pair of pores with the same pore radius and zero for others, resulting in a total penetration probability of ϕ which is the classical assumption usually followed.

Next, we consider steady-state laminar flow through the porous medium, which is idealized as consisting of a bundle of straight capillary tubes. For such a simple model, total flow through the porous medium Q can be simply obtained as [8]

$$Q = -\left(\frac{1}{8\eta}\int_0^\infty r^2 \Omega_A \mathrm{d}r\right)\frac{\mathrm{d}P}{\mathrm{d}x} \tag{4.47}$$

where η is the fluid viscosity, and $\mathrm{d}P/\mathrm{d}x$ is the fluid pressure gradient. In reality, however, pores are not a continuous bundle of capillaries. Moreover, a permeation probability is associated with flow across any section and various other geometrical and scale effects can significantly alter the flow behaviour. The average flow behaviour through a joint of pores of different radii might best be represented by a pore whose radius is the geometric mean of two pore radii. In other words, we assume that the flow through the system of pores of radii r_a and r_b can be represented by an equivalent pore of radius r_{eq} given by

$$r_{eq}^2 = r_a r_b \tag{4.48}$$

This assumption appears to give a more balanced weight to the entire distribution and probably better estimates the permeability compared to the conventional methods that give a rather skewed weightage to the pore distribution [2, 10]. Taking these factors into account, we obtain a modified expression for total flow through the porous media by integrating the flow contributions from all the possible pair combinations of pores of the porous media as

$$Q = -\left(\frac{1}{8\eta}\int_0^\infty\int_0^\infty Kr_{eq}^2\,\mathrm{d}A_a\,\mathrm{d}A_b\right)\frac{\mathrm{d}P}{\mathrm{d}x} \tag{4.49}$$

where K is an unknown parameter that is dependent on the geometrical and scale characteristics of the porous media. Due to the difficulty in the analytical treatment of this parameter, its value has been fixed as unity for the rest of this derivation. Simplifying equation (4.49) and comparing it with Darcy's law gives an expression for the intrinsic permeability of the porous medium as

$$k = \frac{1}{8}\left(\int_0^\infty r\mathrm{d}A\right)^2 \approx \frac{\phi^2}{50}\left(\int_0^\infty r\mathrm{d}V\right)^2 \tag{4.50}$$

where $\mathrm{d}V = n\mathrm{d}A/\phi$ is the incremental normalized volumetric porosity. The only parameter in the above expression is the porosity distribution of the porous medium.

A note of caution is in order when applying equation (4.50) to predict the intrinsic permeability of hardened cement paste, using the MIP method. From a theoretical viewpoint, almost all of the conductivity models emphasize the role of larger pores to some extent. That is, larger pores are the ones that contribute most to the flow if connected continuously. For this reason, it is possible to get erroneous results if, for example, MIP data are used without applying any correction for larger pores. The MIP experimental method itself is not very accurate for larger pore diameters, since the test samples are crushed before testing and might contain inadvertent macro-scale defects. For hardened cement paste samples, the authors recommend that the pores above the first inflection point in the MIP curve or 1000 nm, whichever is greater, should not be used in the analysis. The first inflection point is the point on the MIP curve (where cumulative intruded volume V is plotted against log of the pore radius r) which shows a distinct rise in the volume V of intruded mercury.

4.4.2 *Inconsistency in intrinsic permeability observations*

Ideally, intrinsic permeability is a basic microstructural characteristic and should not be dependent on the fluid used to measure it. However, it has been extensively reported that different values of intrinsic permeability of a porous medium are obtained for different fluids, after applying the density and viscosity normalization. This anomaly can be attributed to the microstructure and pore–fluid interaction. It might occur due to a change in the physico-chemical state of the porous medium or a gradual change in the viscous properties of the fluid being transported, depending on its thermodynamic state or mass transport history. For example, it has been widely reported that even well-hydrated concrete shows a reduction in water permeability over a long period of exposure to moisture [18].

On the other hand, the authors have found that exposure to relatively inert fluids, such as acetone, yields constant permeability or drag coefficients with time. Figure 4.19 shows data on intrinsic permeability of different concrete as reported by various researchers. For this data, permeability measurements for different concrete were taken for gas and water, and converted to intrinsic permeability by applying the density and viscosity normalization. This difference has been attributed to various causes, namely:

1. gas slippage theory [1];
2. plasticity of water (visco-plastic nature of pore liquid);
3. long-range intramolecular force theory [15];
4. yield shear of pore fluid (shearing force acting on pore walls);
5. swelling of hydrates [16];
6. resumption of dormant hydration and self-sealing [16];
7. altered viscosity of pore-water [17].

In addition, experimental problems in measuring extremely small values of water permeability have made it practically impossible to trace the exact cause of the discrepancies in intrinsic permeability as obtained by gas and water measurements. However, it must be pointed out that the permeability values obtained by using gas, oil or even alcohol as the measuring fluid have

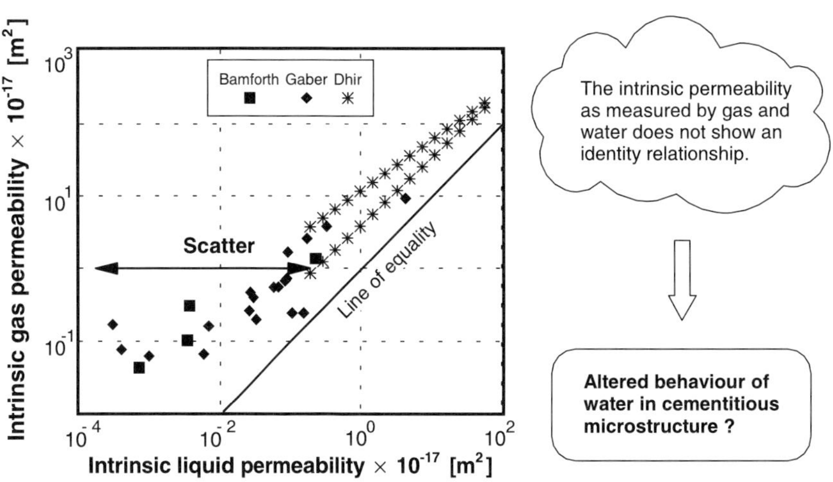

Figure 4.19 Intrinsic permeability of concrete.

traditionally produced more reliable, reproducible and consistent results. It has also been found and reported that usually it takes more than 20 days to obtain steady-state flow conditions in water permeability measurements [18]. Even so, a reduction in water flux or selfsealing behaviour of concrete when exposed to water is also reported [16, 18]. For low W/C ratio mix concrete, the scenario is even grimmer.

For this reason, the authors have tested each of the above-stated causes against a large set of databases of water and gas permeability available in the literature, but none of the theories can explain the discrepancy completely. Furthermore, counter-evidence is also available in the literature, which negates some of the theories, such as the delayed hydration of unhydrated CSH in the presence of water [18]. All of these observations point out the fact that it is the pore water and cementitious microstructure interaction that needs to be properly addressed since, for other fluids, reasonable agreement between observations and predictions by using the rational microstructure-based theories can be obtained. It is believed that the altered nature of pore water, which probably changes the pore water viscosity in a time-dependent manner, is perhaps the most likely cause of large differences in intrinsic permeability as measured by gas and water.

4.4.3 *Modified water conductivity of concrete*

The water present in the cementitious microstructure is far from the ideal condition. Such a nonideal behaviour may result from the dissolution of salts, the long-range forces exerted on the water molecules from the surrounding porous medium, the effects of the polarity of water, or a complex combination of any of these factors. From a thermodynamic point of view, however, the theory of viscosity states that the actual viscosity η of a fluid under nonideal conditions at an absolute temperature T is given by [19]

$$\eta = \eta_i \exp(G_e/RT) \tag{4.51}$$

where G_e is the free energy of activation of flow in excess of that required for ideal flow conditions, η_i is the viscosity under ideal conditions, and R is the universal gas constant. The actual permeability of water can be evaluated if the effect of the nonideal viscosity of water is taken into account in equation (4.50).

Past researchers [20] have reported a viscosity that is about one or two orders higher than the ideal viscosity of water when measured under thin quartz plates (Figure 4.20). The exact physical cause of this phenomenon is not known, but the authors have considered a phenomenological thermodynamic approach to explain this mechanism. When concrete is exposed to water, a transient phase of apparent reduction in the water permeability is observed until it reaches a final value, which is smaller than the permeability expected under ideal conditions. One-dimensional water

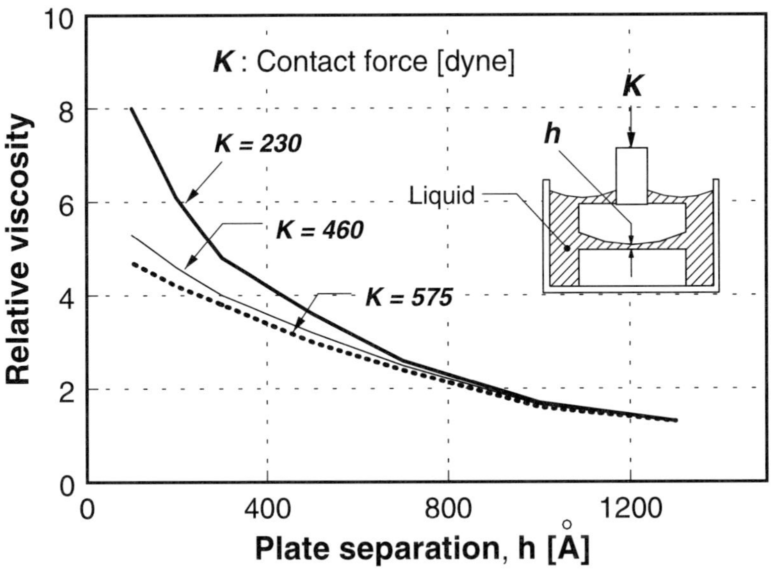

Figure 4.20 Water viscosity and spacing of quartz plates.

sorption experiments in concrete also show a deviation from the square-root law of water absorption after few days [17]. These deviations are not a result of the anomalous diffusion as explained by the percolation theory for porous media where the percolation probability is much lower than critical percolation probability [21]. This is because absorption experiments with alcohol by the authors show a near square-root behaviour, whereas water sorption for the same mortar shows a large deviation from the ideal behaviour (Figure 4.21). It must be noted that drying out the specimens has been shown to restore the initial conditions, which indicates that the time-dependent change of water permeability of concrete is indeed a state-dependent thermodynamic phenomenon.

To account for such time-dependent behaviour, we hypothesize that there exists a pore–water and microstructure interaction which is a time lag phenomenon, bringing about the change in the state of pore water from ideal to nonideal state, and roughly depends on the history of pore humidity. That is, a change in pore humidity brings about a gradual interaction or altered state of the microstructure and the pore–water system, leading to delayed changes in the liquid viscosity and hence observed permeability. This change is not immediate as it would be apparent from one-dimensional water sorption experiment or water permeability experiments. In a nonideal state, the additional energy for the activation of flow, G_e, may be dependent

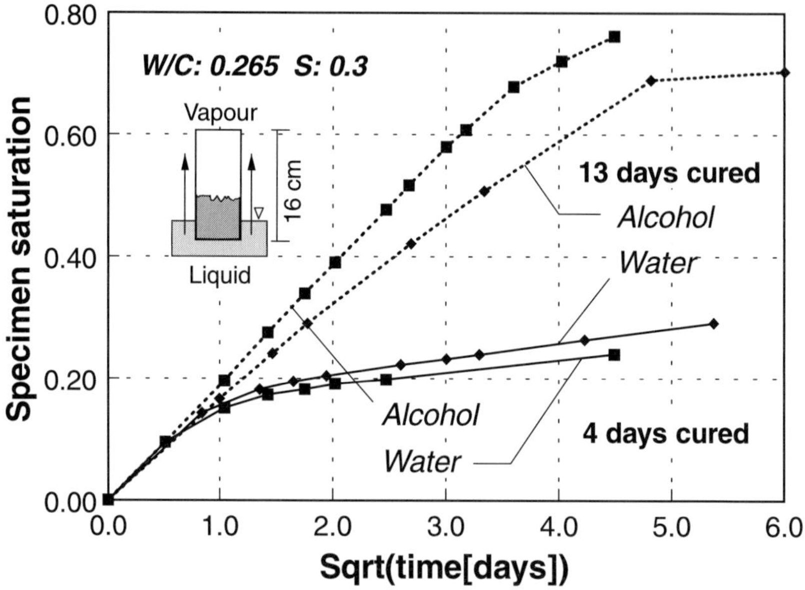

Figure 4.21 Sorption behaviour of alcohol and water in concrete.

on the altered state of the system and can be imagined to bring about a fictitious delayed pore humidity, H_d. In turn, H_d would be dependent on the actual pore humidity, H, history and microstructure characteristics.

In our computational model, this phenomenon is represented by a simple Kelvin chain model (Figure 4.22) where, the dashpot viscosity η_d represents the responsiveness of the microstructure to changes in the actual pore humidity, H (stress analogy), that brings about a change in H_d (strain analogy). Our goal through this model is to obtain the extra energy, G_e, required for the activation of flow at any point and time such that the effective nonideal viscosity of the pore fluid, η, and the effective permeability can be computed. The computational model (Figure 4.23) is given by

$$G_e = G_{max} H_d \qquad \ddot{H}_d + \left(\frac{1+\ddot{\eta}}{\eta}\right) H_d = \frac{H}{\eta}$$

$$\eta = a(1 + bH_d^c) \qquad a = \left[1.59\left(\frac{\phi - \phi_m}{\phi_m}\right) + 0.7\right]^5$$

$$b = 2.5a \qquad c = 2.0 \tag{4.52}$$

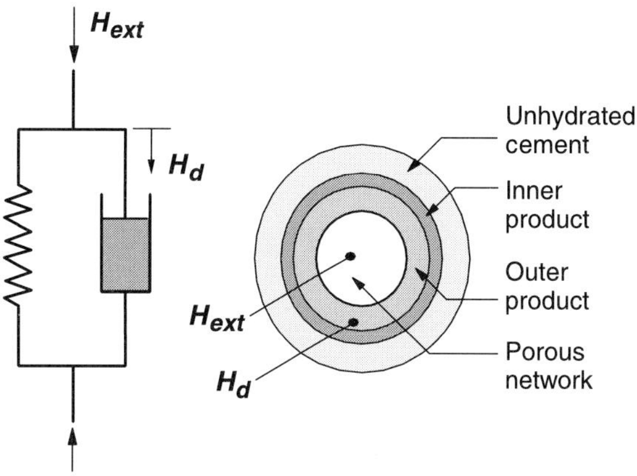

Figure 4.22 The Kelvin chain representation used to compute the pore water viscosity of concrete.

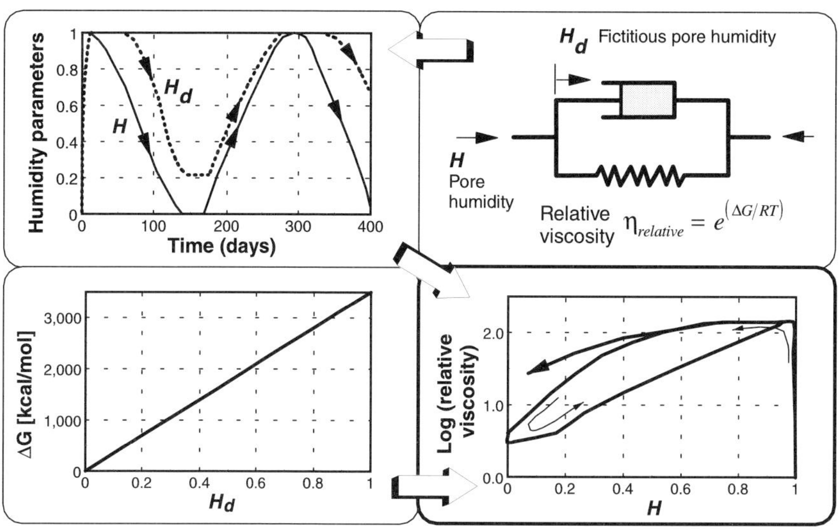

Figure 4.23 Computational scheme to show the relationship of nonideal viscosity of pure water with pore humidity, *H*.

where G_{max} is the maximum additional Gibbs energy for the activation of flow = 3500 kcal/mol (assumed constant for all cases), ϕ is the total suction porosity, and ϕ_m is the porosity accessible by mercury intrusion. The salient point of this model is that the additional energy for activation of flow, G_e, is assumed to be proportional to a fictitious humidity parameter, H_d. Moreover, the responsiveness of the microstructure to an external change represented by η depends on the nature of its porosity distribution. For example, the microstructures that have higher proportions of large pores or capillary porosity are more responsive or react more quickly to changes, and a microstructure with a higher fraction of finer pores reacts slowly to become a nonideal state.

4.4.4 Preliminary verification of intrinsic permeability models

The intrinsic permeability model was verified by comparing the computed permeability value with those obtained from experiments. The experiments were a part of the research project on the long-term performance of concrete in which several kinds of mortars were tested. The W/P ratio (by volume) ranged from 78% to 190%. The sand volume fractions by compaction ratio (S/S_{lim}) ranged from 0.2 to 0.8. Some mortar mixes with high lime volume fractions of up to 17% of the mix volume were also prepared. After sufficient curing, one pair of specimens was exposed to accelerated calcium leaching tests. Gas permeability measurements and pore structure measurements using the MIP method of both the exposed and nonexposed groups of specimens were done. Figure 4.24 shows a comparison of the measured and predicted intrinsic permeability values for many specimens. In the analysis, the porosity distribution lying above 1.1 μm pore radius was neglected for all the specimens due to experimental errors usually associated with large pore radii data in the MIP method. The figure shows a reasonable agreement between the theory and experiment.

4.5 Transport coefficients for unsaturated flows

In most of the concrete structures exposed to the atmosphere, the ingress of moisture and various harmful agents involves the transport of both the liquid and gas phases. For a complete system of modelling concrete performance, we need a transport theory that can deal with the unsaturated conditions of real life concrete structures. In this regard, conductivity of the liquid as well as the vapour phase under arbitrary saturated conditions must be obtained.

4.5.1 Liquid conductivity in unsaturated flows

The discussion in section 4.3 on the permeability of hardened cement paste dealt with the saturated conditions of the paste. The main emphasis was on understanding, in an analytical way, the role that the complex microstructure of cementitious materials plays in the overall transport of fluids. The same concepts can be extended if we take into account the thermodynamic equilibrium conditions which exist in the porous medium as discussed in section 4.2 for an arbitrary state of moisture content. Under unsaturated conditions, only those pores that are completely saturated would contribute to the flow. A knowledge of the contribution to the total flow from these

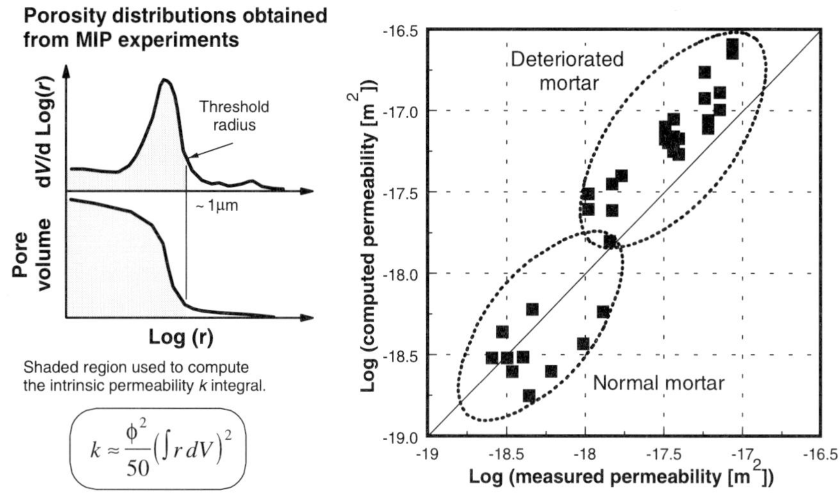

Figure 4.24 Computed and measured intrinsic permeability of mortars. Experimental data from reference [22].

pores would enable us to compute the relative permeability of the porous medium. For a microstructure-based computational system, this means that the intrinsic permeability model equation (4.50) can be used if we know the pore radius, r_c, of the microstructure in which the equilibrated interface of liquid and vapour is created. Obviously, for such a case, the permeation probability should be obtained only across those pores that are completely filled and completely connected, i.e. those pores lying below r_c. For a given pore relative humidity h, r_c can be obtained from equation (4.28). Thus, an expression for unsaturated permeability of water can be obtained as

$$K_l = \frac{\rho\phi^2}{50\eta}\left(\int_0^{r_c} r\mathrm{d}V\right)^2 \tag{4.53}$$

The viscosity term in equation (4.53) corresponds to the time-dependent viscosity of water in a cementitious microstructure and can be obtained from equation (4.52). Although some mechanisms have been proposed to account for the surface flow in the adsorbed water layers, we have neglected the surface flow component completely.

4.5.2 *Vapour diffusivity in unsaturated flows*

At lower humidity, the moisture flux in cementitious materials is dominated by the flux in the vapour phase. The water vapour flow in a pore is governed by mechanisms that depend upon the vapour pressure or relative humidity and radius of the pore, r. For very small pores whose radius is of comparable size to the mean free path length of a water molecule (l_m), an overall reduction in the observed flux occurs due to hindered paths and a greater number of collisions. This phenomenon is known as the Knudsen effect [19] and is dependent on the Knudsen number, N_k, which is the ratio of l_m to the effective pore diameter $2r_e$.

$$N_k = \frac{l_m}{2r_e} \tag{4.54}$$

where r_e is the actual pore radius, r, minus the thickness of adsorbed layer of water, t_a. Using the modified Fick's law, the flux of vapour phase, q_v^r, in a single pore of effective radius, r_e, can be obtained as

$$q_v^r = \frac{\rho_s D_o}{1 + (l_m/2r_e)} \frac{\partial h}{\partial x} \tag{4.55}$$

where ρ_s is the saturated vapour density, D_o is the vapour diffusivity in a free atmosphere. As the saturation of the porous medium increases, fewer paths are available for vapour to propagate along. Therefore, only unsaturated pores should be considered when evaluating total vapour flux through the porous medium. Furthermore, tortuosity effects should be considered to take into account the actual path over which the loss of vapour pressure gradient takes place. At any arbitrary relative humidity, the effective vapour flux, q_v, can therefore be obtained as

$$q_v = \frac{\rho_s \phi D_o}{\Omega} \int_{r_c}^{\infty} \frac{\mathrm{d}V}{1 + N_k} \frac{\mathrm{d}h}{\mathrm{d}x} \tag{4.56}$$

where $\Omega = (\pi/2)^2$ accounts for the tortuosity in three dimensions. The integral in equation (4.56) takes into account the dependence of vapour diffusivity on pore-structure saturation.

4.5.3 *Computational model of moisture conductivity*

The integrals of equation (4.53) and (4.56) are too complicated to be put into practical computational use. Usually, real-life numerical schemes involving finite elements, or other methods, require that conductivity coefficients be computed quite frequently, especially since the problem is nonlinear. For practical purposes, we have simplified the conductivity models discussed above so much so that they can be implemented easily into regular finite-element routines. For a porosity distribution given by equation (4.39), the liquid conductivity, K_l, can be simply obtained by inserting the computational microstructure model of the cementitious matrix into equation (4.53) as

$$K_l = \frac{\phi_{cg}^2}{50\eta}(A_{cp} + A_{gl})^2$$

$$A_{cp} = \frac{\phi_{cp}}{\phi_{cg}} \frac{\{\exp(-B_{cp}r_c)(-B_{cp}r_c - 1) + 1\}}{B_{cp}} \qquad \text{Sly. for } A_{gl} \tag{4.57}$$

Similarly the vapour conductivity is obtained as

$$K_v = \frac{\rho_v \phi D_o}{(\pi/2)^2} \frac{1-S}{1 + l_m/2(r_m - t_m)} \left(\frac{Mh}{\rho RT}\right) \tag{4.58}$$

where t_m corresponds to the thickness of an adsorbed layer of water in a pore of radius r_m, and r_m is the average size of unsaturated pores expressed as a geometric mean. All other symbols have their usual meaning. The above expressions can be readily computed without involving time consuming numerical integration schemes over a complete porosity distribution. The conductivity curve for a typical pore distribution is shown in Figure 4.25, which also shows the reduction in conductivity under long fully saturated conditions. Unfortunately, the dynamic nature of concrete permeability dependent upon the duration and conditions of moisture exposure is a relatively unexplored area in the field of concrete science.

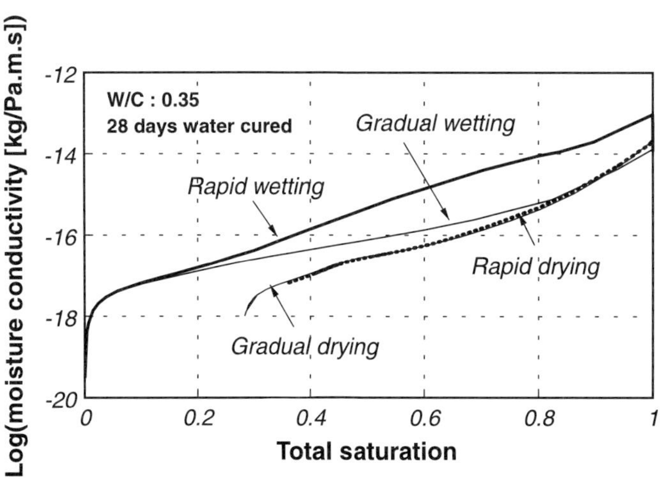

Figure 4.25 Simulated moisture conductivity curves for a typical pore distribution.

4.6 Varying microstructure in the moisture transport formulation

In section 4.2.2, the influence of moisture loss due to self-desiccation and a change in the microstructure on the moisture transport process was briefly discussed. In this section, we focus on a more rigorous analysis by considering the effect of dynamically varying microstructure on the transport behaviour. The changes in the microstructure could be brought about by hydration or any other mechanisms, such as calcium hydroxide leaching or even carbonation. Our main concern will be the determination of the rate of change of the unit water content of concrete as influenced by a change in the microstructure. The change of microstructure will be considered not only in the terms of the change in total porosity but also the changes in the shape of the porosity distribution itself.

Based upon a multicomponent division of porosity, the unit water content of concrete can be expressed as a linear summation of water content of interlayer, gel and capillary porosity components. The change in total water content can therefore be expressed as

$$\frac{\partial \theta_w}{\partial t} = \sum \phi_i S_i \frac{\partial \rho}{\partial t} + \rho \left(\sum \phi_i \frac{\partial S_i}{\partial t} + \sum S_i \frac{\partial \phi_i}{\partial t} \right) \tag{4.59}$$

where ϕ_i and S_i denote the porosity and degree of saturation of the ith porosity component (i.e. interlayer, gel and capillary components). In the case of concrete under normal conditions, the change in pore water density can be neglected due to the very small compressibility of liquid water. The second and third terms in equation (4.59) are the primary contributing terms to the rate of change of the total concrete water content. Generally, in a static microstructure, only the change in pore water pressure (or pore humidity) would result in a corresponding change in the degree of saturation as governed by the thermodynamic equilibrium conditions (section 4.3).

However, in a dynamic microstructure, the degree of saturation will also be influenced significantly by the change in the shape and size of the pore radii. For example, under the conditions of thermodynamic equilibrium and similar pore humidity, a finer porosity distribution will contain more condensed water than a coarser distribution. Therefore, the rate of change of pore distribution parameters that define the porosity distributions of gel and capillary components should be included in equation (4.59). In this treatment, the interlayer properties are assumed to be constant throughout the ageing process of concrete. Therefore, the degree of saturation of the interlayer is essentially a function of the pore humidity or pore water pressure only (equation 4.38). The gel and capillary porosity components are represented by the pore structure parameters ϕ and B (section 3.2.5). The rate of change of porosity of these two components, $\mathrm{d}\phi_i/\mathrm{d}t$, can be obtained directly from the microstructure development model described in Chapter 3, which is based upon the average degree of hydration of reacting powder materials. The last term in equation (4.59) can therefore be resolved in a direct stepwise manner by simultaneously applying the microstructure development model and the multicomponent cement heat of hydration model to the moisture transport model.

To obtain the rate of variation of the degree of saturation of the gel and capillary components (dS_i/dt), we express the saturation as a function of both the pore water pressure, P, and a pore structure parameter, B. Since, the capillary and gel distributions are similar as regards the physical and analytical treatment, we consider a representative porosity distribution that is similar in analytical description to the gel and capillary components. The steps used for this representative distribution can be similarly applied to the gel and capillary components in equation (4.59). The pore water in a typical cementitious microstructure exists in the condensed and the adsorbed states. Therefore, the rate of variation of the total degree of saturation in the representative porous medium is a summation of the rate of change in condensed and adsorbed water, i.e.

$$\frac{\partial S_T}{\partial t} = \frac{\partial S}{\partial t} + \frac{\partial S_{ads}}{\partial t} \tag{4.60}$$

where S_T is the total degree of saturation of the representative porosity distribution, S is the pore water volume fraction in the condensed state, and S_{ads} is the pore water volume fraction in the adsorbed state. Based on the descriptions of the hysteresis nature of condensed moisture and consideration of adsorption, these terms could be expressed as

$$S = S_C + S_{ink}(S_C) \qquad S_{ads} = k(P)U(P, B) \qquad S_C = S_C(P, B) \tag{4.61}$$

where S_C is the volume fraction of the primary condensed water, S_{ink} is the volume fraction of pore water existing due to the inkbottle effect (section 4.3.3), k is the the adsorbed water film thickness, and U is the exposed internal pore structure surface area per unit pore volume. It is important to note the functional dependencies in equation (4.61). Applying these definitions to equation (4.60) results in

$$\frac{\partial S_T}{\partial t} = \left(A\frac{\partial S_C}{\partial P} + U\frac{dk}{dP} \right)\frac{dP}{dt} + Q$$

$$Q = A\frac{\partial S_C}{\partial B}\frac{dB}{dt} + k\frac{dU}{dt} \qquad A = 1 + \frac{dS_{ink}}{dS_C} \tag{4.62}$$

From equations (4.62) it can be seen that in a pore pressure based formulation ((equation (4.16)), the sink terms, Q_P, would contain several additional terms resulting from a consideration of the dynamic nature of the microstructure. In effect, for a microstructure becoming finer, equation (4.62) states that a suitable amount of moisture should be deducted to maintain the moisture mass compatibility conditions. Whereas, for a pore structure getting coarser, a suitable amount of moisture would be released to maintain the thermodynamic compatibility conditions. Of course, for a static microstructure, this formulation would again reduce to the usual moisture transport formulation in a porous medium.

4.7 Summary and conclusions

The moisture transport process occurring in concrete is intriguing, both as a natural phenomenon and as a phenomenon of importance for real engineering applications, such as the durability of concrete under natural environments. This chapter reviewed this important physical process which is of concern to all engineers and theorists dealing with the durability related issues of concrete. Several aspects were clarified through a step by step approach where fundamental properties, such as the hydrated microstructure, were identified as key parameters in the development of the model. It was realized that a general multiphase formulation is necessary to describe adequately the liquid and gas transport phenomena in concrete porous structures. Although, the overall moisture transport process can be apparently reduced to Fick's second law, the underlying mechanisms incorporate a number of varied physical processes. These include, for example, the complex nature of the random porous network where moisture conductivity is not only dependent on the moisture potential but also on the history of moisture potential experienced by the hydrated microstructure. Similarly, a unified adsorption condensation model combined with hysteresis explains the isotherm and moisture retention behaviour of the microstructure.

It can be concluded that the concrete microstructure is the fundamental material property defining moisture transport processes. To this end, quantitative relationships between concrete microstructure and hydraulic properties were envisaged. Computer simulated microstructural networks can quite successfully capture the physics of random geometry. Based on the ideas derived through these methods and supported by theoretical models of random geometry, simple expressions for moisture conductivity coefficients were proposed. As for the moisture retention characteristics, overall microporosity was subdivided into capillary, gel and interlayer porosity. The interlayer water concept as initially proposed by Feldmann and Sereda explains the large amounts of moisture retained by concrete under normal environmental conditions. Furthermore, a theoretical model was proposed to account for the inkbottle hysteresis effects observed in capillary and gel pore systems. The gradual sealing behaviour of even well-hydrated concrete to moisture under prolonged moisture exposure was phenomenologically explained by the history dependent nonideal viscosity of water where additional free energy is required for the activation of flow. It is hoped that this phenomenon will be investigated in detail in future by the research community for enhancing modelling of concrete performance.

References

[1] Bamforth, P.B., The relationship between permeability coefficients of concrete using liquid and gas, *Magazine of Concrete Research*, 1987, **39**, 3–11.

[2] Reinhardt, H.W. and Gaber, K., From pore size distribution to an equivalent pore size of cement mortar, *Materials and Structures*, 1990, **23**, (133), 3–15.

[3] van Breugel, K., Simulation of hydration and formation of structure in hardening cement-based materials, Ph.D thesis submitted to Delft Technological Institute, Netherlands, 1991.

[4] Neville, A.M., *Properties of Concrete*, Elsevier Science, Amsterdam, 1991.

[5] Quenard, D. and Sallee, H., Water vapour adsorption and transfer in cement based materials: a network simulation, *Materials and Structures*, 1992, **25**, 515–522.

[6] Brunauer, S., Emmet, P.H. and Teller, E., Adsorption of gases in multimolecular layers, *Journal of the American Chemical Society*, 1938, Vol. 60.

[7] Hillerborg, A., A modified absorption theory, *Cement and Concrete Research*, 1985, **15**, 809–816.

[8] Feldman, R.F. and Sereda, P.J., A model for hydrated Portland cement paste as deduced from sorption-length change and mechanical properties, *Mater. Constr.*, 1968, **1**, 509–519.

[9] Mehta, P.K. and Manmohan, C., Pore size distribution and permeability of hardened cement paste, in the proceedings of the 7th International Congress Chemistry of Cement, Paris, 1980, Vol. III, VII-1/5.

[10] Nyame, B.K. and Illston, J.M., Relationships between permeability and pore structure of hardened cement paste, *Magazine of Concrete Research*, 1981, **33** (116), pp. 139–146.

[11] Garboczi, E.J., Permeability, diffusivity and microstructural parameters: a critical review, *Cement and Concrete Research*, 1990, **20**, 591–601.

[12] Katz, A.J. and Thompson, A.H., Prediction of rock electrical conductivity from mercury injection measurements, *Journal of Geographical Research*, 1987, **92**(B1), 599–607.

[13] Halamickova, P. *et al.*, Water permeability and chloride ion diffusion in Portland cement mortars: relationship to sand content and critical pore diameter, *Cement and Concrete Research*, 1995, **25** (4), 790–802.

[14] Chaube, R.P., Shimomura, T. and Maekawa, K., Multiphase water movement in concrete as a multi-component system, in Proceedings of the 5th RILEM International Symposium on Creep and Shrinkage in Concrete, Barcelona, E&FN Spon, London, 1993, 139–144.

[15] Luping, T. and Nilsson, L.O., A study of the quantitative relationship between permeability and pore size distribution of hardened cement pastes, *Cement and Concrete Research*, 1992, **22**, 541–550.

[16] Dhir, R.K., Hewlett, P.C. and Chan, Y.N., Near surface characteristics of concrete: intrinsic permeability, *Magazine of Concrete Research*, 1989, **41**(147), 87–97.

[17] Volkwein, A., The capillary suction of water into concrete and the abnormal viscosity of the pore water, *Cement and Concrete Research*, 1993, **23**, 843–852.

[18] Hearn, N., Detwiler, R.J. and Sframeli, C., Water permeability and microstructure of three old concrete, *Cement and Concrete Research*, 1994, **24**, 633–640.

[19] Welty, J.R., Wicks, C.E. and Wilson, R.E., *Fundamentals of Momentum, Heat and Mass Transfer*, 3rd edn, John Wiley & Sons, New York, 1984.

[20] Peschel, G., The viscosity of thin water films between two quartz plates, *Matrx. et Constr.*, 1968, **1**(N6), 529–534.

[21] Stauffer, D. and Aharony, A., *Introduction to Percolation Theory*, Taylor & Francis, London, 1992.

[22] Saito, H. *et al.*, Deterioration of cement hydrate by electrical acceleration test method, *Proceedings of the JCI*, 1996, **1**, 969–974.

[23] Badmann, R., Stockhausen, N. and Setzer, M.J., The statistical thickness and the chemical potential of adsorbed water films, *J. Coll. Interf, Sci.*, 1981, **82**(2), 534–542.

Chapter 5

Concrete: a multicomponent composite porous medium

- Multicomponents of concrete composite
- Interfacial zones and their influence on transport properties
- Moisture transport formulation in multicomponents
- Local moisture exchange among components
- Case study to demonstrate the multicomponent concept

5.1 Introduction

Transport properties of concrete have generally been analysed by treating concrete as an isotropic, homogeneous and uniform porous medium [1, 2]. In this approach, the physical characteristics of various components of concrete, such as aggregates, cement-paste matrix, aggregate cement paste interfaces and bleeding paths, are lumped into a single representative porous medium. This approach can analytically treat most of the cases of practical interest of concrete made of normal mixes and low porosity aggregates. However, it fails to consider the moisture transport behaviour in special concrete, e.g. one that contains high-porosity lightweight aggregates. A different analytical approach is also needed for the cases where, due to high aggregate contents, a distinct aggregate cement matrix interface might be present and enhance the rate of mass transport.

It is the aim of this chapter to identify various components of concrete relevant to moisture transport and combine the moisture transport behaviour in these components into a unified moisture transport theory of concrete. This treatment will not only enable us to consider rationally the different components of concrete but also help us to identify the conditions in which a single uniform porous medium assumption can be applied with confidence to analyse the moisture transport behaviour in concrete. The previous chapter discussed the theoretical aspects of moisture transport modelling in concrete by treating it as a uniform porous medium. This chapter extends those concepts to multicomponents present in concrete. Depending on the microstructural characteristics and distributions of porosity in each of these components, the overall moisture transport behaviour in the concrete might be influenced significantly. It is the aim of this study to incorporate these additional systems of moisture movement into the generic moisture transport formulation of concrete.

5.2 Multicomponents of concrete

The modelling of concrete performance identifies three distinct media existing in concrete through which moisture mass transport can take place in liquid and gaseous phases. These three basic media of mass transport will hereafter be referred to as the components of concrete. These components are: (a) the hardened cement paste (HCP) matrix; (b) the aggregate cement-paste matrix interface and bleeding paths; and (c) the aggregates. These three basic media of moisture transport are schematically represented in Figure 5.1. A brief discussion of each of these components is given below.

5.2.1 *Hardened cement-paste matrix*

Under usual conditions, this component is responsible for the bulk of the mass transport through concrete and provides most of the concrete bulk porosity. It constitutes the various products of hydration and includes the so called gel and capillary pores. The capillary or larger pores, existing primarily between products of hydration, transport most of the bulk flow at high relative humidity (due to the presence of condensed liquid), whereas gel pores play an important role at low relative humidity when most of the flow may occur in the vapour form. The pores in this component usually follow a very tortuous path and constitute a complex network of random

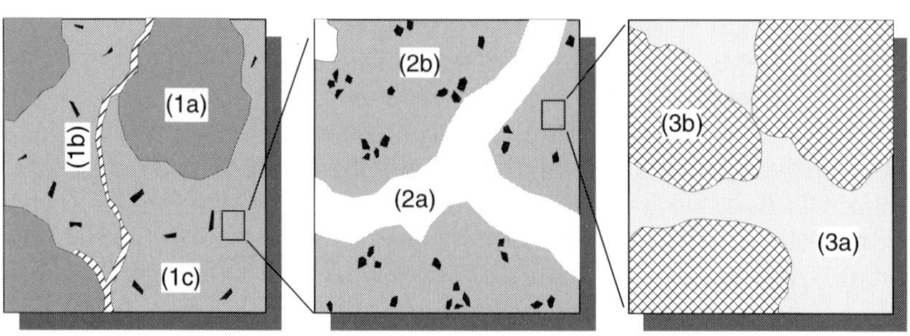

(1a) *Aggregates* (2a) *Capillary pores* (3a) *Gel pores*
(1b) *Channels* (2b) *Hydration products* (3b) *Hydrates*
(1c) *Cement paste*

Figure 5.1 The multicomponents of concrete.

interconnections. The overall tortuosity of flow path may also increase with the addition of aggregates.

5.2.2 Aggregate matrix interfacial zones and bleeding paths

This component includes the large pores existing at the aggregate and bulk matrix interfaces as well as some additional settlement or bleeding paths that might be formed due to insufficient compaction in large water-to-powder ratio mix concrete and/or segregation of water from cement paste mix. The water-to-powder ratio near the aggregate surfaces is usually larger than the bulk water-to-powder ratio, primarily because of the so called *wall effect*, which essentially means that cement powder particles near a wall-like surface cannot pack as efficiently as in the bulk medium, leading to a lower powder density near the aggregate surfaces [3]. This effect, coupled with the one side restricted growth of powder particles during hydration [4], leads to a zone of coarse microstructure and high porosity near the aggregate surface, whose thickness depends on the ratio of the size of the aggregates to the powder particles. For the smaller of the fine aggregates, the wall effect might be practically nonexistent as these particles are stereologically similar to the powder particles, whereas large aggregates would have an interfacial zone of uniform thickness primarily dependent on the powder particle size distribution. The thickness of interfacial zones has been reported to be lying in the range 15–50 μm [3, 5]. The probability that the interfacial zone would form a completely connected path or percolate through the concrete is dependent on the aggregate volume fraction and grading. An analytical treatment of the wall effect and its effect on the zone of higher local porosity near an aggregate surface can be found in reference [3]. The effect of the volume fraction and the grading of aggregates on the probability of percolation of interfaces is discussed later in this book using computer simulation models.

Insufficient compaction and bleeding may also lead to the formation of additional coarser paths. Usually, the total volume fraction of the pores contained in this component is very small. However, owing to their large size, these paths may have subtle influence on the permeation characteristics, such as the diffusivity and permeability of concrete. For example, under fully saturated wet-face conditions of concrete, a large amount of liquid might be carried through this porous network and transferred to the hardened matrix pores and aggregates, thus increasing the rate of moisture transfer within the concrete and thereby aiding the deterioration of the concrete.

5.2.3 Aggregates

The aggregate pores constitute a distinct range of pore-size distribution. Since aggregates are usually dispersed inside the paste matrix, they do not constitute a continuous porous path. Their role in the moisture transport can be idealized as a buffer zone which either stores or releases the moisture during interaction with the surrounding paste matrix. It is believed that no net movement of moisture occurs through these pores, or it is very small compared to the amount of moisture carried through channel or cement paste components. Thus, only local moisture transfer is assumed to take place between the aggregates and the surrounding paste matrix or channels or both. Obviously, this component needs to be considered for moisture transport analysis only when a significant amount of concrete porosity exists in aggregates, i.e. when high porosity aggregates, such as artificially produced lightweight aggregates, are used in the concrete mix. The additional components responsible for mass transport in concrete besides the HCP described above would generally change the pattern and behaviour of moisture transport in the concrete. The difference, as compared to a HCP system only, may vary depending on the distribution characteristics of these additional components.

For example, Figure 5.2 shows the experimental result of water absorption into a long, one-dimensional mortar bar specimen. The plotted lines correspond to the samples made with increasing volume fractions of the sand content (considered as aggregates). It can be observed that the amount of water absorbed initially decreases and then increases with increasing sand content. These results can be interpreted in the following way. First, the addition of a small amount of sand increases the tortuosity of water movement paths and some absorption of water may also take place into the aggregates, thus lowering the rate and amount of water absorbed. In addition, for a low water-to-powder ratio (W/P) mix, aggregates may also provide some degree of stiffness to the overall hardened matrix, thereby reducing macrodefects, such as bleeding paths or channel structures, as compared to the neat cement-paste matrix.

However, at high sand contents, a large number of aggregate-to-aggregate contacts might result in the formation of some directly connected pathways of interfacial zones, through which moisture transport can be very rapid. Increase of W/C ratio also indicates an increase in the rate of water transport into the concrete. It can be primarily attributed to the well-known fact that a high-water content mix produces a concrete of relatively coarser microstructure. Moreover, at very high aggregate contents, the large aggregate matrix interfacial zones or channels may form a continuous path throughout the concrete thereby enhancing the rate of moisture ingress, compared to the increase that can be attributed to the coarser HCP matrix microstructure alone.

5.3 Equilibrium moisture distribution assumption

The problem of moisture transport in concrete as a multicomponent porous medium is quite different from the approach usually taken for a single

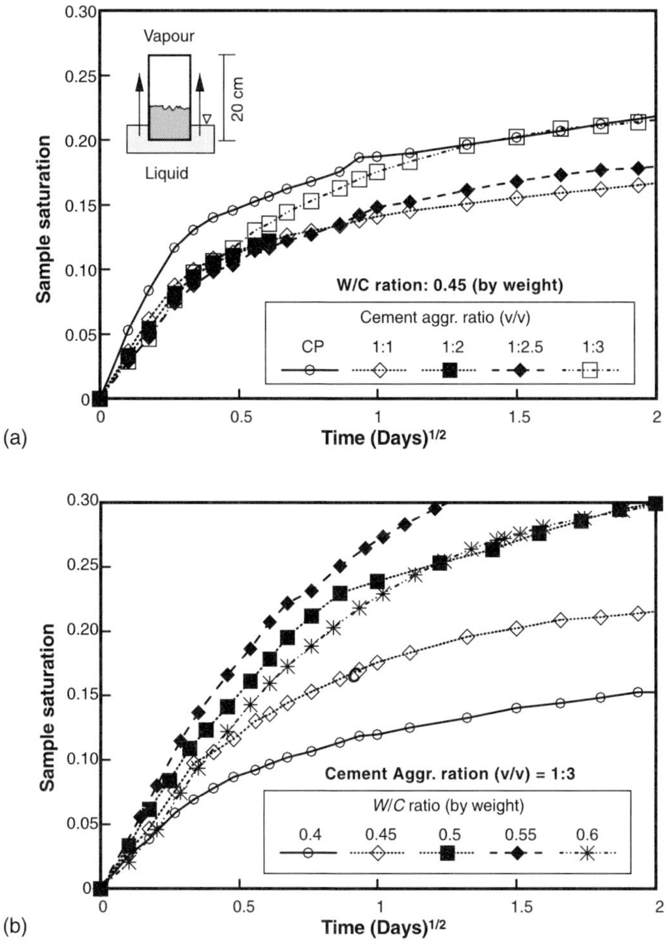

Figure 5.2 The influence of *W*/*C* ratio and aggregate content on water sorption behaviour in mortar prisms: (a) effect of *W*/*C* ratio; (b) effect of aggregate volume fraction.

homogeneous porous medium. For example, apart from the overall mass transport through each of the components, a dynamic intercomponent interaction also takes place that needs to be considered to define the mass transport behaviour completely. Usually, the sorption–desorption behaviour of a porous medium is described by assuming a steady-state thermodynamic condition of multiphases in the porous network. This implies that the distribution of liquid and vapour phases inside the pores is similar to that attained under steady-state equilibrium conditions after infinite time. Consider a very small sample of the porous medium exposed to rapid wetting conditions. Under these conditions, transient stages of moisture distribution in the porous network would be represented as shown in Figure 5.3. The moisture distribution in such a porous network would be as shown in Figure 5.3 after a long time as the system reaches equilibrium conditions. The usual assumptions of equilibrium distribution of moisture for such an infinitesimal element may indeed be true if the moisture redistribution or, more explicitly, the transfer of moisture from coarser pores to finer pores in the porous medium occurs much quicker than the transport of moisture across the representative distance depending on the size of the element of the porous medium.

A generic method that can deal with transient stages of moisture redistribution inside the porous medium would be to treat it as a multiporosity medium where each porosity group consists of the pores exhibiting similar hydraulic properties, such as conductivity and diffusivity. For such a case, the equilibrium assumption of moisture distribution can be applied to each porosity group. Furthermore, taking into account the interaction with other porosity groups, the overall transport behaviour can be obtained. This is especially important if the porosity groups differ significantly in terms of the rate and behaviour of moisture transport through them. For example, aggregate pores constitute such a distinct porosity group wherein the state of moisture might not be in thermodynamic equilibrium with the surrounding HCP matrix, hence leading to localized moisture transfer.

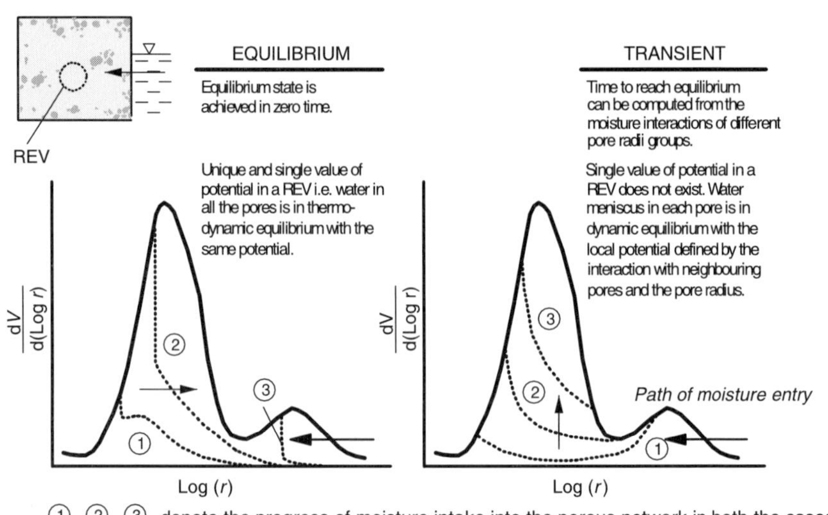

Figure 5.3 Transient stages of moisture distribution in a porous network.

For the case where the channels form a percolated path across the concrete, the channels also need to be considered separately in the system of transport analysis, since their transport characteristics are quite different from the bulk paste. In essence, to get an accurate description of the transport behaviour in a multicomponent porous medium, each component should be dealt with separately by considering intercomponent interactions. This implies that a single value of the thermodynamic potential in a REV of concrete does not exist. For example, in the modelling of concrete microstructure, the state of moisture in a REV is described by the independent thermodynamic potentials of moisture in the three basic components of concrete identified earlier, which are HCP matrix, channels and aggregates. This is opposite to the usual method of analysis of transport in a porous medium, where a single value of potential is adopted by treating the porous medium as a single homogenous porosity system. Consideration of the nonequilibrium moisture distribution assumption as described above may also aid in identifying the range in which the usual single potential assumptions can be used, by applying parametric analysis.

Evidently, it can also be deduced from the discussion so far that for very high intercomponent mass transfer rates, behaviour of a multicomponent porous medium can be represented by a representative single component homogenous porous medium.

5.4 Percolation of aggregate matrix interfaces

The amount of aggregate volume fraction and aggregate porosity play important roles in deciding the overall transport behaviour and hence durability oriented performances of concrete. For high-porosity aggregates, the influence may be in terms of the local moisture transport with the neighbouring HCP matrix. Furthermore, interfacial zones are formed between the different types of aggregate and the cement-paste matrix due to the wall effect and one-sided growth of the cement particles. It has been widely reported that the porosity within the interfacial zone (especially near aggregate surfaces) can be up to three times the porosity of the bulk matrix [3, 4]. Digital computer simulations by Bentz and Garboczi [4] also support this experimental fact. Provided that the aggregate particles are absorbent enough, the interfacial zone will be less porous and its microstructural properties will be more or less similar to the bulk matrix. In most cases, the interfacial zone will be weaker than either the matrix or aggregate and provide easy passage for the transport of degrading agents, such as gases and liquids, thereby influencing the overall transport behaviour in concrete. In this regard, the role of aggregates from the viewpoint of the aggregate matrix interfacial zone needs to be examined further.

The impact of interfacial zones on the transport properties of concrete depends not only on the chance of their continuous interconnection throughout the concrete but also on the microstructural characteristics of the interface itself. For example, for concrete mixes containing high fine powder material admixtures, such as slag and fly ash or highly absorbent aggregates, the microstructural properties of the interface may not be significantly different from the bulk paste. Also, isolated aggregate matrix interfaces should not significantly affect the bulk transport properties of the paste. At low aggregate volume fraction in the composite, such as self-compacting concrete [6], the interfaces would be isolated and the bulk transport properties would be primarily dependent on the HCP micro-structure. At sufficiently high aggregate volumetric percentage in the composite, the interfaces will start to overlap to form a continuous connected path of high porosity through the composite. Such a state can be called the state of *percolated interfaces* in the composite. Such a state of the composite may influence the transport properties so as to reduce the overall service life of the structure, since the durability of a concrete structure is closely related to its transport properties.

5.4.1 Hard core soft shell computer model

The percolation problem of interfaces in a composite has been considered in the past using analytical and computational methods [5, 7]. Bentz and Garboczi [4] used a numerical computer model called the *soft shell hard core* approach to compute the probability of percolation of interfaces for arbitrary volume fractions and grading of the aggregate. The model treats aggregates as impenetrable hard cores, which are concentrically surrounded by penetrable soft shell interfaces (Figure 5.4). In this computer model, aggregate particles with decreasingly small sizes are packed randomly in a mortar sample of given size, such that aggregate boundaries do not overlap with each other until a predetermined aggregate volume fraction in the mortar has been achieved. Afterwards, soft shells or interfaces are placed

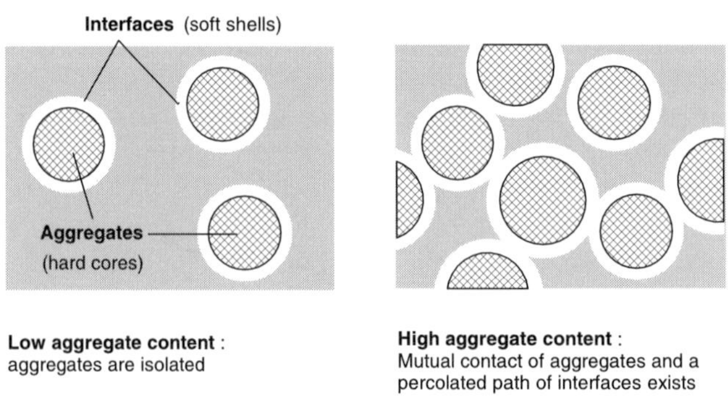

Figure 5.4 Schematic representation of hard core soft shell model.

concentrically around each particle. For this continuum representation of mortar, percolation probability or the probability that there exists at least one continuously connected path of interfaces throughout the composite is determined by applying a *burning algorithm* [5]. The volume fraction of aggregates at which such a percolation of interfaces occurs can be called the *critical volume fraction* of the aggregates. As the aggregate volume fraction in the mortar is increased, a larger number of interface contacts occur. This effectively leads to a larger number of interfaces becoming connected to the initial percolating backbone of interfaces and subsequently, at high enough aggregate volume fractions, all the interfaces would be percolating. Using this computer simulation model and reverse data analysis of MIP results on mortars of increasing sand volume fractions, an interface thickness value of about 15 μm was suggested.

In the hard core soft shell model, a constant thickness of aggregate matrix interface was assumed. This might be true if the aggregate particle size is very large compared to the cement particles, but it may not give a true representation when the composite contains a large volume fraction of finer aggregates. Interface thickness dependence on aggregate particle size has also been reported in the literature [3, 8]. For this reason, the authors carried out parametric continuum computer simulations of hard core soft shell models to study the percolation behaviour in a composite with various assumptions regarding the interface thickness and its dependence on the aggregate particle size. Moreover, an approximate quantitative estimate of the composite transport properties as a function of aggregate volume fractions was also obtained. The methodology adopted by the authors is roughly similar to that adopted in reference [4] and will be described in brief here.

For a given particle size grading and volume fraction of aggregate in the composite, particles were randomly placed in a unit cell (1 cm^3) of the composite, with larger particles placed first, so that no overlap of particles occurred. Afterwards, soft shells of interfaces of predetermined thickness were placed around the particles concentrically. The thickness of the interface for each case study was based upon the model assumptions to be followed. Periodic boundary conditions were adopted to avoid any spurious effects resulting due to the finite size of the computer model. This means that any particle extending outside any face of the cell would intrude into the cell from the opposite face. A typical computer model contained 6000–15 000 aggregate particles in the unit cell of the composite (mortar). Upon completion of the random packing of particles, clusters of particles connected to each other through soft shell interfaces were identified [7]. A cluster here is defined as a group of particles or hard cores in which a continuous path exists through their corresponding interfaces or soft shells by virtue of mutual overlapping. For example, in a case where all the particles are isolated, the number of clusters would equal the number of particles in the composite, since no overlap of interfaces occurs. On the other hand, at complete percolation, all of the particles would belong to a single cluster only. The aggregate volume fraction, when at least one cluster spanning the whole of the composite cell exists (left to right or top to bottom), is termed the *critical aggregate volume fraction*, G_{crit}. The cluster concept is similar to the one adopted in percolation problems of random lattices. It helps in accurately computing the volume fractions of interfaces that are part of the percolating backbones. The approach of direct volume fraction computation of percolating clusters, adopted by the authors, seems to be more accurate than the grid point count method [4, 5].

The computer model was verified by running test cases for a test sand size distribution and similar interface thickness as described in reference [4]. The results were comparable in terms of the critical sand volume fraction and also the variation of interface volume fraction percolated as a function of sand volume fraction. The important task in these computer simulations is to identify an approximate value of G_{crit} to serve as a guiding value for practical use and check its effect on the macroscopic transport properties, such as conductivity. Moreover, by using these computer simulation methods, the trends in the relationship between the interface thickness and the sand particle size in a mortar composite can be investigated. Of course, this relationship might also be dependent on numerous other factors, such as curing temperature, cement powder particle characteristics, etc., but the generic tendency will be similar for all such cases, since the stereological characteristics of the constituents of the mortar composite will be similar.

5.4.2 Model results and discussion

The continuum computer model of a composite mortar was used to study the percolation and conductivity behaviour of a composite based upon different models of interface thickness. In the first case, a constant interface thickness and monosized sand particle distribution was considered. For this case, interface percolation curves were obtained for different interface thicknesses. Comparing this case with the experimental results on conductivity of mortar specimens, an appropriate interface thickness corresponding to the critical sand volume fraction can be obtained and important features of the nature of variation of interface thickness with the sand particle size can be observed. In the second case, a grading of sand particles was assumed and the interface percolation curves were obtained as a function of the interface thickness to particle radius ratio and sand volume fraction. The third case involves studying the interface percolation behaviour of a real composite material made from the sand grading given in reference [4] and the differences observed when using the interface thickness assumption. As pointed out earlier, an interface thickness of 10–15 μm has been suggested by Bentz [4]. Considering this as a reference value for the interface thickness, and from the observations of the three cases described above, it may be possible to formulate a tentative model of interface thickness that is dependent on the aggregate particle size when all other factors are constant.

Figure 5.5 shows the results of computer simulations for a 1 cm^3 mortar specimen, specified as in the first case above. The graph shows the volume fractions of interface that percolate across the composite mortar as a function of the sand volume fraction. Here, the interface thickness is considered in terms of the ratio of interface thickness, t, to the sand particle size, r. Below the critical sand volume, aggregate matrix interfaces are isolated. At the critical sand volume, percolation of the interfaces takes place. However, only a fraction of the total interface volume in the composite is connected to the percolating backbone. As sand volume is increased in the composite, a larger fraction of these interfaces becomes connected to the percolating backbone and, ultimately, all of the interfaces constitute a percolating system, whence percolating interface volume fraction is unity. In Figure 5.5, the sand particles are all assumed to be of the same size. Furthermore, the effect of interface thickness is considered in terms of the t/r ratio, since it gives a more fundamental insight into the percolation behaviour of a monosized particle system, independent of the size scaling effects. As can be observed from the simulation results, a relative increase in the interface thickness compared to the particle size leads to an earlier percolation of the interfaces. This is because the chances of overlap in

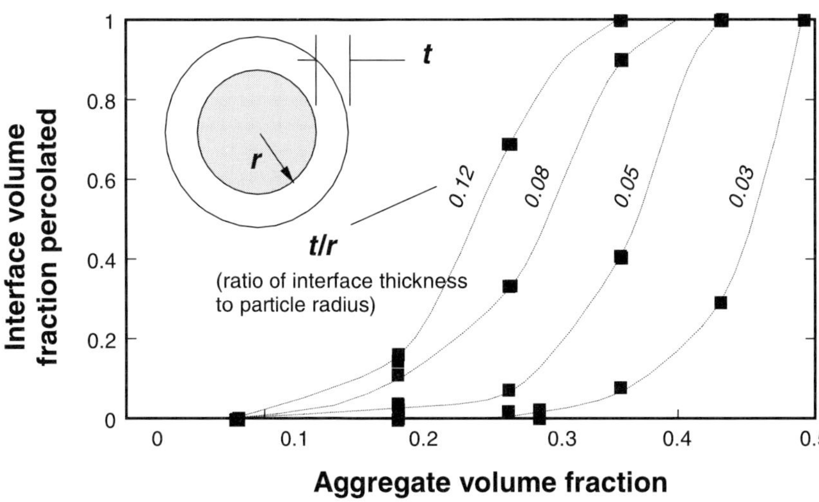

Figure 5.5 Computed percolation curves of a composite containing spherical, equally sized aggregate particles.

interpenetrable soft shells increase with the increase of t/r ratio leading to an earlier percolation.

The influence of interface percolation on transport properties, such as gas permeability of a composite, was studied by Ishida and Maekawa [9] by means of experiments done on model mortars. In the experiments, glass beads of uniform sizes were used as a replacement for sand particles. This guaranteed the same size and uniform properties of the particles in the composite. Subsequently, gas permeability data of the mortars of a similar water-to-cement ratio and curing conditions were obtained for a number of glass bead volume fractions. Similar tests were repeated for different sizes of glass bead whose diameters ranged from 0.05 mm to 6 mm. In this way, the effect of the size and amount of glass beads on the permeability, and hence of the percolation properties on the mortar composite, could be systematically measured. The details of the experimental methods and mix proportions can be found in reference [9]. To infer the interface percolation characteristics from such data sets analytically, the overall conductivity of the composite must be expressed as a function of the percolated volume fractions of the interface. In addition, a knowledge of the conductivity of interfaces and the bulk matrix is also required.

In its most simple form, the overall conductivity of a composite normalized with respect to the volume of the bulk matrix can be expressed as the averaged conductivity of the interfaces and the matrix, where the averaging is done with respect to their relative volume fractions. Expressed mathematically, if k_{in} and k_{mx} denote the permeability of the interfaces and the matrix, V_{in} and V_{mx} denote the relative volume fractions of the interfaces and the matrix respectively at an arbitrary stage of percolation, such that $V_{in} + V_{mx} = 1$, and G denotes the aggregate volume fraction, then the bulk conductivity of the composite k_b normalized with respect to the matrix volume can be obtained as

$$k_b = V_{in}k_{in} + V_{mx}k_{mx} \tag{5.1}$$

where it is assumed that the interfaces and bulk matrix behave independently as regards the transport behaviour. Since we are trying to interpret the permeability data obtained by applying external pressure gradients, this assumption appears to be reasonable. Such an assumption would not be correct for fluid flows driven by internal capillary forces or thermodynamic potentials, since internal local redistribution of mass might take place. Also, it has to be noted that even at complete percolation, V_{in} would be appreciably smaller than V_{mx}. Since k_{in} is usually at least one order larger than k_{mx} due to the higher porosity and coarse microstructure of the interfaces, the overall permeability k_b may be affected significantly, once the percolation of interfaces takes place.

Since k_{in} and k_{mx} can be assumed to be similar for composites prepared with similar water-to-powder ratios, equation (5.1) suggests a one-to-one relationship with the computer simulation model of percolation of interfaces. In other words, in Figure 5.6, which shows the relationship of relative gas permeability with aggregate volume, the aggregate volume fraction at which a sharp increase in permeability is observed can be approximately taken as the critical aggregate volume fraction. The relative permeability in this figure is the ratio $K/K_{0.2}$ where K is the actual permeability of mortar at a specified aggregate volume fraction, normalized with respect to the volume of cement-paste matrix. Here, $K_{0.2}$ is the normalized permeability at $S/S_{lim} = 0.2$, where S is the aggregate volume fraction in the mortar mix, and S_{lim} is the limiting filling capacity of aggregate in a unit volume ($\sim$0.6). The relative permeability parameter has been chosen here, since it gives a quick estimate of the aggregate volume fraction at which percolation might be occurring. The t/r (interface thickness to particle radius) ratios of 0.05 and 0.03 cannot successfully fit the observed data of permeability of mortars with 1.0 mm and 0.4 mm glass beads. Therefore, t/r ratios for mortars prepared with 1.0 mm or 0.4 mm diameter glass beads should be in the vicinity of 0.02. Moreover, mortar made with 0.04 mm size glass beads shows no increase in permeability, therefore the interface in such a case might be extremely thin or nonexisting. For very large aggregate particle size, the interface thickness should be nearly constant since, considering the wall effect, large aggregate surfaces

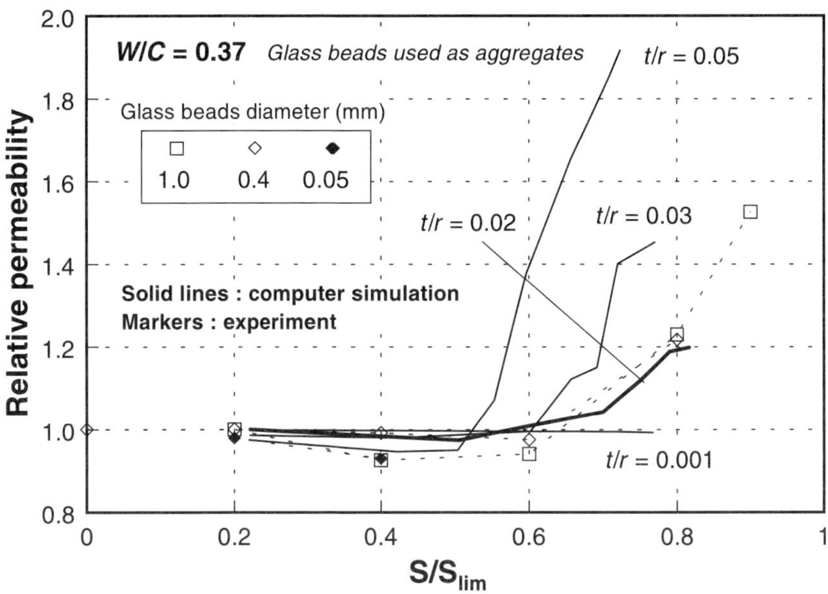

Figure 5.6 Experimental and simulated relationships of relative gas permeability (normalized with bulk matrix volume fraction) with relative aggregate volume fractions.

are similar to an infinite wall. This discussion forms the basis for a particle size dependent interface thickness model explained next.

5.4.3 *Particle size dependent interface model*

Summarizing the discussion of the previous section, it appears that the interface thickness depends on aggregate particle size as shown in Figure 5.7. In other words, although the interface thickness is the highest for the largest particles, the incremental t/r ratio shows a peak somewhere for intermediate size of aggregate particles. It is understood that while the ultimate thickness of the interface will depend on the properties of composite, such as the water-to-cement ratio, the variation of the t/r ratio as a function of aggregate particle radius will depend on the cement powder particle characteristics. This result is interesting since it not only negates the usually held opinion about constant interface thickness, but also shows that interface thickness might be, in fact, dependent on the aggregate particle size itself, within a certain range. Obviously, for very narrow grading of aggregates, the interface percolation behaviour of the composite using this model will be similar to that obtained using the constant interface thickness assumption. The result has immediate implications for problems of practical interest. It explains that the risk of interface percolation can be avoided, even for high aggregate content mixes, by choosing a suitable grading of aggregates depending on the water-to-powder ratio and powder particle characteristics.

For example, Figure 5.8 shows experimental gas permeability measurements as a function of natural river sand volume fraction in a typical mortar. The grading of sand is as given in Table 5.1. It can be observed that even for

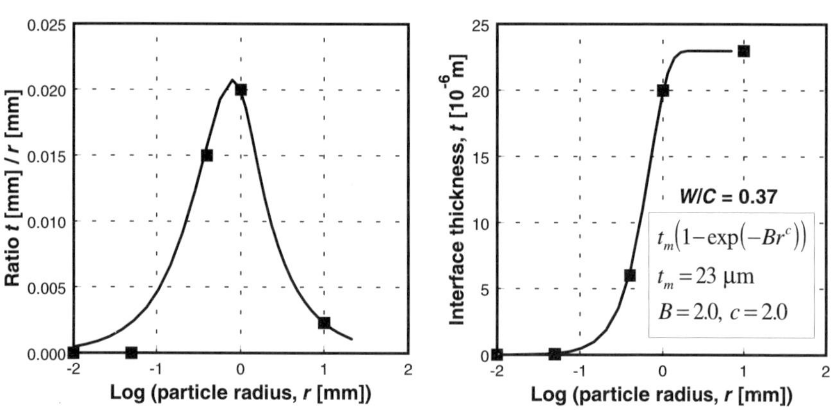

Figure 5.7 Aggregate particle size dependent interface thickness model proposal. The squares denote the approximations obtained by reverse analysis using experimental data.

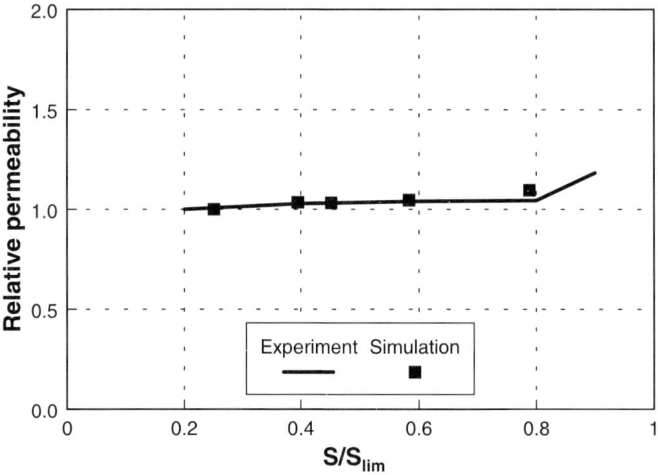

Figure 5.8 Simulated and measured relationship of relative permeability with aggregate volume fraction (normalized with its limiting filling capacity S_{lim}). A particle size dependent interface model is used.

quite high sand volume fractions approaching 50%, no apparent interface percolation takes place. Figure 5.8 shows the computed conductivity if a particle size dependent interface thickness model as shown in Figure 5.7 is adopted and shows a good conformance with the experimental observation.

This probably explains the reason for the very low permeation properties of selfcompacting high-performance concrete, which contains a large amount of fine aggregates. Moreover, due to the low water-to-powder ratio, the absolute thickness of interfaces for the largest of sand particles may be quite small. As a further validation of the proposed model, revised computer simulations were done on a model mortar containing sand of gradings as given in Table 5.2 [5].

Table 5.1 Natural river sand grading (see Figure 5.8)

Sieve size (mm)	*% passing*
5.0	100.0
2.5	91.0
1.2	66.0
0.6	33.0
0.3	13.0
0.15	5.0
0.09	2.1

Table 5.2 Sand size distribution used in the computer model (see Figure 5.9)

Sieve size (mm)	*% passing (mass)*
4.75	100.0
2.36	94.4
1.18	64.6
0.6	41.4
0.3	19.0
0.15	0.0

The interface percolation characteristics for this case are compared to the results given in the reference [5]. It can be seen that the particle size dependent interface thickness model can also explain the percolation behaviour in a similar way to the constant interface thickness model (Figure 5.9). However, the improved interface modelling takes into account the effect of size of sand particles on interface thickness. The largest thickness of the interface in this case is 20 μm, which can be compared to the 15 μm figure obtained in reference [5]. This also explains several disagreements in the data reported by various researchers on the approximate thickness of the aggregate matrix interfaces. In concluding this section, we note that it is difficult to assign some absolute values for the thickness of aggregate matrix interfaces observed in cementitious composites. In general, the interface thickness will be dependent on the size of aggregate particles. This fact alone may influence the percolation properties of such composites when measured with respect to increasing aggregate volume fractions. Moreover, the cement powder particle distribution, curing conditions and especially aggregate grading may have significant additional effects on the interface percolation probability, which would decide the overall transport properties of the cementitious composite, such as concrete.

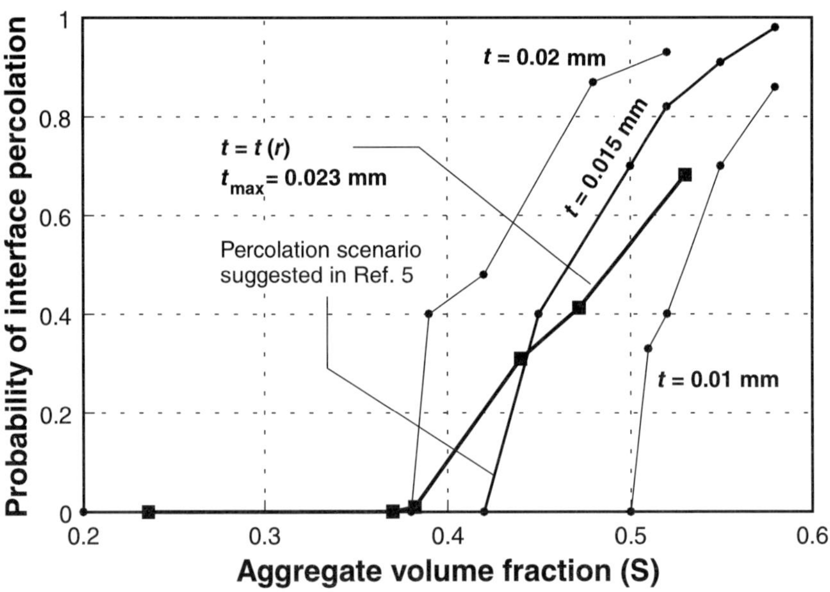

Figure 5.9 Comparison of interface percolation probability curves obtained using size dependent interface thickness model and constant thickness (as assumed in reference [5]).

5.5 Moisture transport formulation in a composite

The moisture transport in concrete, treating the concrete composite as a uniform isotropic, homogenous porous medium, has been discussed in detail in Chapter 4. In the development of the transport model, flow contributions from vapour and liquid phases were unified in the same model. Also, an assumption regarding the thermodynamic equilibrium of different phases was made. In other words, it was assumed that various transportable phases in the composite exist in an equilibrium state distribution inside the complex porous network of the composite. These assumptions may be valid if the aggregate matrix interfaces or channel structures do not percolate across the composite, or when the aggregates are practically impervious so that their porosity does not constitute a significant amount of the total composite bulk porosity. These conditions might exist for most normal concrete, however, to consider a generic treatment of moisture transport, the transport in interfaces and aggregates as well should be integrated into a unified moisture transport formulation.

To consider the moisture transport behaviour in a concrete composite as a multicomponent system, the three basic components, bulk hardened cement-paste matrix, channels or interfaces and aggregates (section 5.2), need to be considered. It is assumed that the thermodynamic equilibrium assumption holds good for each of these porosity components considered individually. Thus, each of these components can be treated as an effective homogenous porous medium in itself or each of the porosity components can be viewed as a continuum. For this reason, the moisture transport model originally developed for an isotropic and homogenous porous medium can be applied to each of the porosity components individually. Also, for a complete description of the problem, the mutual interactions of these components between themselves needs to be considered. The bulk transport behaviour of the composite can be obtained by summing the transport behaviour of its components.

5.5.1 *Mass and momentum conservation*

The assumptions of thermodynamic equilibrium as well as other conditions that were adopted in the modelling of moisture transport in cementitious materials will be adopted here. Of the three basic components of concrete, the channel (interface) component is similar to the bulk matrix system, the only difference being higher porosity and relatively larger pore distribution; therefore, analytically, it can be treated in a similar way to the cement paste pore system (uniform porous medium).

However, aggregate porosity differs due to the nature of moisture transport modelled through them. Essentially, the modifications required in the formulation described in the previous chapter are minimal. It is basically the mass balance equations that require a reconsideration. Furthermore, since no movement of moisture is assumed to take place through the aggregates across a representative elementary volume (abbreviated to REV), the momentum balance of moisture for the aggregate component is not required. Accompaning the moisture transport in individual components of concrete will be a mutual interaction, in terms of the moisture exchange. The mass transport equations for these components will be coupled therefore with some interaction terms. The interaction terms define the intercomponent rate of moisture exchange within a REV and can be called the *local mass transfer* rate terms. The term *local* is used since the rate of moisture exchange is supposed to be dependent on the thermodynamic state of moisture, such as the relative humidity in individual components at any given location and time. Therefore, in the consideration of overall moisture balance in each of the components, the respective mass exchange rates due to interaction with other components must be included.

In the development below, all differential equations are written over a REV of interest, which is assumed to be very small compared to the specimen size and quite large compared to microscopic scale. A continuum approach is adopted in the modelling of concrete performance. For example, aggregates are modelled as uniformly and equally distributed in the REV according to their volume ratio, a similar approach will be applied later for the development of a local mass transfer model. It is assumed that the mass of vapour in a REV can be neglected compared to the mass of liquid and that the total gas pressure is constant everywhere. We also assume that the liquid phase does not interact with the surrounding solid cementitious mass, so that no chemical fixation or permanent loss of moisture takes place within a REV.

It is assumed that the deformation field in the porous medium is uncoupled to the moisture transport problem, so that the moisture transport will be independent of the deformation fields in the composite material. A general subscript and superscript notation is used. A subscript denotes the phase of interest. For example, l indicates the liquid phase, whereas v denotes the vapour phase. Total water present in the REV within a component is represented by subscript w. A superscript denotes the component in the composite of interest. Cement paste matrix, channels and aggregate components are abbreviated as *cp*, *ch* and *ag* respectively. With these conditions, the appropriate moisture balance equations for the HCP matrix, channels and aggregates can be obtained using general laws of mass conservation. The term *moisture* is used to represent the liquid and vapour phases of water collectively.

Moisture in the channels and cement-paste matrix

The rate of change of moisture in channel and cement-paste matrix components of the concrete in a given REV will be a consequence of the flow of moisture in the liquid and vapour forms into the REV due to potential gradients and phase transitions of one phase to other phase. Moreover, there will be a local exchange of moisture within the components leading to a redistribution of moisture locally. Therefore, mass balance of liquid moisture in the cement paste matrix, considering it as an independent porous component, yields

$$\frac{\partial \theta_l^{cp}}{\partial t} + \operatorname{div} q_l^{cp} + \upsilon^{cp} + R_{cp.ag} - R_{ch.cp} = 0 \tag{5.2}$$

where θ_l^{cp} is the volumetric liquid water content in kg/m^3 of total concrete, i.e. $\theta_l^{cp} = \rho_l \phi^{cp} S_l$, and ϕ^{cp} is the bulk porosity of the HCP matrix, i.e. the pore volume of cement-paste matrix per unit volume of the concrete composite. Also, ρ_l is the liquid density, S_l denotes the degree of saturation of the porous network of HCP matrix with liquid water, q_l is the flux of liquid water per unit area of the concrete, υ^{cp} is the rate of phase transition from liquid to vapour phase, and $R_{cp.ag}$ and $R_{ch.cp}$ denote the rate of moisture exchange per unit volume of concrete between *cp* and *ag*, and *ch* and *cp* components, respectively. We have a similar expression for channel structures as

$$\frac{\partial \theta_l^{ch}}{\partial t} + \operatorname{div} q_l^{ch} + \upsilon^{ch} + R_{ch.ag} + R_{ch.cp} = 0 \tag{5.3}$$

with all the terms having similar meaning as in the case of the cement-paste matrix.

It should be noted that equation (5.3) would be included in the overall moisture transport problem of the concrete composite only if the channels form a percolated path across the composite. The porosity, ϕ^{ch}, in such a case would correspond to the porosity existing in the percolated interfaces per unit concrete volume. When a percolated or continuously connected path of channels does not exist (section 5.3), concrete can be effectively considered as aggregate and HCP matrix components only and the channel component need not be considered.

The mass balance of moisture existing as the vapour phase in the channel and hardened cement-paste matrix yields

$$\frac{\partial \theta_\upsilon^i}{\partial t} + \operatorname{div} q_\upsilon^i - \upsilon^i = 0 \tag{5.4}$$

where i denotes either of the *cp* or *ch* components, and θ_υ is the mass of moisture present in the vapour phase in component i per unit volume of concrete. This quantity is usually very small in magnitude compared to the liquid water content θ_l^i. Therefore, the total moisture content (in absolute mass terms) of a component in the composite can be approximately taken as the mass of liquid phase of moisture only. The term q_v represents the flux of moisture across the composite in component i in the vapour form. Considering that the total gas pressure is assumed to be constant, it is the vapour pressure gradient and diffusive flux of water vapour that will be responsible for the movement of the vapour phase in the porous microstructure.

Moisture in the aggregates

The role of aggregates can be unified in the same framework of moisture transport modelling using the concept of local moisture transfer described earlier. It can be easily observed that usually aggregates form porosity regions that are isolated from each other. Thus, aggregate porosity, unlike hardened cement-paste matrix and interfacial zones, does not form a continuously connected path across the composite and any change in its moisture content would result from the interaction with other continuously connected components only. The aggregates can be idealized as kinds of reservoir zones that either release or store moisture locally due to the interaction with the surrounding cement-paste matrix and interfaces. Thus, we have

$$\frac{\partial \theta_l^{ag}}{\partial t} = R_{cp.ag} + R_{ch.ag} \tag{5.5}$$

where θ_l^{ag} is the volumetric liquid water content of aggregates per unit volume of concrete, such that $\theta_l^{ag} = \rho_l \phi^{aq} S_l$, and ϕ^{ag} is the porosity of the aggregate relative to the unit volume of concrete, i.e. it is the volume of aggregate pores in a unit volume of concrete. Mathematically, $\phi^{ag} = G\phi_{abs}$, where G is the volumetric fraction of aggregates in the composite, ϕ_{abs} is the absolute porosity of the aggregate, and S_l denotes the degree of water saturation of aggregate pores. The terms $R_{cp.ag}$ and $R_{ch.ag}$ denote the rate of moisture exchange with the cement-paste matrix and channels respectively per unit volume of concrete. The phase transition term, v, has been neglected since mass of vapour in the aggregates is considered to be negligible.

5.5.2 *Reduction to simplified classical form*

To obtain the total mass transport behaviour in concrete, equations (5.2)–(5.5) need to be solved simultaneously. The basic item of interest in this formulation is the moisture content of each component. Total moisture content of concrete can be obtained by a simple summation of the moisture content of each component. In the basic mass conservation problem defined above, we have in all seven primary variables (including v) and only five conservation equations. It is assumed that the remaining terms, such as flux, q, and local mass exchange terms, could be expressed as a function of the primary variables. The constitutive law defining the phase transition terms, v, of each component must be developed, so that for a given problem the above equations provide a unique solution. While it is possible to derive the moisture flux models and local moisture exchange rates based upon the microstructural characteristics of components and thermodynamic equilibrium conditions, a generic model that defines the rate of phase transition between liquid and vapour phases remains to be found. A few models defining this relationship have been proposed in the field of soil science [10], unfortunately, they are primarily empirical and at best qualitative in nature.

An attempt has been made to redefine the problem such that the phase transition terms are not required in the overall formulation and at the same time basic features of the model are retained. The model is restated in terms of the total moisture contained in individual components. That is, we are now concerned with the conservation of total moisture instead of individual phases, such as liquid water and vapour, separately in each component. This is achieved by summing the conservation equations of liquid and vapour phases and neglecting the mass of vapour present in a REV but retaining the flux contribution terms. This manipulation leads to a cancelling out of the phase transition terms and we obtain a mass conservation equation for total moisture in a component. The assumption seems to be reasonable enough since the mass of liquid water in a REV would be many times more than the vapour mass, which can be therefore neglected. Using this redefinition, we have the reformed moisture balance for *cp*, *ch* and *ag* components as (i.e. adding equation (5.2) and (5.3) to (5.4),

$$\begin{aligned}
\frac{\partial \theta_w^{cp}}{\partial t} &= -\text{div}\, q_w^{cp} - R_{cp.ag} + R_{ch.cp} \\
\frac{\partial \theta_w^{ch}}{\partial t} &= -\text{div}\, q_w^{ch} - R_{ch.cp} - R_{ch.ag} \\
\frac{\partial \theta_w^{ag}}{\partial t} &= R_{cp.ag} + R_{ch.ag}
\end{aligned} \tag{5.6}$$

where θ_w^{cp}, θ_w^{ch} and θ_w^{ag} denote the total moisture content of *cp*, *ch* and *ag* components respectively, expressed in kg/m^3 of concrete. Here q_w^{cp} and q_w^{ch} denote the flux of total moisture in the *cp* and *ch* components and includes the moisture flux contributions from both the liquid and vapour phases. Once again, it is stressed here that the channel component needs to be considered only when it constitutes a percolated path across the concrete composite. In nonpercolated composites, we are essentially left with only two primary unknowns, θ_w^{cp} and θ_w^{ag} (equations (5.6)).

The classical constitutive law defines the moisture flux q in a porous medium as [2, 10]

$$q = -D(\theta)\nabla\theta \tag{5.7}$$

where $D(\theta)$ is the so-called diffusivity parameter of the porous medium and is usually expressed as a nonlinear function of the moisture content θ of the porous medium. Since it is the only parameter that defines the transport characteristics of the porous medium, we find many studies devoted to analytical description of this relationship. However, the suggested relationships have generally been empirical in nature [11]. Furthermore, in the realm of the empirical models, it appears that no consensus exists on the nature of the D–θ relationship. This has led to several experimental measurements of diffusivity of concrete under various exposure conditions, which show that indeed diffusivity is a complex parameter and cannot be simply specified as a function of moisture content. Factors, such as the age of concrete, exposure history, maturity and temperature, among others, have a direct bearing on the transport properties of concrete. The empirical models of diffusivity cannot take into account all these factors.

For these reasons, and a certain level of arbitrariness involved in choosing the values of the so-called material parameters of these models, the problem in the previous chapter of momentum balance of moisture in a concrete porous network was reworked. It was found that by making certain assumptions, it is indeed possible to arrive at a similar form of the moisture flux equation, equation (5.7). The only difference being that through this approach the diffusivity function of a porous medium can be analytically evaluated in a nonempirical way, once the microstructural properties of the porous medium, its exposure history and the thermodynamic state of moisture inside the porous network are known. The details of this development of conductivity and diffusivity models have been described in the previous chapter. In brief we state that it is possible to obtain equation (5.7) from general considerations of multiphase dynamics under the assumptions of constant total gas pressure in pores and a linear relationship of viscous drag force with velocity of liquid moisture. The diffusivity $D(\theta^i_w)$ of component i (cp or ch) can be evaluated from the general concepts of adsorption of vapour molecules, surface tension, thermodynamic equilibrium and Hagen–Poiseuille flow characteristics through a random network of micropores. In a compact form $D(\theta^i_w)$ can be expressed as

$$D(\theta^i_w) = \left(\frac{\rho RT}{Mh} K_l(\theta^i_l) + \rho_v D_a f(\theta^i_w)\right) \frac{\partial h}{\partial \theta^i_w} \tag{5.8}$$

where R is universal gas constant, T is absolute temperature, M is the molecular mass of water, h is the relative humidity in the REV of component i, K_l is the permeability of liquid and depends on the liquid content (degree of saturation) of the component, ρ_v is vapour density, and D_a is the vapour diffusivity in free atmosphere. The factor $f(\theta_w^i)$ accounts for the effect of tortuosity and varying moisture content on vapour diffusion. The slope of the relative humidity with moisture content is obtained from the isotherms of individual component. An extensive discussion of these parameters can be found in sections 4.3–4.5 which discuss the basic constitutive models related to the moisture transport in concrete.

5.6 Local moisture transport behaviour

Associated with the diffusive and bulk movements of moisture through the components of concrete, there exists an associated phenomenon, namely, the rearrangement and redistribution of moisture among the components of concrete composite locally. Thus, the exchange of moisture may take place between aggregates, channels and continuous hardened cement paste as shown in Figure 5.10.

An accurate formulation of the mechanisms for this local moisture transfer needs to be found. In this book, however, guided by the general nature of moisture transport in a porous medium, the pressure potential level of moisture in each of the components in a given REV is taken as a key parameter controlling the intercomponent moisture transfer. The formulation of the local moisture transfer can be obtained by incorporating a difference in these potentials.

In the most simplistic scenario, the rate of intercomponent exchange can be taken to be linearly proportional to the difference of liquid pressure potentials of interacting components. Alternatively, this result can be obtained as follows. Consider a system of two conducting components in contact with each other, as shown in Figure 5.11. Here δ_1 and δ_2 represent the average spatial

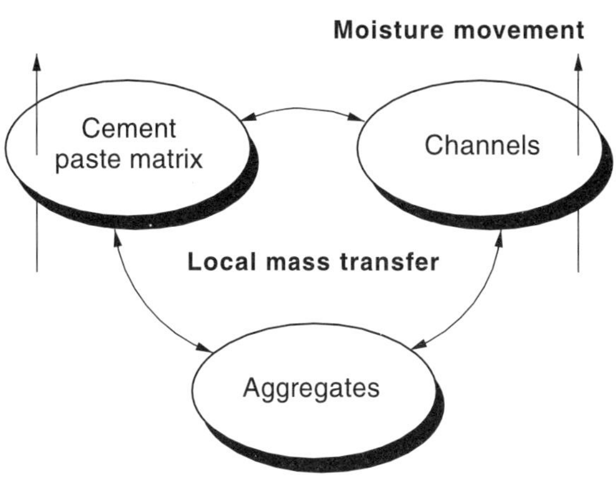

Figure 5.10 The local moisture exchange concept.

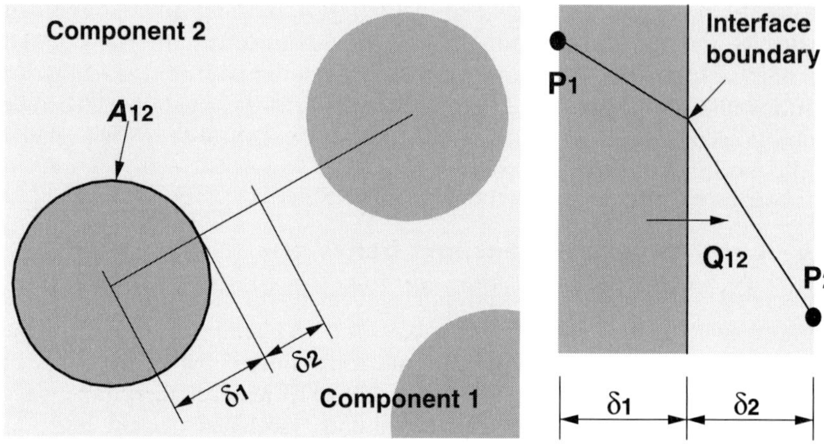

Figure 5.11 Formulation of local moisture transfer model: component 1 = suspended sphere, component 2 = surrounding media.

dimensions of the components 1 and 2 in a unit volume of the porous medium. If P_1 and P_2 represent liquid pressures at the centroids of components 1 and 2 respectively, the moisture exchange rate R_{12} from component 1 to 2, in a unit volume of the overall solid matrix can be obtained as

$$R_{12} = q_{12}A_{12} = -\frac{2A_{12}}{\sum_{i=1}^{2}(\delta_i/K_i)}(P_2 - P_1) \tag{5.9}$$

where A_{12} represents the actual area of contact per unit volume between the two components, q_{12} is the corresponding flux, and K_i represents the unsaturated permeability of the ith component. It has to be noted that a linear variation of liquid pressures P_1 and P_2 is assumed in the individual components considered above. Moreover, this formulation is based upon quasi steady-state assumptions. Expressing the rate of mass transfer as linearly proportional to the difference in pressure potentials, the local mass transfer coefficient α_{12} can be obtained as

$$\alpha_{12} = \frac{2A_{12}}{\sum_{i=1}^{2}(\delta_i/K_i)} \tag{5.10}$$

In this work, three major components of moisture mass transport have been identified. Thus, the local mass transfer coefficients for the three interacting pairs, namely, $\alpha_{cp.ag}$, $\alpha_{cp.ch}$ and $\alpha_{ag.ch}$,will be derived next. These derivations should be treated at best as qualitative in nature since a large amount of uncertainty is involved in exactly computing the actual areas of contact between different components. Moreover, the accuracy of length scales, δ, for different components is unclear. Therefore, many assumptions and simplifications are inherently involved in the derivation of these coefficients. Nevertheless, such an exercise will give us an idea of the order of magnitudes of these coefficients and related sensitivities of the various parameters involved.

5.6.1 Aggregate and fine porosity matrix

Aggregates can be modelled as spheres of the same radius, r. If the volumetric concentration of aggregates in the total hardened concrete is G, s is the average distance between the outer surfaces of two spheres and G_o is the limiting filling capacity of aggregates. Then, we have

$$G = G_o/(1 + s/2r)^3 \tag{5.11}$$

Average separation s represents the length scale of a fine porosity matrix. The area of contact between two components in a unit volume will be $3G/r$. Therefore, the local mass transfer coefficient is obtained as

$$\alpha_{cp.ag} = \frac{3G}{r^2}\left(\frac{1}{2K_{ag}} + \frac{(G_o/G)^{1/3} - 1}{K_{cp}}\right)^{-1} \tag{5.12}$$

Generally, permeability of the finer porosity component is very small compared to the permeability of aggregate pores. Also, the actual effective area of contact may be smaller than the estimate obtained from the above expressions. Thus, a parameter, β_1, is introduced which represents the apparent effectiveness of the moisture transfer across actual contact areas. Using this and the assumption that $K_{cp} \ll K_{ag}$, $\alpha_{cp.ag}$ becomes

$$\alpha_{cp.ag} = \frac{3\beta_1 G K_{cp}}{r^2\{(G_o/G)^{1/3} - 1\}} \tag{5.13}$$

5.6.2 Fine porosity matrix and interfaces/channels

Coarser porosity paths are modelled as circular cylindrical paths with the same radius, r. With similar symbols and assumptions as earlier, the local moisture transfer coefficient $\alpha_{cp.ch}$ can be obtained as

$$\alpha_{cp.ch} = \frac{2\beta_2 G K_{cp}}{r^2\{(G_o/G)^{1/2} - 1\}} \tag{5.14}$$

where β_2 represents the effectiveness of moisture transfer across the area of contact between fine and coarse porosity paths. The value of the limiting packing factor G_o can be taken as 0.91 in this case. The length scale, r, for coarse porosity paths has been reported to be of the order of 10–50 μm [3, 5]. A tentative value of 20 μm has been assumed.

5.6.3 Interfaces/channels and aggregates

It is extremely difficult to ascertain the area of contact of these two components. When a complete percolation of the interfacial zone occurs, the contact area will be of a similar magnitude to the external surface areas of aggregates. At any intermediate stage, it will be approximately proportional to a fraction of the total surface area of aggregates that are in the percolating backbone. Also, the permeability of both the components can be assumed to be roughly similar. Thus, we have,

$$\alpha_{ag.ch} = \frac{3\beta_3 G}{r(r/2K_{ag} + r_{ch}/K_{ch})} \tag{5.15}$$

where G is the volume fraction of aggregates, and β_3 is the fraction of the effective area of contact between the interfaces and aggregates. The approximation that $K_{ag} \approx K_{ch}$ and $r_{ch} \ll r$ (radius of aggregates), yields

$$\alpha_{ag.ch} = (6\beta_3 G K_{ch})/r^2 \tag{5.16}$$

From the expressions for local moisture transport coefficients, it can be observed that the local mass transfer rate is quite sensitive to the average spatial dimensions of aggregate and interface components. A linear sensitivity can be expected for other parameters involved in the above model. It must be restated here that the transport coefficient models discussed above are quite primitive in nature, considering the limitations of our understanding of the complex microstructure formed in a hydrated sample of the concrete and the difficulty in the analytical description. The coefficients discussed above help us to make informed and educated guesses about the moisture transport in a multicomponent system.

5.7 Simulations of moisture transport

The system of partial differential equations (equations (5.6)) describes the governing equations of moisture transport in a nondeformable, isothermal composite porous medium. Compared to the formulation for an isotropic and homogenous porous medium, the multicomponent formulation is complex and involves a higher degree of nonlinearity due to the pressure potential based local mass exchange terms. To solve this system of equations, a coupled finite-difference numerical solution scheme has been adopted (Figure 5.12) to simulate the moisture transport behaviour in a

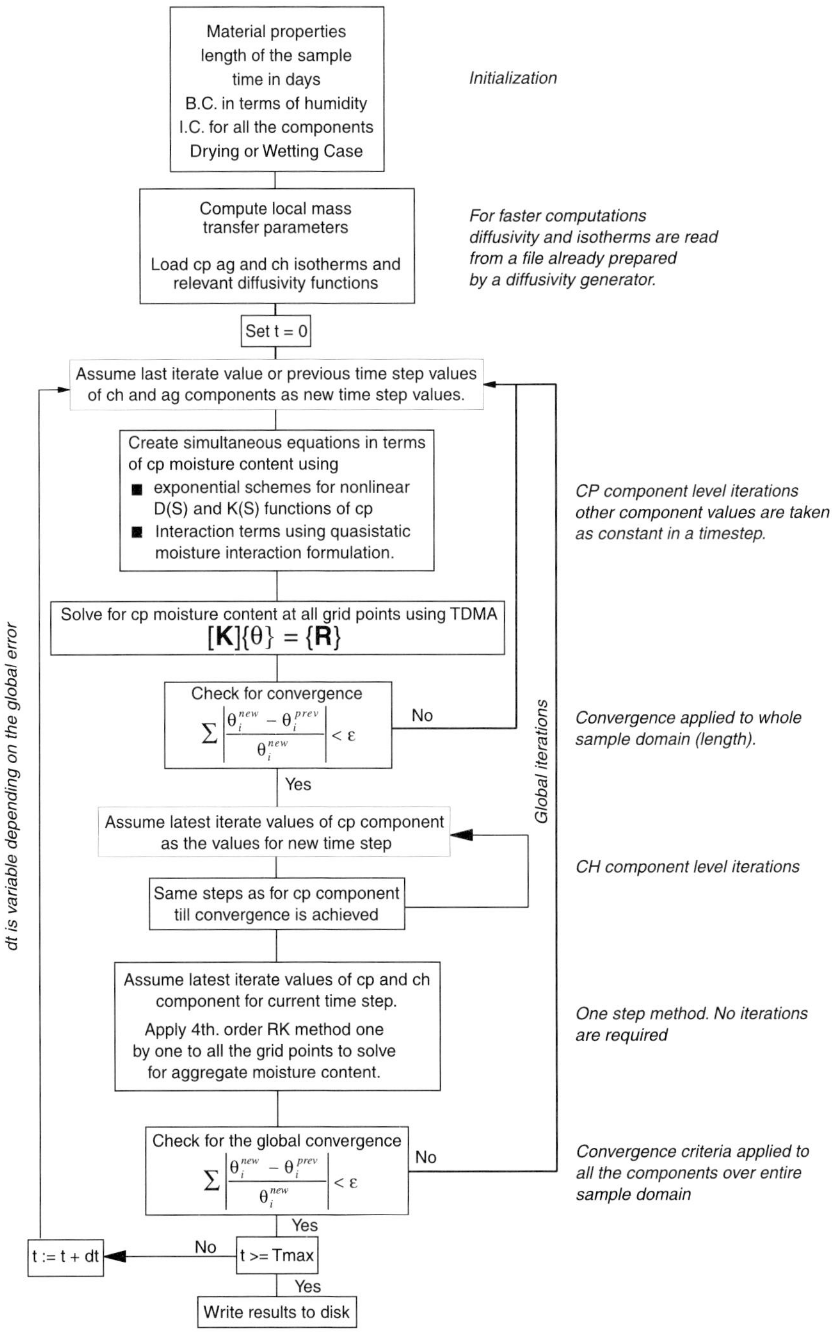

Figure 5.12 The coupled finite-difference scheme of solutions.

composite. The intercomponent moisture transfer models are based on coefficients derived in the previous section. Through these simulations, the role of each component on the moisture transport process as a whole is clarified. Parameters for which a numerical parametric study is done are (a) local mass transfer coefficients to measure their sensitivity and (b) the effect of aggregate volume fractions and aggregate porosity. Both the drying and wetting processes have been included. The initial and boundary conditions adopted in the simulations are shown in Table 5.3. Perfect moisture transfer across the surface has been assumed. That is, the surface emissivity parameter used is very high compared to the diffusivity D of the bulk matrix.

5.7.1 Sensitivity of local moisture transfer coefficients

As discussed in section 5.6, the length scale for the individual components as well as the permeability of the cement paste component is very important in the computation of the intercomponent moisture transfer rate coefficients. At this moment, only an approximate guess can be made. Another way to deal with this difficulty is to do an inverse analysis. That is, from a given set of experimental data, choose the values that can predict a larger set of experimental data. Once a pore distribution is selected, a more or less accurate guess for permeability can be made. The only parameters that need investigation are length scales (δ_i) and effectiveness of the area of contact between the various components (β_i). Some approximate estimate for δ_i is possible based on the results of past researchers. The parameters that are left are β_i. A sensitivity analysis for β_1, β_2 and β_3 is done with experimental comparison and an estimate for these coefficients is made. The length scales required are radius of the aggregates and radius of the coarser porosity zone, tentatively assumed to be 2 mm and 20 μm, respectively.

First, let us examine the behaviour of the water sorption when simulated numerically. Analysis of the wetting case for different values of β_1, β_2 and β_3 from 0 to 1 is shown in Figures 5.13 (a)–(c). The moisture passes through both the interfacial channels and the paste matrix with pores. At the same

Table 5.3 Initial and boundary conditions adopted in the simulation

Process	*Initial condition*	*Boundary condition*	*Length*
Wetting	$RH = 1\%$ for all the components	Face 1: $S = 1$ for all the components Face 2: Atmospheric relative humidity = 90%	$L = 5$ cm for (a) $L = 20$ cm for (b)
Drying	$S = 1$ for all the components	Atmospheric relative humidity = 50%	$L = 5$ cm for (a) $L = 20$ cm for (b)

time, the moisture will be supplied from the channel to the paste matrix provided that the diffusivity of the channel system is predominant. Second, a similar analysis of the drying case for different values of β_1, β_2 and β_3 is shown in Figures 5.13 (d)–(f). The point of discussion here is the averaged overall diffusivity as a composite.

For the wetting case, it is seen that increasing β_1 for aggregate–cement paste interaction gradually increases the rate of sorption (Figure 5.13(a)–(c)). But, the total normalized saturation and the rate of sorption decreases with the addition of aggregates due to the addition of nonparticipating porosity of the aggregate component. As the aggregate versus cement paste interaction is increased, the aggregate porosity starts participating in the moisture transport by virtue of the local mass transfer. Therefore, the rate and level of the normalized sorption increase gradually from the minimum value but, even at maximum interaction, this value is less than the value for the case when there are no aggregates. Needless to say, this effect will be significantly smaller for normal aggregates. Since the normal aggregate porosity is around 2%, the contribution of aggregate porosity to the total concrete porosity would be very small compared to the contributions from the hardened cement mass. Overall sorption is not very sensitive to the β_1 parameter for normal aggregates.

Figure 5.13(e)–(f) shows the effect of β_2 (cement paste–channel interaction) on the rate of sorption. It can be seen that the overall rate of sorption is very sensitive to β_2. The rate of sorption increases rapidly with increasing values of β_2. This is due to the fact that ingress of moisture into the concrete through channel systems is very rapid owing to its high diffusivity. A rapid interaction with the cement paste component transfers a large amount of moisture to the paste matrix, increasing the apparent rate of sorption. It has to be noted that higher values of β_2 approximate closely to the usual assumption of equilibrated moisture distribution inside the REV of a porous network. It is difficult to understand why β_2 should be very small for the case when interfaces are closely dispersed and form a percolated path, as in a typical concrete containing a high aggregate volume fraction. Therefore, we can infer that when interfaces form a continuously connected path, the net effect of rapid moisture suction can also be represented by a single matrix system, which has a higher apparent diffusivity than the real diffusivity of a cement-paste matrix in concrete.

Aggregate–coarser porosity interaction shows a similar trend to that observed for *cp–ch* interaction. If the aggregate porosity is small, the sensitivity of β_3 on the total normalized weight gain curve would not be large. It has to be noted that, for β_3 value greater than 0.01, the results are similar for all the subsequent higher values of β_3. That is, even a 100 times increase in β_3 does not affect the overall result very much. This result shows the importance of percolation. That is, once a continuous connected interaction path of aggregates and channels is established, a change in the effective area of

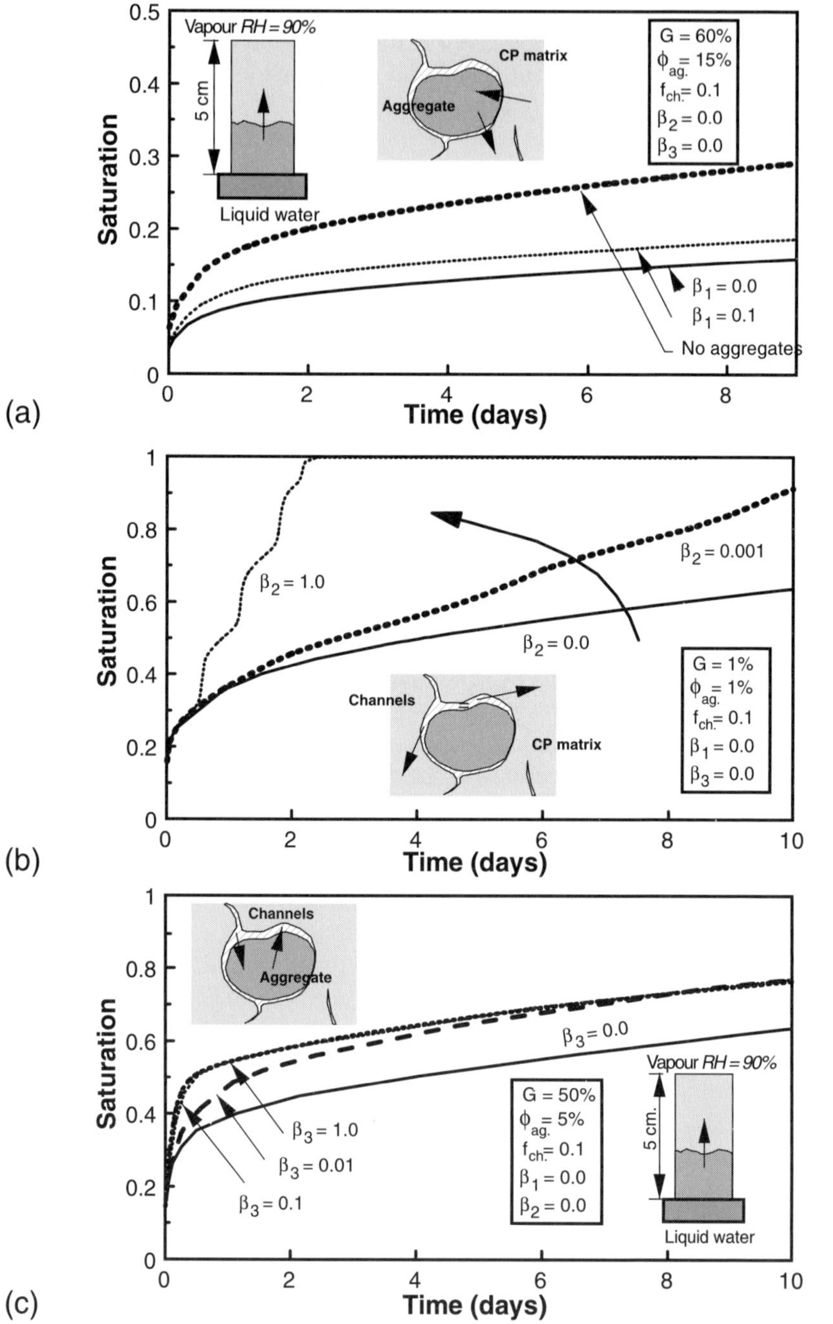

Figure 5.13 (a)–(c) Parametric study for local mass transfer coefficients (wetting phenomenon).

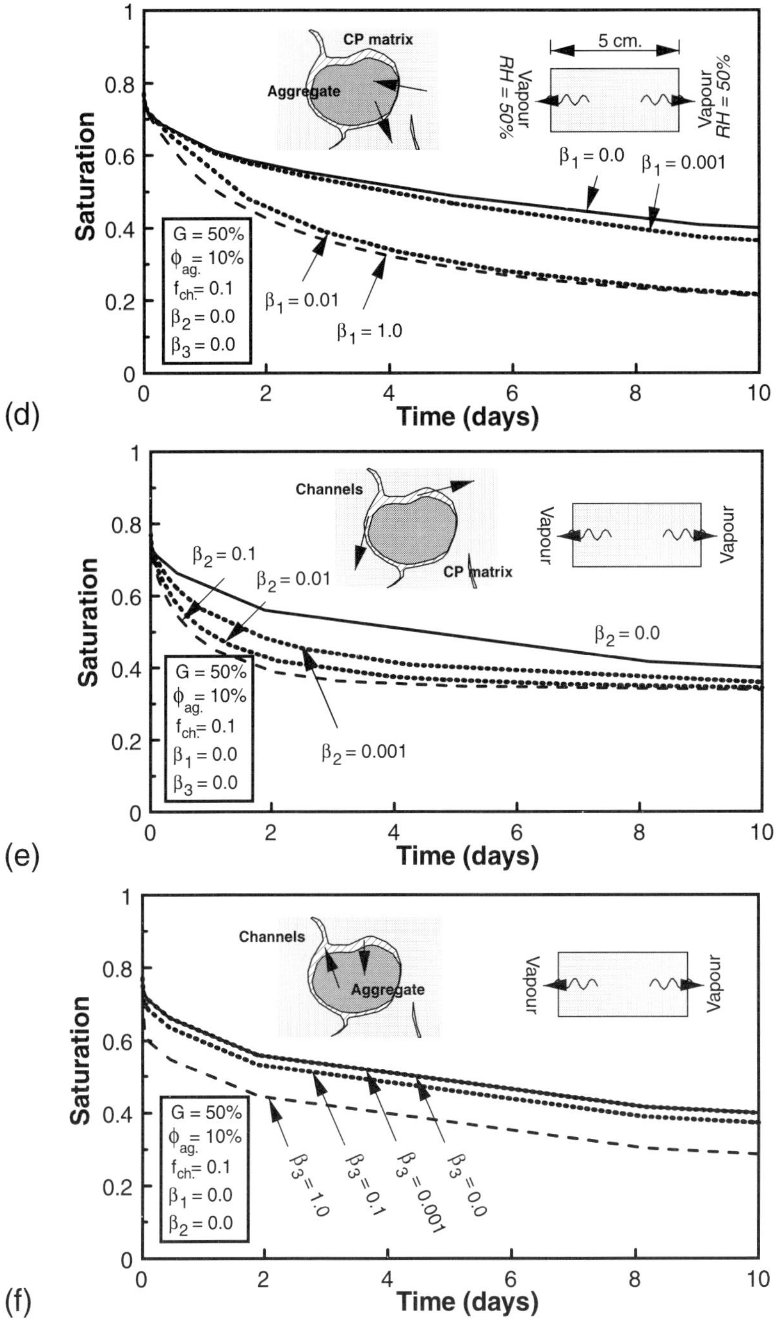

Figure 5.13 (d)–(f) Parametric study for local mass transfer coefficients (drying phenomenon).

contact does not cause significant change in the overall sorption behaviour. Thus, establishment of a continuous connected interaction cluster of aggregate and coarser paths (which may occur at some critical aggregate content) is important and it will be critical for the longer life of concrete structures. In this regard, the computer simulation models that describe the probability of connectivity of the channel–aggregate path with increasing volume fraction of aggregates are indispensable.

Similarly, for the drying process, it can be seen that all the interaction constants follow a similar trend in sensitivity as for the wetting case, although the magnitude of variation is decreased for the channel–aggregate and channel–cement paste constants.

5.7.2 *Influence of aggregate volume fractions and porosity*

It was recognized in the previous section that, after some critical aggregate content for a given W/C ratio and aggregate size distribution, a continuous connected channel–aggregate interaction path will exist. It is not clear at this stage how this variation can be formulated in a generic analytical manner. Tentatively, for sensitivity analysis, a curve obtained by computer simulation methods as shown in the Figure 5.5 is chosen, that gives the probability of percolation of channel–aggregate interfacial paths with increasing sand volume fraction. In general, this curve will be dependent on the grading of sand particles and to some extent on the W/C ratio. The volume fraction of percolating interface components is also obtained by similar computer simulation methods. From these considerations, the aggregate–channel interaction constant β_3 is multiplied by the probability curve of Figure 5.5 to represent realistic interaction constants for a given aggregate content.

The material parameters assumed for the simulations of moisture transport are shown in Table 5.4. Simulation results for increasing aggregate contents that automatically incorporate the effect of the interfaces are

Table 5.4 The material parameters assumed in the simulation of the effect of aggregate content on moisture transport

Aggregate content (G)	*Interface percolation probability*	*Total interface volume fraction*	*Composite porosity*	β_1	β_2	β_3
0.3	0.0	0.0	0.183	1E−3	0.0	0.0
0.5	0.8	0.011	0.141	1E−3	1E−3	1E−3
0.7	1.0	0.024	0.097	1E−3	1E−3	1E−3

Basic porosities: Basic aggregate porosity = 0.027; Basic cement matrix porosity = 0.25; Basic interfacial zone porosity = 0.45. Diffusivities: Interfacial zones = 3E−7 m^2/s; hcp matrix: computed from a typical microstructure distribution.

shown in Figure 5.14 for wetting and drying processes. It is implicitly assumed in the simulation that addition of aggregates does not change the basic pore structure of the fine porosity region or cement-paste component.

From Figure 5.14, it is observed that small additions of aggregate do not affect the overall normalized sorption behaviour. In fact, for cases with large aggregate porosity, up to some cement saturation threshold, a reduction in the normalized sorption may be observed due to the effect of the nonparticipating nature of aggregate pores in the sorption process. Beyond the critical aggregate volume fraction, the larger aggregate content

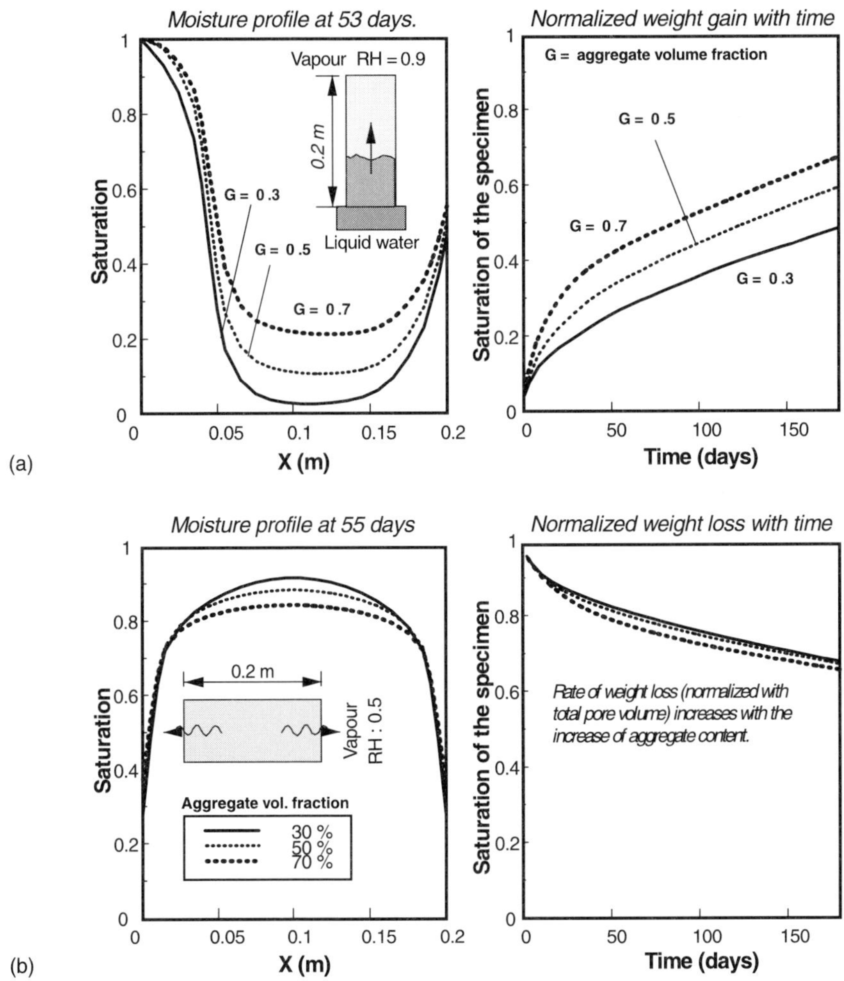

Figure 5.14 Simulations of the influence of aggregate volume fraction on the wetting and drying process of concrete: (a) moisture profiles and weight gain curves for wetting phase; (b) moisture profiles and weight gain curves for drying phase.

increases the rate of sorption several fold due to the presence of interfaces (section 5.4.2). Also it can be seen that the overall sorption characteristics will not change much for an already percolating concrete where percolation is occurring due to interfaces or channel structures.

Figure 5.14 also shows the effect of aggregate content on the drying behaviour of a one-dimensional specimen. As expected, a larger aggregate content will mean faster drying of the concrete since moisture is carried by the channels to the drying surface. However, the effect is not so pronounced as in the wet sorption case, since most of the moisture is carried by channels in the vapour form. In the model, this effect is manifested by the exponential decrease of moisture diffusivity of channel components. In a more exact analysis, an exponential decrease in the rate of moisture transfer will also

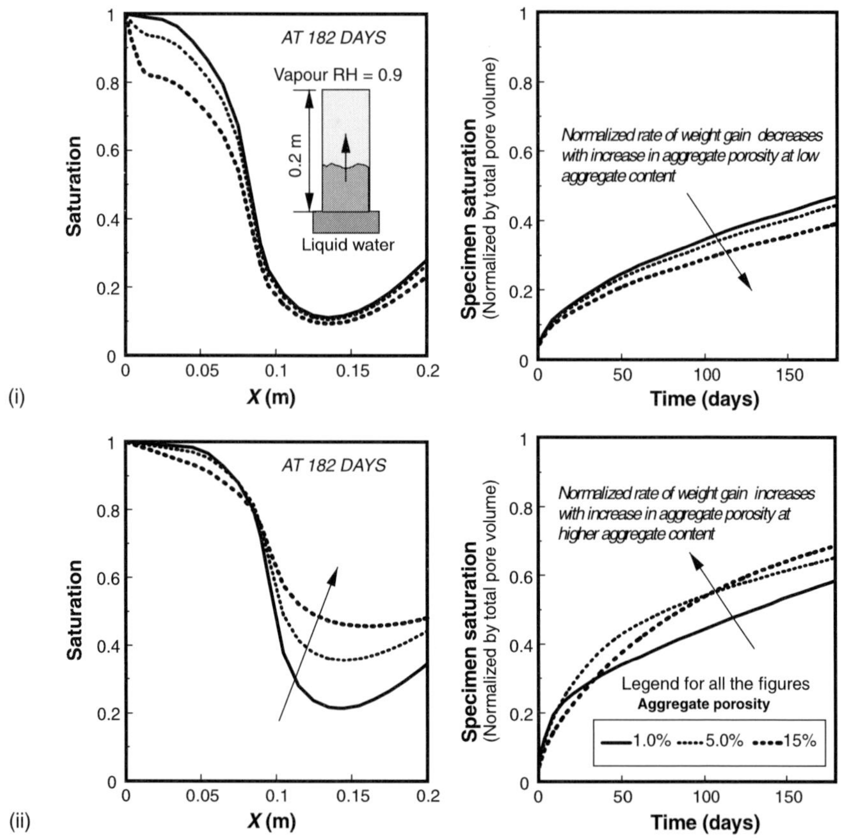

Figure 5.15 (a) Influence of aggregate porosity on moisture transport (wetting): (i) moisture profiles and weight gain curves for aggregate content of 0.3; (ii) moisture profiles and weight gain curves for aggregate content of 0.7.

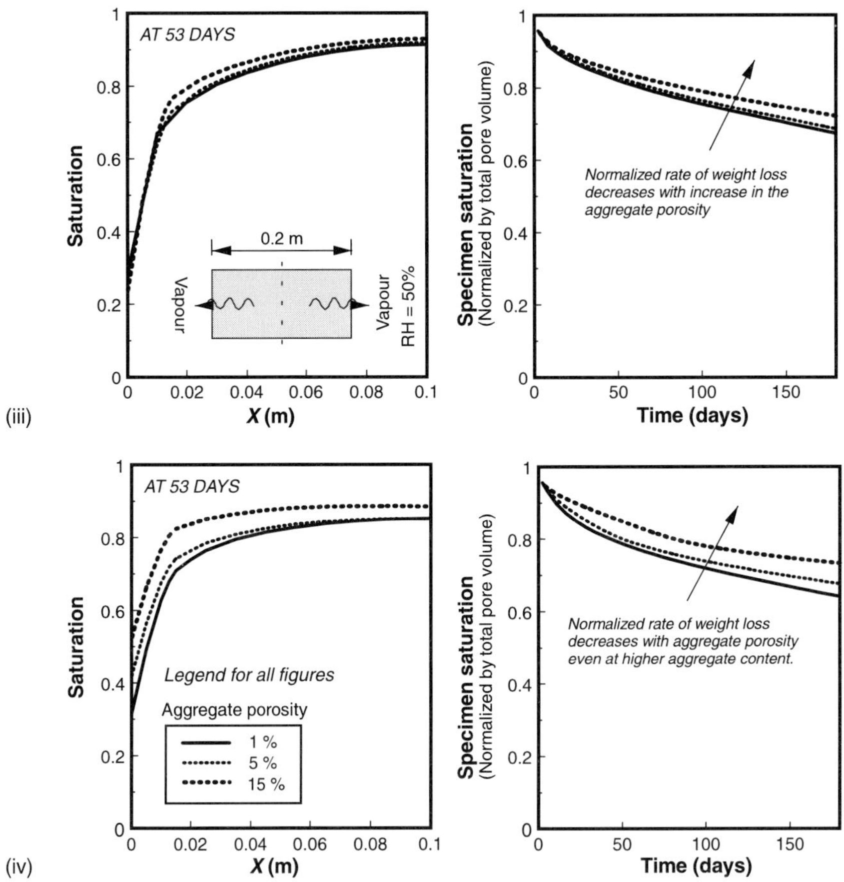

Figure 5.15 (b) Influence of aggregate porosity on moisture transport (drying). (iii) moisture profiles and weight loss curves for aggregate content of 0.3; (iv) moisture profiles and weight loss curves for aggregate content of 0.7.

take place, since the interaction constant combines the effect of moisture conductivity of both the interacting components.

Effect of increasing aggregate porosity

Figure 5.15(a) shows the effect of aggregate porosity on wet face sorption. An initial decrease in the normalized sorption is computed due to the discrete aggregate porosity that does not participate in the sorption effectively. If there are no channels then the normalized weight gain with time for increasing aggregate porosity would always be less than or equal to the sorption curve when no aggregates are present in the hardened cement matrix system. Also, the rate of sorption in the cement paste component will be less than the rate of sorption when there are no aggregates.

Drying behaviour has a much more pronounced effect on the desorption behaviour in terms of the spatial distribution of moisture. It can be seen from Figure 5.15(b) that steeper moisture gradients are observed for increasing aggregate porosity. In the internal regions, the cement component is kept wet with the moisture lost from the aggregates. This leads to steep moisture gradients near the drying surface. The steeper moisture gradients would mean risk of cracking near the drying surfaces during the early drying periods. In terms of the normalized moisture weight loss, the desorption rate as well as the desorbed amount decreases with the aggregate porosity. Thus an optimal balance between the risk of surface cracking and reduced moisture loss should be decided upon in the mix design.

5.8 Drying shrinkage analysis of lightweight aggregate concrete

Concrete subjected to the environment from the green stage to the posthardening period undergoes a complex history of deformations. Deformations related to moisture movement in the concrete, temperature variations and creep continue to take place even in the absence of externally applied loads. The mechanisms of the deformations related to nonstructural loads are still not well understood. The multicomponent concept introduced in this work can be a useful tool for analysing such deformational fields and related phenomena (e.g. hydration, etc.), and during the curing period and hardening of the concrete, when due account has to be made for the special properties of the composite's components (e.g. the high porosity of aggregates). As an example of the application of the multicomponent concept, we shall consider a simplified case of drying shrinkage occurring in already hydrated and hardened mortars where due account is taken of the aggregate component present in the hardened matrix. Through this exercise, we aim to demonstrate the versatility and usefulness of the multicomponent concept. This is also the first stage of the hygrothermal physics and structural mechanics linkage for structural performance evaluation discussed in Chapter 8.

5.8.1 Two-dimensional analysis of drying

Most of the experiments related to the measurement of unrestrained drying shrinkage strain are done on long prisms with constant cross-sectional area. As an approximation to a full three-dimensional analysis, a two-dimensional analysis would meet the requirements, provided that the specimens to be analysed are long enough so that the edge effects do not cause any considerable change in the solutions. The normal ratio of length to width (or height) for such specimens in various experiments [8, 12] ranges between 4 and 5. Thus as an approximation, edge effects are assumed to be negligible (Figure 5.16).

5.8.2 Drying shrinkage strains

It is now well known that shrinkage mechanisms are related to the moisture movement inside the concrete pores [8, 12]. Therefore, to predict the unrestrained drying shrinkage strains when the moisture distribution inside the concrete is known, it is important to have a knowledge of the mechanisms that unify the moisture distribution and drying shrinkage stresses into one concept. The method tentatively adopted in this study for the prediction of unrestrained drying shrinkage is outlined in Appendix A. The method is based upon capillary stress mechanisms. Unrestrained drying shrinkage strains are obtained by this method, once the liquid pressure potential (or RH) inside the concrete pore space is known.

5.8.3 Mass transport equations and numerical scheme

As a further simplification of the problem, in the drying shrinkage strain analysis, only aggregate and hardened cement components have been considered. That is, the interfaces between the cement and aggregate components have been eliminated in this analysis. Using the assumption of isotropic conditions for cp and ag components we have

$$\begin{aligned} \frac{\partial \theta_{cp}}{\partial t} &= \frac{\partial}{\partial x}\left(D_{cp}\frac{\partial \theta_{cp}}{\partial x}\right) + \frac{\partial}{\partial y}\left(D_{cp}\frac{\partial \theta_{cp}}{\partial y}\right) - \alpha_{cp.ag}\left(P_{cp} - P_{ag}\right) \\ \frac{\partial \theta_{ag}}{\partial t} &= \alpha_{cp.ag}(P_{cp} - P_{ag}) \end{aligned} \tag{5.17}$$

The shrinkage strain, which primarily occurs due to the loss of water from the cement component is computed by taking the average saturation of the cement component in the element as shown in Figure 5.16. As the drying proceeds, more and more cement pores are emptied. Pores near the surface dry out more than the pores lying in the interior parts of the element. An average saturation of the cement component in the element can be obtained by integrating the cement-paste component saturation of each control volume of the element. Using the average cement-paste degree of saturation and the corresponding isotherm, the average relative humidity of the cement-paste component can be computed. This value of the averaged

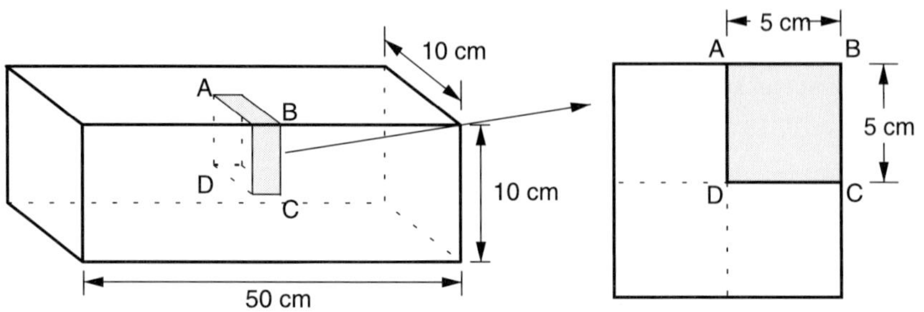

Figure 5.16 Two-dimensional element considered in the analysis of the drying shrinkage.

relative humidity is used to computed the average shrinkage strains of the specimen using the method outlined in Appendix A. It must be noted that this scheme is quite simplistic in nature and only an approximate estimate of the shrinkage strains can be obtained under the assumptions made. But, considering that the size of the specimen is small (cross-sectional dimension 4 cm $\times$ 4 cm) and the complexity involved in solving the general force equilibrium equations for all control volumes, this scheme may be justified. Here, no shrinkage cracking accompanying stress release is considered. If we try to trace the post-shrinkage cracking deformation, tension softening with fracturing energy release has to be considered in nonlinear structural analysis.

An assumption of a uniform boundary condition over the entire length of the specimen is made. The boundary conditions adopted are similar to the ones used in the previous section. A typical value of surface moisture emissivity E_b is taken as 5×10^{-8} for numerical analysis. Equation (5.17) is discretized using a fully implicit type finite-difference method in the x-y and time domains. The finite-difference grid incorporating 'half' and 'quarter' control volumes used in the analysis is shown in Figure 5.17. The resulting simultaneous equations for N^2 grid points have been solved using the implicit scheme.

5.8.4 Numerical simulations and verification

The experimental work of Kokubu *et al.* [13] done on lightweight aggregate concrete to measure the weight loss and drying shrinkage strains was

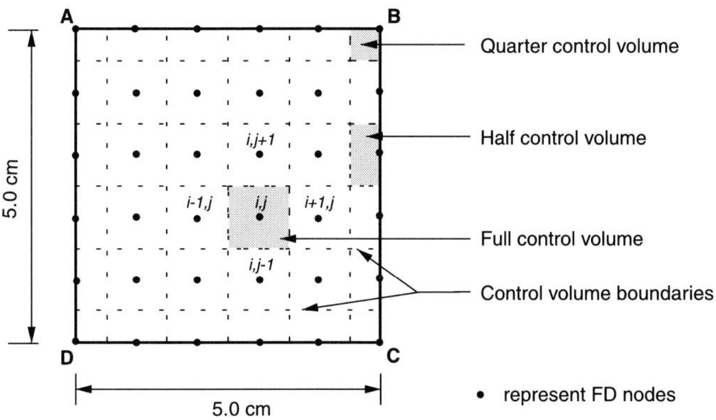

Figure 5.17 Two-dimensional finite-difference grid adopted in the simulation.

referred to for the verification of the multicomponent formulation. The experiments were conducted for the same mix proportions of concrete but varying aggregate porosity. The goal of these experiments was to observe the systematic effect of aggregate porosity on the overall transport behaviour and drying shrinkage strains. A comparison of the computed and measured shrinkage strains and shrinkage weight loss relationship is shown in Figure 5.18, which shows a good qualitative agreement between the computed and measured trends.

In the numerical simulations, the basic cement-paste matrix microstructural properties and local mass transport coefficients were kept constant for all the cases. This is justifiable since the W/C ratio was kept constant during the experiments. Thus, the difference in behaviour arises only from the interaction of the cement-paste matrix with the lightweight aggregates. From Figure 5.18, the following can be observed. It is seen that the shrinkage strains measured during the later stages of weight loss have similar slopes with respect to weight loss. That is, this behaviour is identical for all the cases irrespective of aggregate porosity. Also, the point of rapid rise of shrinkage strain with weight loss shifts towards the right as the aggregate porosity is increased. Thus, the drying shrinkage strain arising due to the moisture loss from cement porosity appears to be an invariant.

We know that an interaction takes place between the aggregate–cement paste components in terms of moisture transfer. It would require the same humidity throughout a given control volume to produce equilibrium. Owing to the nature of moisture isotherms of aggregates, this would mean that aggregates would have to lose a large amount of moisture to reach

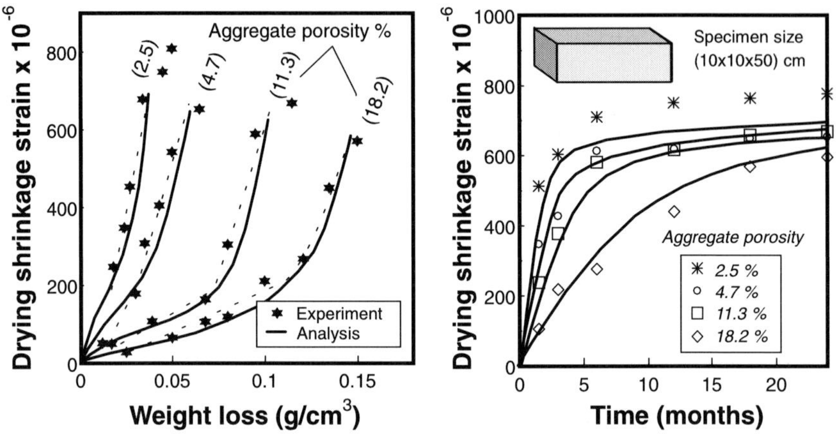

Figure 5.18 Computed and measured weight loss versus drying shrinkage relationships for lightweight aggregate concrete.

thermodynamic equilibrium with the surrounding cement pores (Figure 5.19). Since aggregates are discretely distributed in the matrix, the only way aggregates can do this is by unloading some of their moisture to the surrounding environment, i.e. cement-matrix pores. During the initial phase of drying, most of the cement-matrix pore system is kept wet by the moisture lost from aggregates. This in turn would mean little overall drying shrinkage strain, as the bulk of the contribution of shrinkage strain comes from the drying of cement-matrix porosity. Therefore, very little drying shrinkage will be observed until aggregates have released most of their moisture to the surrounding cement-matrix porosity by virtue of the equilibrium requirements as defined by their respective isotherms.

The condition of moisture interaction formulated in terms of the difference of pressure potentials of pore water of interacting components properly treats this phenomenon and defines the conditions of equilibrium correctly. The computed trends are similar to the one observed experimentally. For the same reason, the degree of saturation of components cannot be taken as the fundamental driving factor. In the previous analysis, interfaces were assumed to be absent. However, incorporation of interfaces would not change the fundamental nature of the drying behaviour. There would not be any appreciable shrinkage arising solely due to the moisture loss from interfaces. Therefore, the addition of this component would simply shift the origin of shrinkage strain–weight loss curves on the positive weight loss axis by an amount corresponding to the porosity contained in the channel regions.

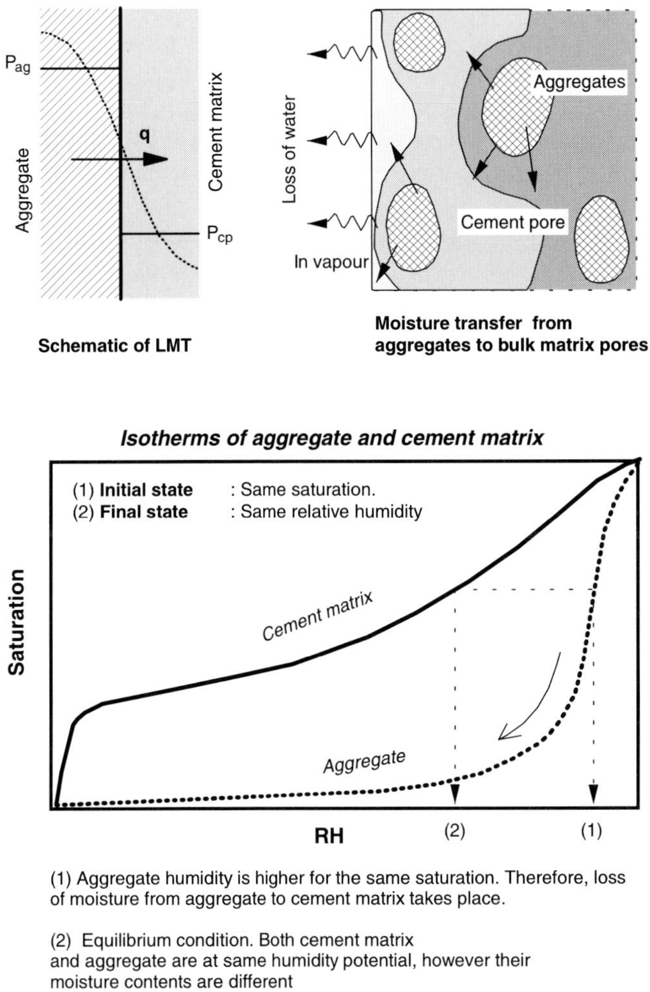

Figure 5.19 Local moisture transfer.

5.9 Summary and conclusions

Concrete is a composite material. Hence, to deal with the moisture transport process in such a material, the major components of porosity must be duly accounted for. The multicomponent formulation considers the hardened cement-paste matrix, aggregates and the interfacial zones as the primary components to describe the moisture transport behaviour. This formulation can incorporate the interactions among these components, in the form of quasistatic moisture interaction terms. Computer simulation methods suggest that aggregates play a crucial role in defining the role of coarser interfacial regions in moisture transport behaviour. For any given aggregate grading and *W/C* ratio, there exists a critical aggregate volume fraction at which the interfacial zones form a continuously connected path across the concrete. This critical aggregate volume fraction usually lies above 45%. The results of permeability and computer simulations suggest that the self-compacting high performance concretes typically do not contain a percolated path of interfacial zones, giving them higher durability. To a large extent, the *W/C* ratio defines the basic pore structure of a hydrated cement-matrix system. The interface zone characteristics are dependent on the aggregate porosity as well as the *W/C* ratio. For normal mix proportions (high water content and large aggregate volume fraction), interfacial zone characteristics are crucial in defining the overall transport characteristics of the hardened and hydrated concrete. In this regard, the critical aggregate volume fraction for a given grading must be obtained.

Regarding the moisture transport formulations, it appears that concrete can be treated as a uniform and homogenous porous medium as long as aggregates used in the mix are inert with respect to moisture. For high porosity aggregates, a multicomponent consideration is necessary. The explicit consideration of interfacial zones in the transport formulations may not be generally necessary. This is because, for low aggregate contents, the interfacial zones are discrete and unconnected, and therefore hardly affect the transport behaviour. On the other hand, at high aggregate contents, when a percolated path of interfacial zones exists, the transport characteristics are primarily decided by the moisture transport in the interfaces and moisture redistribution to the surrounding bulk matrix. If local redistribution is fast enough, the net effect of interface percolation is an apparent increase in the bulk conductivity of the composite. This is especially true in a concrete composite since a randomly dispersed structure of aggregates ensures that the interfaces are uniformly and closely distributed, thereby leading to a rapid local moisture transfer. Thus, a uniform porous medium assumption can still be applied to a percolated composite but with apparent transport coefficients that correspond neither to the interfaces nor the bulk cement paste matrix. Of course, if crack-like structures are present that are not closely dispersed, then a multicomponent approach must be adopted.

References

[1] Hall, C., Water sorptivity of mortars and concretes: a review, *Magazine of Concrete Research*, 1989, **41** (147), 51–61.

[2] Dhir, R.K., Hewlett, P.C. and Chan, Y.N., Near surface characteristics of concrete: intrinsic permeability, *Magazine of Concrete Research*, 1989, **41** (147), 87–97.

[3] van Breugel, K., Simulation of hydration and formation of structure in hardening cement-based materials, Ph.D thesis submitted to Delft Technological Institute, Netherlands, 1991.

[4] Bentz, D.P. and Garboczi, E.J., Percolation of phases in a three dimensional cement paste microstructure model, *Cement and Concrete Research*, 1991, **21**, 325–344.

[5] Synder, K.A. *et al.*, Interfacial zone percolation in cement aggregate composites, *Interfaces in Cementitious Composites*, J.C. Maso (Ed.), E&FN Spon, London 1990.

[6] Okamura, H., Maekawa, K. and Ozawa, H., *High Performance Concrete*, Giho-do Press, Tokyo, 1993.

[7] Stauffer, D. and Aharony, A., *Introduction to Percolation Theory*, Taylor & Francis, London, 1992.

[8] Chaube, R.P., Shimomura, T. and Maekawa, K., Multiphase water movement in concrete as a multi-component system, in Proceedings of the 5th RILEM International Symposium on Creep and Shrinkage of Concrete, Barcelona, E&FN Spon, London, 1993.

[9] Ishida, T. and Maekawa, K., Study of the effect of aggregate content on the permeability of concrete, *Proceedings of JSCE*, Annual Conference, 1994, 1020–1021.

[10] Connell, L.D. and Bell, P.R.F., Modeling moisture movement in revegetating waste heaps–I. Development of a finite element model for liquid and vapour transport, *Water Resources Research*, 1993, **29**(5), 1435–1443.

[11] Hall, C., Water sorptivity of mortars and concretes: a review, *Magazine of Concrete Research*, 1989, **41**(147), 51–61.

[12] Shimomura, T. and Maekawa, K., Analysis of the drying shrinkage behaviour of concrete using a micromechanical model based on the micropore structure of concrete, *Magazine of Concrete Research*, 1997, **49**(181), 303–322.

[13] Kokubu, M. *et al.*, Problems of lightweight concrete, *Concrete Library*, *JSCE*, 1969, No. 24, 1–13.

Chapter 6

A simulation model of early age development in concrete: DuCOM

- The early age development phenomenon
- DuCOM: the finite-element computational coupling of hydration and structure formation phenomenon with moisture transport in concrete
- Sample scenarios of use of DuCOM for practical use
- Case studies for the evaluation of DuCOM

6.1 Introduction

In Chapter 2 on the quantification of concrete durability, a proposal to quantitatively estimate the performance of concrete under generic conditions was discussed. It was shown that the strength and microstructure development of cementitious materials are the most basic and crucial factors that control the various mechanisms affecting the long-term performances of concrete. In this chapter, the authors seek to establish a rational computational framework for the durability analysis of concrete structures by quantitatively evaluating these parameters. The method involves a dynamic coupling of cement hydration, moisture transport and microstructure formation models discussed in previous chapters into a finite-element computational program. As a verification target, several case studies are considered which include, for example, the effect of different curing conditions on the strength gain, moisture loss and the microstructure development.

6.2 Coupled computational formulations

The system of analytical models describing the microstructure development, hydration and moisture transport is complex and furthermore inter-related. The only practical way of making good use of such a system is to combine these separate formulations into a unified computational framework. In fact, the use of the hydration model alone can predict the achieved performance of an isolated lump of concrete material, but this is not applicable to structural concrete actually inside a structure with nonuniform fields of temperature, moisture and degree of hydration. The hydration model without any coupling with moisture transport cannot fully contribute to engineering practices on concrete structures under natural environments. However, the unification of concrete modelling would enable us to study the overall early age development phenomenon under generic conditions from a material science point of view. Also, from an engineer's view point, the computational approach would enable us to apply these models to real-life structures and to study the effects of mix proportions, curing conditions and environments, etc., on concrete performance and associated overall structural durability. This process would help the concrete community in making more informed guesses in the performance-based design scheme of long-term serviceability and safety.

For computational integration, finite-element based methods are adopted because of their versatility. Another background point is that the structural nonlinear analysis under static as well as dynamic actions is coded extensively for structural safety and serviceability assessment based on a finite-element method. As Chapter 1 states, the authors aim to combine concrete performance assessment with time, i.e. the main theme of this book, and verification of structural functions.

The entire theoretical formulation of microstructure formation, moisture transport and hydration phenomena is integrated into a finite-element based computational program named DuCOM (durability model of concrete). The method is applied both in time and space domains to obtain the solutions for primary variables, i.e. pore water pressure P and temperature T, by mass and energy conservation. As a part of the framework, solutions for the internal variables are also obtained for the development of pore structure in terms of microstructure distribution and porosity of various phases, the average pore water content, degree of hydration of individual mineral components and the strength of the paste. The input required in this scheme is initial mix proportions, powder material characteristics (density and mineral compositions), initial temperature, the geometry of target structures in terms of finite-element geometry and the boundary condition to which the structure will be exposed during its life. A schematic representation of the overall computational scheme is shown in Figure 6.1.

For the numerical method of solution, the conservation equations for temperature (energy) and pressure (moisture mass) can be restated in a general Poisson form as

$$\alpha \frac{\partial X}{\partial t} - \mathrm{div}(D \nabla X) + Q = 0 \qquad (6.1)$$

where the specific capacity, conductivity and other parameters for both the cases are obtained according to Table 6.1 which summarizes various material modelling discussions so far.

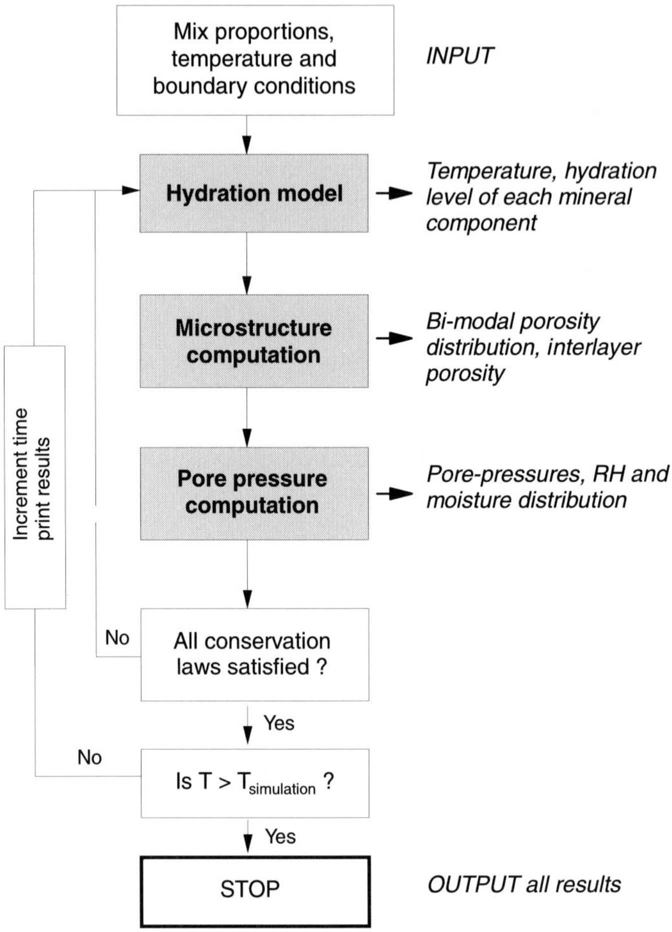

Figure 6.1 DuCOM computational scheme.

Mass conservation is simply extensible to chloride, carbon dioxide, oxygen and other ingresses related to steel corrosion. The dual system of mass and energy conservation gives another interpretation of the word DuCOM: (DUal system of COncrete). A standard Galerkin procedure is used to obtain the space discretizations of the system of partial differential equation given by equation (6.1). Since the temperature and pressure fields are inherently coupled in this system, we have adopted an alternate staggered method of solution. In this scheme, temperature and pressure fields are obtained alternatively in given steps of time, until complete convergence is achieved. This alternate scheme of solution has been

Table 6.1

Variable	*Transport parameters*		
	Specific capacity, α	*Conductivity, D*	*Sink term, Q*
P	$\phi\rho\ \partial S/\partial P$ Path dependent moisture isotherms [kg/Pa m^3]. Eqns (4.40), (4.20)	$K_v + K_l$ Random geometry of pores and Knudsen vapour diffusion [kg/Pa m s] Eqns (4.57), (4.58)	$Q_{hyd} + \rho S\ \partial\phi/\partial t$ Water combined due to hydration; bulk porosity change effects[a] [kg/m^3 s] Eqns (3.32), (4.17)
T	~ 641 [kcal/K m^3] (constant)	$\sim 4.75 \times 10^{-4}$ [kcal/K m s] (constant)	CH Multi-component heat of hydration model of cement[a] [kcal/m^3 s] Eqns (3.34)

[a] denote the parameters that require the knowledge of porosity distribution.

recommended for coupled systems in general, since it leads to many interesting possibilities of applications, for example:

1. completely different methods could be used in each part of the coupled system;
2. independent codes dealing efficiently with single systems could be combined;
3. parallel computation with its inherent advantage could be used;
4. efficient iterative solvers could be developed in systems with the same physics.

Perhaps, the computational time might increase in such cases, but stable convergence is guaranteed as compared to the direct simultaneous solution schemes. Due to the alternate solution schemes, the finite-element discretizations can be illustrated for the variable X and can be identically applied to T and P. By applying a one-step time discretization to equation (6.1), the following system of equations can be obtained

$$[\mathbf{C} + \mathbf{K}\theta\Delta t]\mathbf{X}_{n+1} = [\mathbf{C} - \mathbf{K}(1-\theta)\Delta t]\mathbf{X}_n + \mathbf{f}_{n+\alpha}\Delta t \tag{6.2}$$

with the usual meaning of symbols. We use **C**, **K** and **f** matrices as

$$\mathbf{f}_{n+\alpha} = -\theta\mathbf{f}_{n+1} - (1-\theta)\mathbf{f}_n$$

Volume integrals:

$$\mathbf{C} = \int_V \alpha\mathbf{N}^T\mathbf{N}\,\mathrm{d}V \qquad \mathbf{K} = \int_V \mathbf{B}^T\mathbf{D}\mathbf{B}\,\mathrm{d}V \qquad \mathbf{f} = -\int_{\mathbf{V}} \mathbf{N}^T Q\,\mathrm{d}V \tag{6.3}$$

Surface integrals:

$$\mathbf{C}=0 \qquad \mathbf{K}=h_X\int_V \mathbf{N}^T\mathbf{N}\,\mathrm{d}S \qquad \mathbf{f}=-h_X X_s\int_S \mathbf{N}^T\,\mathrm{d}S$$

The algebraic system of equations (6.2) and (6.3) is solved by using the skyline substitution method. Convergence of the system is based on the limitation of relative errors of **P** and **T**, which is stricter than the usual total limitation criteria. Moreover, for stability of the solution and spurious oscillation removal, especially under rapid change situations and at the very beginning of the solution procedure, a complete diagonalization of **C** based on mass lumping parameters [1] has been adopted. Also, for guaranteed stability, θ is usually taken as $\frac{2}{3}$ for time discretizations. In the computational program, boundary conditions can be specified as known values of the primary variable or in terms of a convective condition, where the ambient value of the variable and convective transfer coefficients as h_X should be specified.

For example, in the case of moisture transport, boundary conditions could be specified as a given pore pressure head directly or in terms of the ambient relative humidity at any given time, i.e.

$$\begin{aligned} P &= P_s \\ q_s &= -E_b(h-h_s) \end{aligned} \tag{6.4}$$

where q_s represents the flux of moisture into the porous media at the surface. Also, P_s is the specified pore pressure head, and h_s is the environmental humidity corresponding to a pore pressure of P_s. A value of 10^{-5} m/s for surface moisture emissivity coefficient, E_b, is used to represent the usual convective moisture transfer conditions at the surface.

DuCOM is a full three-dimensional computer program and useful for any shaped structure. For practical cases, we often encounter a nearly one-dimensional flow condition with respect to both mass and heat, and the development of the performance of the concrete skin below the surface is of great interest to engineers. Then, an adaptation of DuCOM for the one-dimensional cases is freely available to anyone connected to the World Wide Web as briefly described in the Appendix B. The quality assessment of cover concrete for slabs and walls with a two-dimensional wider extent can be made. The drying shrinkage crack risk assessment of very massive concrete structures close to the surface is possible within one-dimensional restraint. The quality of concrete with and without mechanical defects serves as a main parameter for durability performance assessment of concrete structures.

In the computational framework, the only basic input required are mix proportions, the properties or type of cement and powder materials, the geometry of the structure, the initial casting temperature and the boundary conditions specified in terms of history of exposure of the structure to the environment. All other parameters are intrinsically computed based upon micromodels of material behaviour. For example, the critical parameters required to evaluate various transport coefficients in the moisture transport formulations are the pore distribution parameters B_{cp} and B_{gl}, and the total porosity of interlayer (ϕ_{lr}), gel (ϕ_{gl}) and capillary (ϕ_{cp}) components. In the course of simulations, these parameters are actually obtained as an output of the degree of hydration dependent microstructure development model. For a fully matured concrete, however, these can also be estimated approximately from the experimental measurements of porosity and pore distributions as obtained by MIP methods. The interdependency of seemingly different physical phenomena during the hardening stage of concrete can be therefore well appreciated in such simulation methods.

6.3 Model features and accuracy considerations

This section aims to investigate the effect of macroparameters usually referred to in concrete material science and engineering on the hydration and structure development phenomena. The results are discussed in terms of the relationships between intrinsic material models and, in the process, accuracy of such intrinsic parameters is questioned. Experimental results available in the literature or in this book are also used for verification with systematically arranged parameters.

6.3.1 Effect of chemical composition of cement

The chemical composition of cement and the effect of different powder materials are considered in the heat of hydration model of powder materials as discussed generally in Chapter 3 and in detail in Chapter 7. If the concrete is isolated in regard to mass and energy release, moisture transport model and pore structure formation have nothing to do with the temperature change of the entire concrete. This is done by adopting heat generation models based on Arrhenius' law of chemical reaction.

In a pioneering work, Suzuki *et al.* [2] applied the temperature-dependent model to represent the hydration phenomenon and thermal fields for assessing initial defects in massive concrete structures. At this stage, the cement hydration process was simply treated as being averaged as a single chemical reaction. Thus, the basic parameters, such as the relationship of basic heat generation rate with the accumulated heat and average thermal activity, had to be obtained experimentally on a case-by-case basis. When the kind of cement and/or mix proportion with pozzolan additives was changed, the adiabatic temperature rise measurement had to be performed again to deduce the average activation energy and reference heat liberation rate.

Later, Kishi extended the same concept to each mineral reaction of cement clinkers including their interactions. Here, it is not necessary to conduct tests to identify the heat properties of cement, but it is simply required to input the weight ratio of the cement mineral compounds reported in the mill sheet of cement products by manufacturers. DuCOM installs this multicomponent hydration model applied to Portland cement and mixed powders with pozzolans as discussed in Chapter 7.

The relationships of reference heat rate and the thermal activities with accumulated heat of each mineral are the material parameters of this model that are fixed intrinsically. These intrinsic parameters have been obtained from numerous sets of inverse data analysis. Furthermore, the intercomponent interaction models, which make use of the results obtained from stoichiometric chemical balance of hydration reactions can consider the effect of alkalinity and organic admixtures. Figure 6.2 shows a comparison of the computed and measured degrees of hydration and temperatures for different cement types. The Bogue chemical composition of cement used in these computations is given in Table 6.3.

Alite-rich Type III cement shows early development of strength, whereas belite-rich Type IV cement which has a very low heat generation rate shows the development of strength and hydration over a longer period of time, ultimately giving higher strength than any other cement.

6.3.2 *Effect of cement fineness and particle size distribution*

The effect of cement fineness is, in general, to increase the rate of hydration since a larger surface area of the cement grains is available for surface reactions. The current hydration model does not directly consider the effect of cement fineness. However, the microstructure development model makes

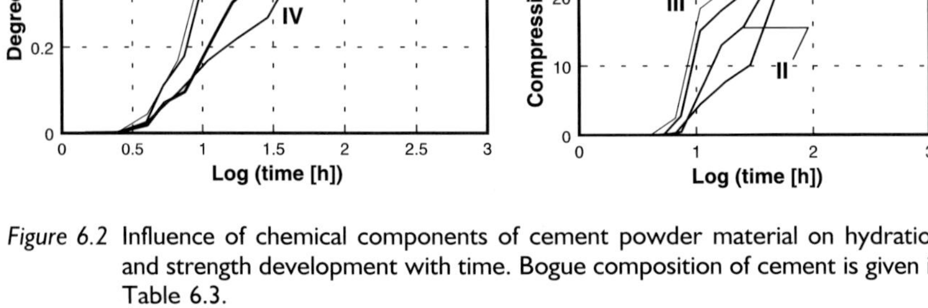

Figure 6.2 Influence of chemical components of cement powder material on hydration and strength development with time. Bogue composition of cement is given in Table 6.3.

Table 6.2 Bogue composition of cement types used in computation

	C_3S (%)	C_2S (%)	C_3A (%)	C_4AF (%)
Type I	51.4	22.6	11.1	7.9
Type II	41.6	34.4	5.4	13.2
Type III	60.0	13.5	8.9	8.1
Type IV	24.0	51.5	4.9	11.6

use of this information in deciding the cement grains' packing efficiency. This affects the structure development process. Therefore, only under those conditions when microstructural characteristics can influence the hydration process (e.g. low relative humidity curing), can we observe the effect of cement fineness. Thus, the current model would not show any difference if the curing was performed under fully saturated conditions. In reality, however, some effects of cement fineness have been reported [3]. The effect of particle size distribution is not considered in the formulations at the current stage since the powder phase is assumed to be monosized. Figure 6.3 shows the effect of cement fineness and particle size distribution on the degree of hydration in the simulation model and, as expected, does not show

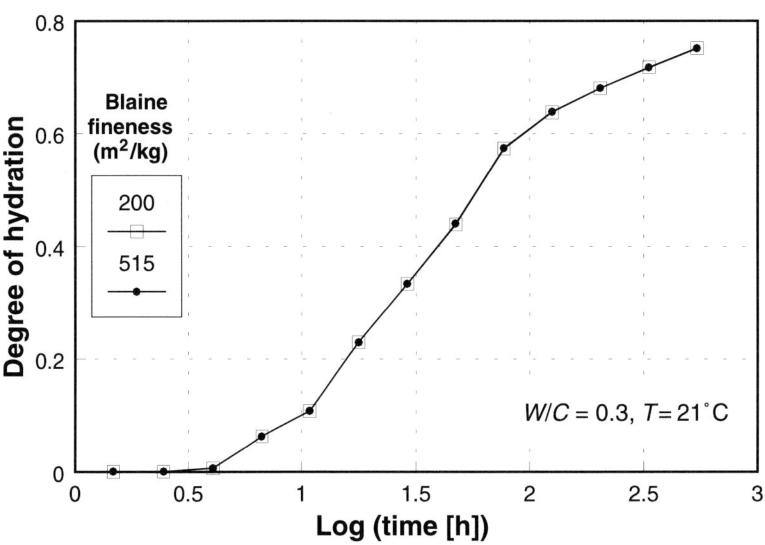

Figure 6.3 Computed influence of cement fineness on the progress of hydration.

any difference. A refinement in the current model to consider the cement fineness effect is desired.

6.3.3 *Effect of water-to-cement ratio*

The overall hydration level that can be achieved in a concrete mix increases with the water-to-cement ratio (W/C ratio). The W/P ratio is one of the most important factors influencing the early age and strength development process as well as the mechanics of deformable fresh concrete. In the models of this book, the influence of the W/P ratio on the development of concrete can be primarily attributed to the stereological aspects as well as the heat of hydration models of concrete. These are briefly explained below.

STEREOLOGICAL ASPECT

The water-to-powder ratio directly influences the stereological aspect of microstructure formation by affecting the mean particle-to-particle spacing. At a high W/P ratio, the mean free space available for expansion of mineral particles will be large, resulting in a lower specific surface area. In the model, a mean particle diameter for the powder mixture is assumed to compute the free space.

The total free space (voids) also accounts for the bulk of capillary porosity. If a capillary/gel porosity model is assumed to be a major factor in the porosity of the total hydrated mass, gel porosity is dependent on the weight of powder materials per unit volume of mix. Thus a high W/P ratio means a higher fraction of total porosity residing in the solid hydrated mass (or gel) which would account for the fine porosity distribution. Capillary porosity can be computed thereafter by subtracting the volumes of unhydrated and hydrated mass from the total volume. Therefore, both the content and distribution of pore space has a direct bearing on the W/P ratio.

RATE OF STRUCTURE DEVELOPMENT AND LEVEL OF HYDRATION

In the hydration model, the rate of hydration is dependent on the cluster thickness and free water-to-powder ratio at any instant. Since free water is dependent on the initial W/P ratio, boundary condition and capillary porosity, an implicit interdependence of the rate of structure development on the W/P ratio occurs. This relationship becomes predominant in conditions where structure formation is strongly influenced by the rate of hydration and hence microstructure development. Typical curves of the degree of hydration with time for different water-to-cement ratios are shown in Figure 6.4 for the Bogue compositions of cement given in the figure.

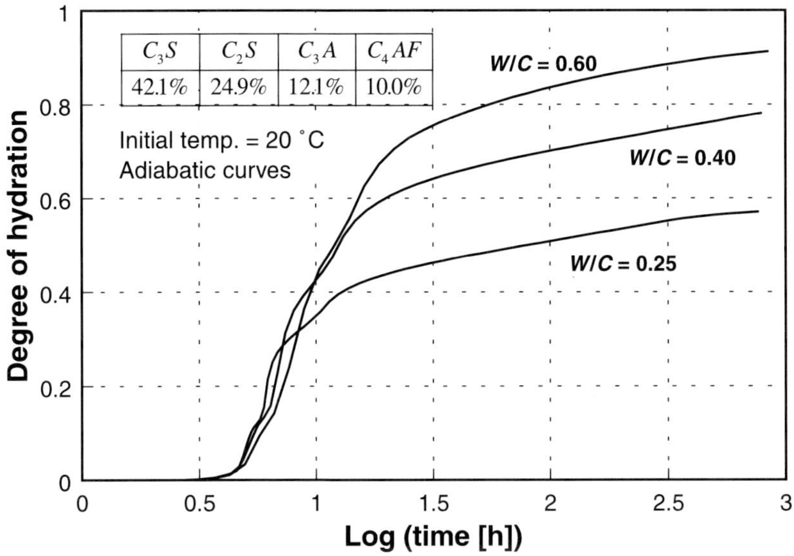

Figure 6.4 Typical progress of hydration curves for different W/C.

6.3.4 *Effect of casting temperature*

Thermodynamic processes and reactions are highly dependent on the phase temperatures [4]. The effect of temperature is directly considered in the heat of hydration model through the Arrhenius' law. Small changes in the casting temperature can give significantly different results. Actually, some production procedures for fresh concrete with cooling have been introduced for massive concrete constructions to avoid thermally induced cracking. Higher temperature does not only result in higher reaction rates and heat liberation but also changes the morphological features of the matrix. Due to higher reaction rates at higher temperatures, the hydration products are produced rapidly and precipitate near the reactant particle surfaces owing to high product concentrations. This produces a coarse microstructure having non-uniform crystals.

The temperature dependence of morphological characteristics of the microstructure is currently not considered explicitly in the microstructure development model. This can be achieved however by considering a temperature history based volume to surface area model of the crystals of hydration products. Of course, higher rates of hydration due to higher temperatures would lead to faster microstructure and strength development.

Since the microstructure development model used in DuCOM is based on the average degree of hydration, these effects are automatically computed.

The effect of curing temperature on hydration is shown in Figure 6.5. Generally, a higher casting temperature leads to a rapid start of hydration due to reduced requirements on the energy barrier according to Arrhenius' law of kinetic reactions. Moreover, the level of ultimate temperature rise in adiabatic heat of hydration measurement experiments appears to be generally constant. For example, Figure 6.5 shows the case of an OPC mix and blended mix (PC + fly ash), where the ultimate temperature rise is about 60 °C and 40 °C respectively from the casting temperature. The rate of temperature rise as well as the ultimate level of temperature can be predicted reasonably in both these cases. The temperature rise depends not only on the thermodynamic properties of the powder materials but on the amount of powder material in the mix. The specific heat capacity of the hydrating mix might control the temperature rise. In general, for concrete, heat capacity appears to increase with temperature and reduce with the bulk weight.

6.3.5 Effect of curing conditions, ambient RH and temperature

Under a generic curing condition that involves form stripping after a stipulated period of time in arbitrary environmental conditions, all the mechanisms of moisture transport, hydration and structure development are active and significantly influence each other. The coupling of hydration and moisture transport occurs by virtue of the amount of free water available in the concrete microstructure. Under sealed or fully wet conditions, moisture

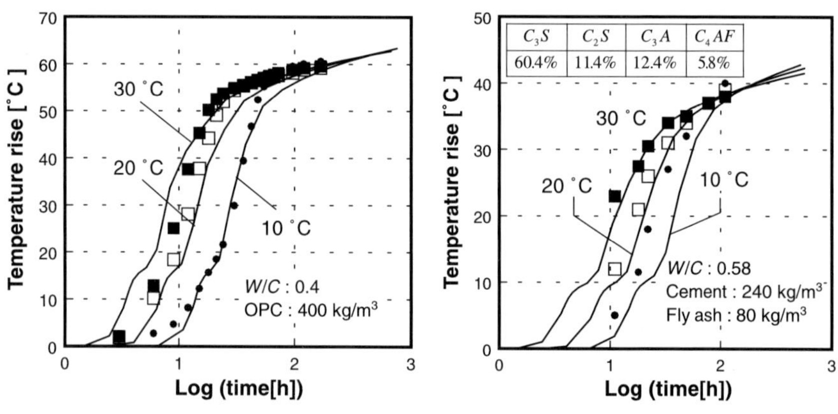

Figure 6.5 Computed and measured temperature rise for different mix proportions, showing the influence of curing temperature on progress of hydration under adiabatic conditions. Lines show computed results.

transport is hardly an issue since there is no moisture loss to the external environment and therefore the free water content never reduces below the critical requirement. A retarded hydration due to lack of free water would produce a coarse microstructure that would lead to even higher moisture loss. The basic rules defining these dependencies are simple and unchanging in nature, since in this case we make use of only a few of the several parameters that DuCOM computes during the course of simulation.

Figure 6.6 shows the simulated effect of curing on strength development and weight loss along different points of a cylinder specimen used for the

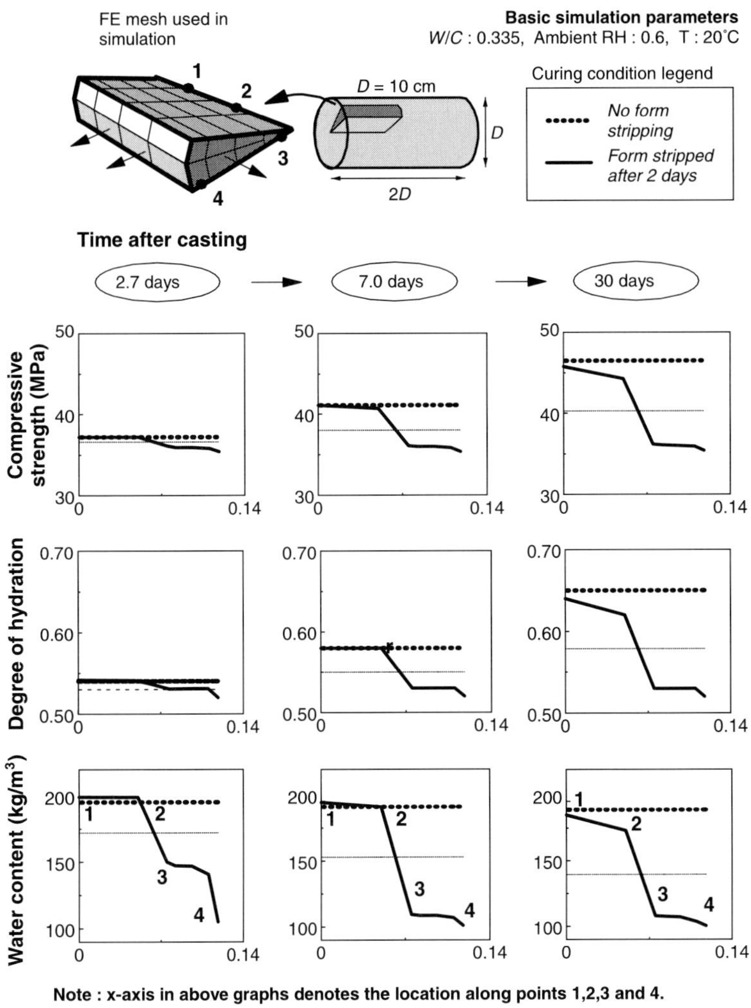

Figure 6.6 Effect of curing on spatial distributions of key material characteristics, such as strength and degree of hydration.

standard strength test. Under moisture sealed curing conditions, most of the hydration process occurs within 2 days of casting. Also a rapid increase of strength gain accompanied by particle-to-particle contact occurs after a few hours of casting. Such processes could be simulated and verified by the finite-element simulation methods. It is clear that a completely sealed condition leads to perfect uniformity of strength, hydration and water content in space. But, early form stripping brings about much poorer concrete performance close to the surface and premature incomplete hydration is seen. When we direct our attention to the strength distribution, a 2–3 times difference in achieved strength is seen. Thus, the averaged compressive strength of the cylinder shown in Figure 6.11 and Figure 2.6 is computed as the average of local strengths. The strength of the core centre is close to that of the sealed condition, since the water loss hardly reaches the core. It implies that the cylinder strength as a mechanical performance of concrete is very size dependent when the concrete is exposed to early drying.

The water content at the corner is much reduced, especially at the early stage of hydration and drying. The associated strength and degree of hydration are also reduced. The location dependent strength is a factor for assessment of the cracking risk of concrete under drying and temperature variation with volumetric change. Furthermore, the poor quality that tends to be developed close to the skin concrete is of great importance for long-term protection against corrosion of steel. The DuCOM analysis can provide the quantitative concrete performance, especially related to the durability of structures. As stated in Chapter 3, the input value of the water-to-cement ratio is not the one specified at the concrete production stage but is that of fresh concrete at the start of hardening. Thus, if the target concrete is likely to exhibit segregation or premature compaction, the estimated substantial mix of fresh concrete has to be input.

6.4 Verifications and practical evaluations

The overall computational system of simultaneous structure formation, hydration and moisture transport was applied to several test cases to check the accuracy and verify the overall moisture transport computational systems. The simulation system was verified by parametric analysis, solved for some ideal cases and lastly applied to predict the experimental observations for various cases. By using these computation methods, the influence of factors, such as the W/C ratio, mix proportions, curing conditions and specimen geometry, etc. on the microstructure and strength development can be rationally studied.

Moreover, plugging in other deterioration models (e.g. carbonation, chloride movement, etc.), long-term durability of concrete structures can also be predicted once the environmental conditions are known. Early age development is the prime area of interest in this book, since it involves the simultaneous occurrence and couplings of hydration, structure development and moisture transport processes described in the previous chapters. Also, the developments during the hydration phase lay the foundations for the long term durability of the concrete.

6.4.1 Cyclic drying–wetting of mortars

The computations were performed to predict the weight loss behaviour and moisture isotherms of mortar samples with time under cyclic drying–wetting conditions. The moisture isotherms were measured by subjecting 1 cm^3 samples of mortar to known repetitive drying and wetting histories. The temperature of the humidity chamber was kept constant during the experiments as shown in Figure 6.7(a). The change of weight measurements provided the drying and wetting paths of these isotherms. Similarly, experimental procedure, data [5] and computed results for another set of data are shown in Figure 6.7(a). In this case, the specimens were put under extremes of cyclic conditions, i.e. vacuum drying and then exposed to near 100% humidity in a cyclic manner. In the computations, these states were taken as 1% humidity during drying and 95% humidity during wetting.

Reasonable agreement can be observed in the rate of moisture weight loss as shown in Figure 6.7(c). The use of hysteresis models in the sorption isotherms improves the agreement as shown in Figure 6.7(b). In practice, for the modelling of concrete performance, alternate humidity generated by the natural environments has to be fairly simulated as a normal condition surrounding structures. Since the moisture migration has much to do with ion transport, the prediction of moisture content is central to the long-term behaviour and performance of concrete.

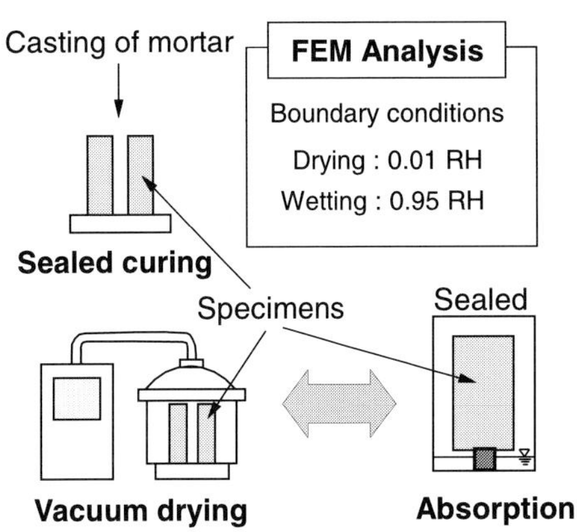

Figure 6.7(a) Cyclic drying–wetting chamber with vacuum pump.

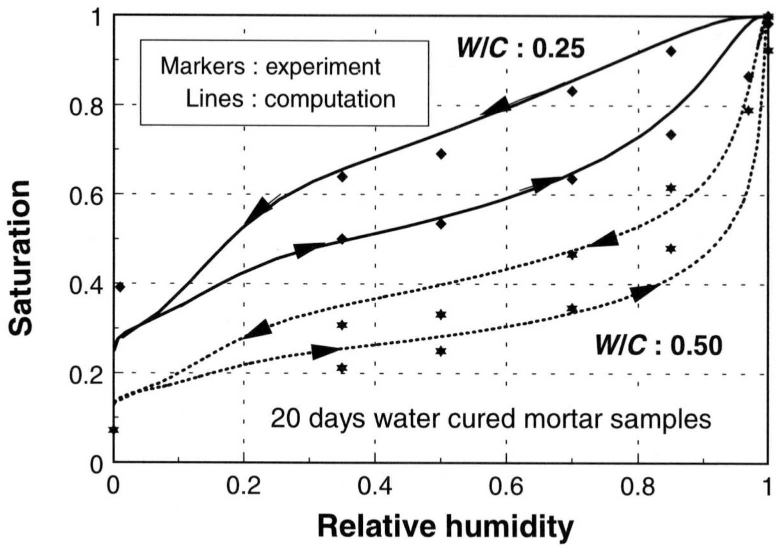

Figure 6.7(b) Predicted and computed moisture isotherms.

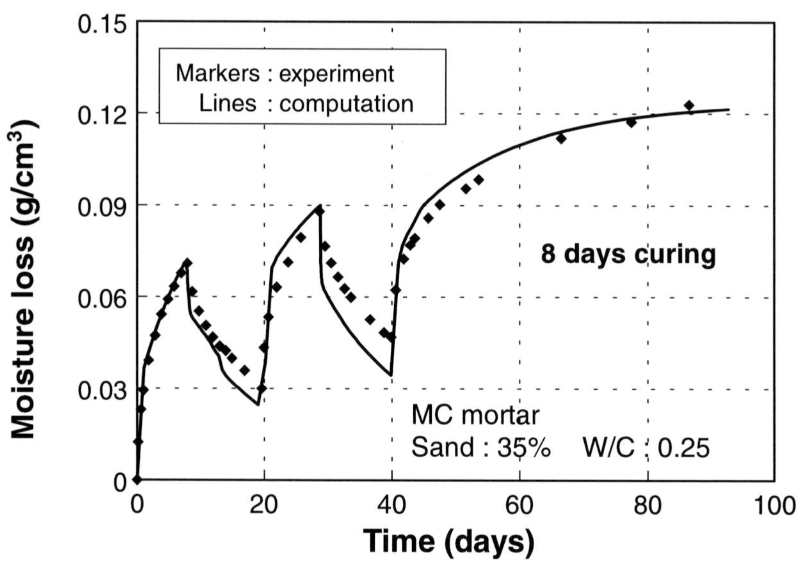

Figure 6.7(c) Prediction of moisture loss under cyclic drying–wetting conditions.

6.4.2 *Water sorption in one-dimensional mortars*

It appears that the computational model of conductivity of water in a cementitious matrix works reasonably well for arbitrary drying–wetting cases of the porous medium and the weight change curves of mortar specimens due to moisture gain or loss can be traced quite satisfactorily with time. Figure 6.8 shows the moisture gain curves for different dry mortar specimens that were exposed to one-dimensional water sorption experiments. This boundary condition is applied to the case of water gain by rainfall after dry weather. The opposite extreme end of a mortar prism is open to the air and free bulk motion of gas evacuated by the water gain from the wet face is assumed. A clear deviation from the square-root law of water sorption can be observed here which is also reasonably predicted by a history dependent water permeability model.

The deviation from a square-root law is exhibited not only by the small W/C mortars but also by large W/C mortars of up to 60% by weight. Of course, for the initial few days, sorption follows a near square-root path, but as the conditions tend to become nonideal a deviation from the expected path of sorption is observed. The deviation is not because steady-state conditions are achieved since the samples used were quite long (16 and 20 cm long). The deviation is primarily due to a retardation in the advancing wetting front of water with time. This phenomenon is clearly exhibited in the moisture profiles measured for a mortar prism that was exposed to wetting on one face and the other faces were open to the ambient relative humidity. The decrease in W/C means denser micropores with less voids and smaller sizes of micropores with higher surface tension. The higher traction forces of the pore water and the greater reaction against water migration result in a

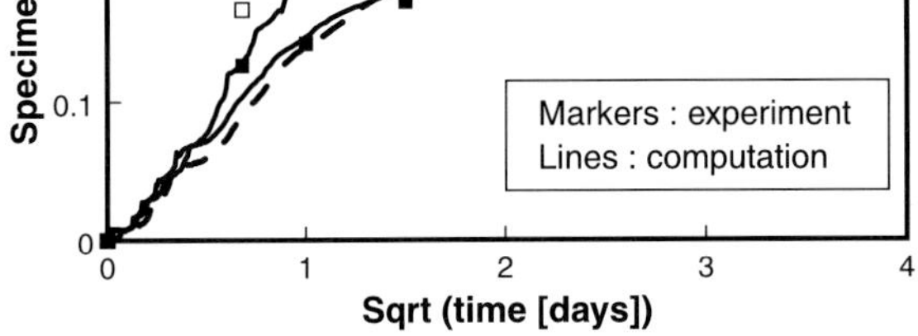

Figure 6.8 Prediction of one-dimensional water sorption in mortar prisms. Sand volume fraction = 0.6 and 0.4 for W/C. All mortar specimens were wet cured for 28 days.

complexity of water sorption as a whole. From the sensitivity analysis shown in Figure 6.8, it can be said at least that the overall resistance to bulk penetration of moisture is much deteriorated by increasing W/C, no matter how small the suction in the micropores might be.

6.4.3 Predictions of pore structure under severe drying

The first case study of pore structure formation and its verification is the comparison of the effect of early drying on microstructure formation and moisture loss behaviours. Specimens made of mortar of 0.265 W/C ratio were exposed to one-dimensional drying in 40% RH and 20 °C after one-day moisture sealed curing. After 8 days of drying, the microstructures were measured, at 1 cm intervals from the exposed surface. The experimental data showed that hydration almost immediately stops near the surface after exposure while the inner parts continue to hydrate, adding to the inside mix water. The microstructure formed in the surface zone is markedly different and coarser compared to the inner zones. A numerical simulation using the same initial and boundary conditions as the experiment shows a similar trend and exhibits an extremely strong coupling between hydration and the moisture transport process at early ages.

It is observed that while no development of microstructure occurs near the surface due to a rapid loss of moisture, the inner zone of the sample continues to hydrate and significant microstructural development takes place. The computed moisture profiles at 2 and 8 days after drying also show reasonable agreement with the measured moisture profiles (see Figure 6.9). From the profile, it is found that approximately 2 cm from the surface the quality of the concrete is much deteriorated, and at depths of more than 2 cm it is nearly the same. As a matter of fact, the size of the deterioration zone is similar to the size of cover concrete expected for protecting reinforcement.

6.4.4 Predictions of weight loss with time in vacuum drying

The second case study is the prediction of weight loss with time for specimens of different mix proportions and cured for different periods. The chemical compounds of the powders, the sand volume and water-to-powder ratios were changed during the experiments. After a stipulated period of curing as shown in Figure 6.10, the specimens were put in a vacuum desiccator and mass measurements were taken with time [6]. Numerical computation using a finite-element scheme as described in Figure 6.1 was performed to simulate the experimental procedure to obtain quantitative estimates of various physical quantities related to the early age development and drying behaviours. The computed results show reasonable agreement not only in terms of the rate of moisture loss, but also in terms of the absolute amount of moisture loss of various mortar specimens (Figure 6.10).

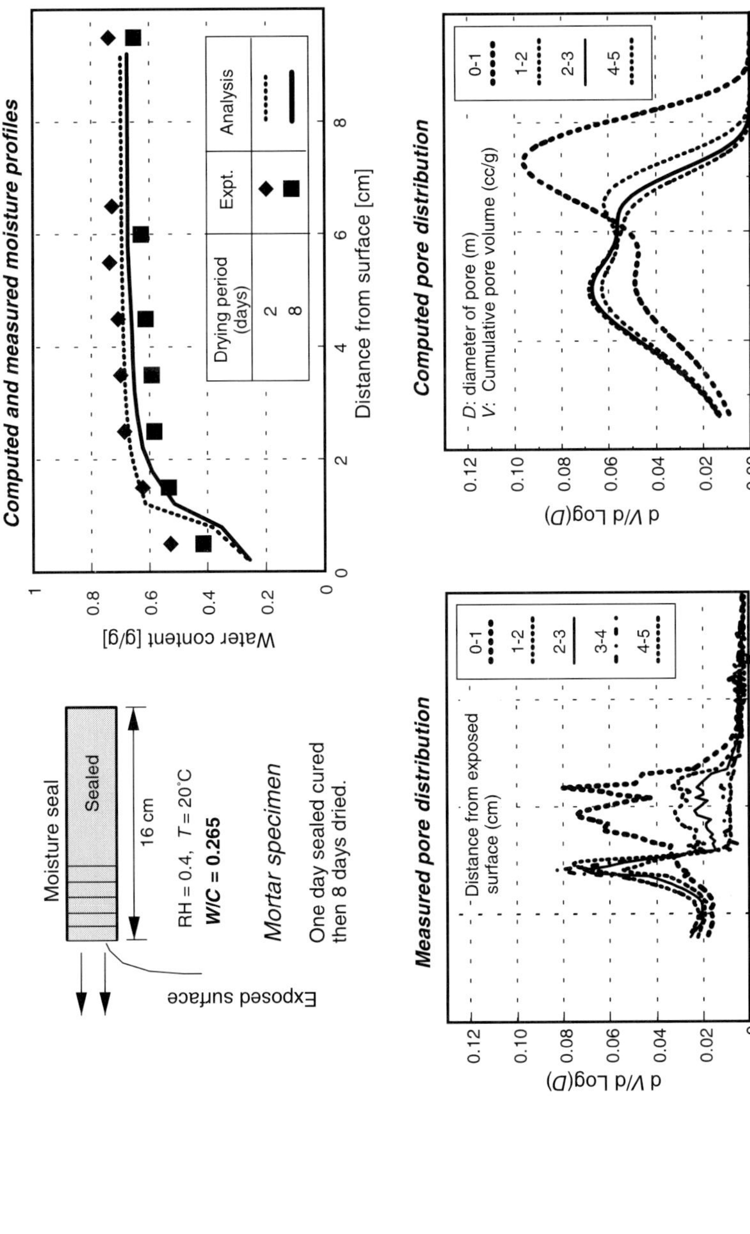

Figure 6.9 Verification of the influence of early age drying on microstructure development.

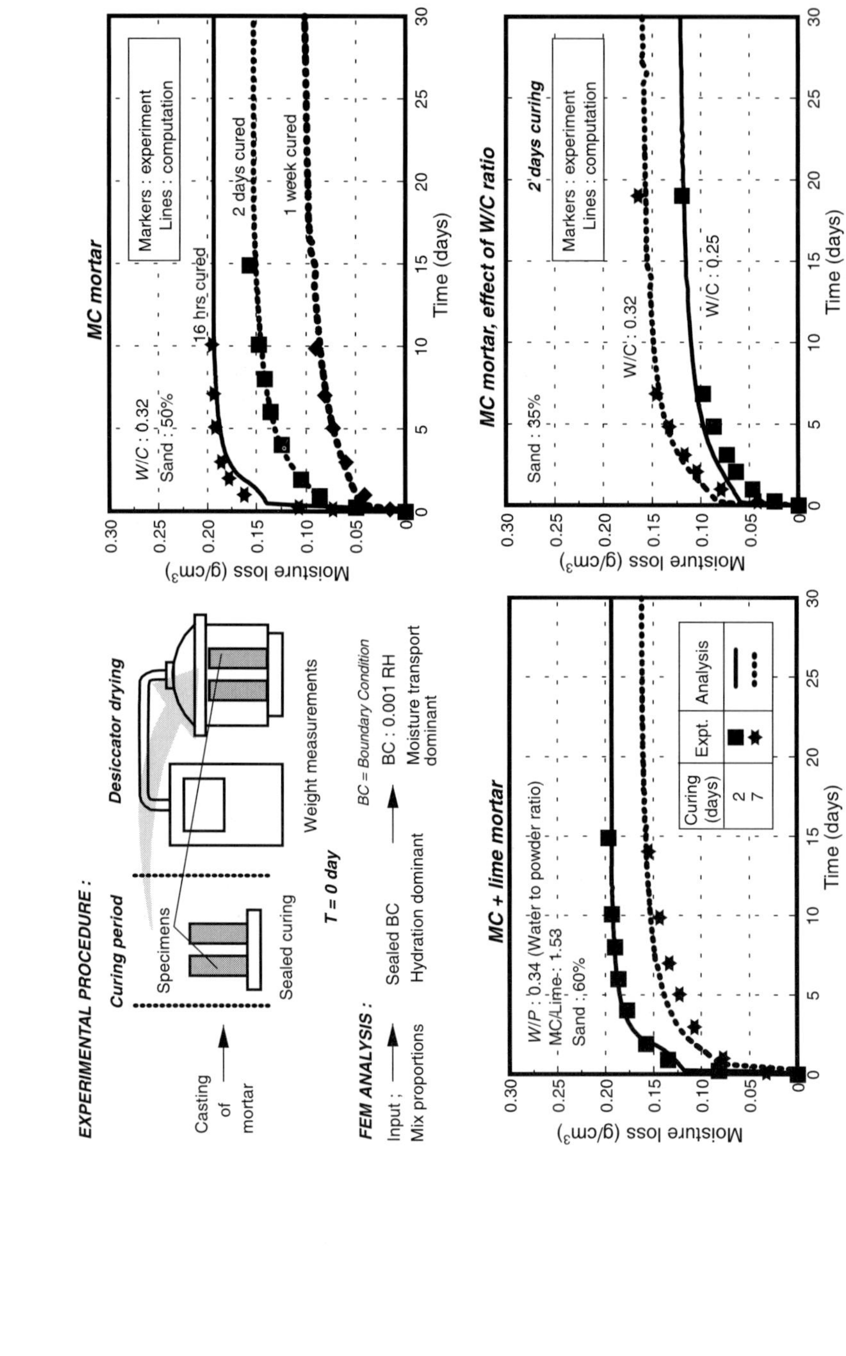

Figure 6.10 Prediction of moisture loss behaviour for early age under severe conditions.

It is clear that the insufficient curing may lose much water and a lower water-to-powder ratio can keep mix water even under a severe drying environment.

However, it must be stated that the agreement of analytical results for other experimental cases, where the mix was prepared with slag and pozzolans as powder material, does not show as good agreement compared to the results of Figure 6.10. Probably the treatment of powder material as single-sized particles tends to overestimate the capillary porosity distribution leading to greater conductivity and moisture loss.

6.4.5 Predictions for the effect of various curing conditions on strength development and weight loss

Three different case studies were conducted to evaluate quantitatively the coupled computational system of moisture transport and hydration. The first case involves the quantitative study of the effect of different curing conditions applied to various mortars, on their strength development as one of mechanical performances. The experimental strength development and moisture loss with time was obtained by the authors for different curing conditions and various mix proportions as shown in Table 6.3. Here, the water-to-cement ratio ranging from 33.5% to 55% and the mineral compounds of the cementitious powders vary widely. The very early form-stripping case was tested by considering slip-form continuous concrete placing in practice and the adequate development of young aged strength is a key factor in deciding the rate of form lifting. We have similar cases for planning the prestressing procedures for bridges and towers on-site.

Figure 6.11 shows a comparison of predicted and experimental compressive strength values at 7 and 28 days for mortars of different mix proportions exposed to various curing conditions as listed in Table 6.3. The computed strength of the cylinder is the average of locally developed strengths as discussed in Figure 6.6. Two marks in each combination of

Table 6.3 Mix proportions and curing conditions [7]

Case	W/C (%)	Air (%)	*Mix unit weight (kg/m³)* Water	Ordinary Portland cement	Medium heat cement	Lime	Slag	Sand	Gravel
MS	33.5	3.5	172	–	513	28	–	828	827
S6	55.8	3.5	172	–	308	17	200	828	827
OP	55.0	4.5	165	300	–	–	–	927	924

Case	Curing condition
SL	Sealed
16	16 hours stripped
2D	2 days stripped
WT	Submerged

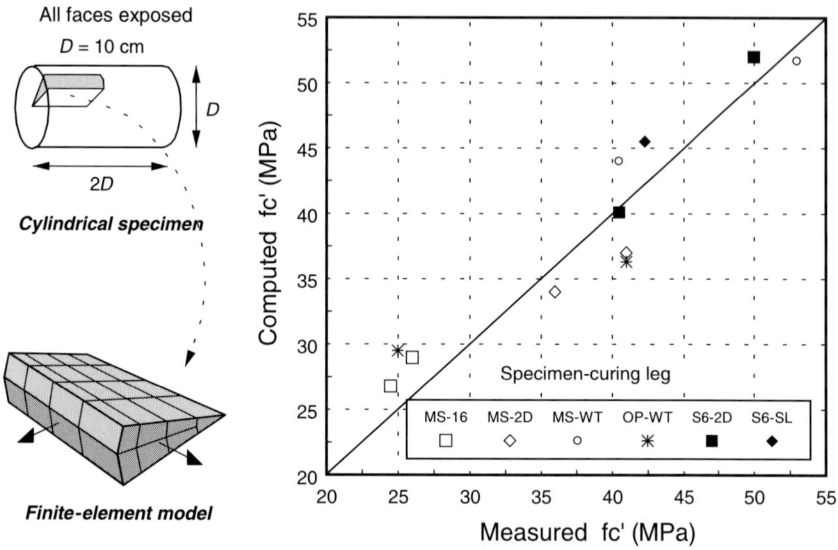

Figure 6.11 Prediction of development of compressive strengths in various mixes under different curing conditions (6 different marks) at different material ages of 7 and 28 days (lower and higher strengths in each mark).

concrete type and curing can be seen in Figure 6.11. The mark of higher strength corresponds to the concrete tested 28 days after production.

Wet curing for more than 2 days reproduces adequate strength development, but the sealed condition leads to the drop of strength since additional water is not supplied for the case of S6, which includes a slag consuming much water than ordinary cement minerals. The most severe curing of 16 hours form stripping creates much lower strength development and an additional strength gain after 28 days can hardly be expected. As is well known, the absolute strength is very dependent on the water-to-cement ratio and type of cement. These values at 7 and 28 days are well predicted and the perfect coincidence is qualitatively shown in Figure 6.11.

6.4.6 Curing period and quality of cover concrete

It is well known that the period of curing necessary for fly ash cement and slag concrete is longer compared to the case of ordinary Portland cement where we expect a concrete performance equivalent to that of OPC concrete. Codes of construction have widely recognized this fact and usually rough guidelines are given for longer curing periods in the case of pozzolan cement concrete (~2–3 times, required for OPC). The need for the longer curing periods is recognized based on the following background.

The rate of hydration of pozzolan-mixed cement concrete is usually lower than in the pure OPC case and, in most cases, the ultimate level of hydration achievable for the OPC case cannot be achieved for pozzolan cement concrete (especially fly ash). Moreover, the pozzolanic cement consumes higher amounts of water during hydration. Thus, early form stripping might result in significant loss of free water in cover concrete, leading to a much poorer quality of cover. Of course, the sensitivity of water loss on the retardation of hydration reactions of pozzolanic cement is dependent on the chemical composition of the powder materials. The aim of this section is to consider the influence of different curing conditions on the quality of cover of pozzolan cement concrete by applying the coupled simulation system. The results are then compared with the corresponding loss in quality of an OPC concrete for similar curing conditions. This comparison will allow us to make a quantitative judgment on the curing conditions to be adopted for pozzolan cement concrete to achieve a given performance level.

To quantitatively assess the performance level of cover concrete compressive strength and concrete tightness in terms of intrinsic permeability are evaluated in the simulation. The water-to-powder ratio is kept constant at 0.5 by weight and one mix type of pure OPC and two pozzolanic mixes containing OPC and fly ash in varying proportions (80 : 20 and 60 : 40) are considered. The curing conditions adopted are form stripping at 16 hours, 2 days and 7 days, perfectly sealed curing without any moisture ingress and wet curing with a free water supply. The performance level criteria, such as permeability and strength, are then computed for each of the mix types at the end of a specified curing condition and period. Figure 6.12 shows the computed results of these quality indicator parameters for different curing conditions (and periods).

As can be expected, there is a huge difference in the cover concrete quality indicator levels for 16 hours curing and sealed curing for all concretes. In most of the cases, based upon the design performance level that is desired, suitable curing conditions can be adopted. What is more interesting in these figures is that the relative gain in the quality levels for different curing conditions depends on the material of the concrete. Also, additional curing periods might be required for some pozzolan concrete to gain the same absolute level of performance as OPC concrete. For example, if requisite quality levels of cover concrete tightness and strength are fixed at some absolute level, OPC mix achieves these levels faster than pozzolan cement concrete containing large amounts of fly ash. To gain the same performance level, the curing periods required for 60 : 40 mix may be up to twice as large as those for OPC concrete (Figure 6.12). But, the longer curing periods depend on the powder material composition as well. For example, the 80 : 20 mix shows similar performance levels to those of the OPC mix, since the fineness effect of fly ash in this case is counterbalancing the reduced degree of hydration effect.

This study also shows that the estimates of curing requirements for different combinations of arbitrary cement concrete and performance levels can be obtained by applying the concepts of coupled transport and

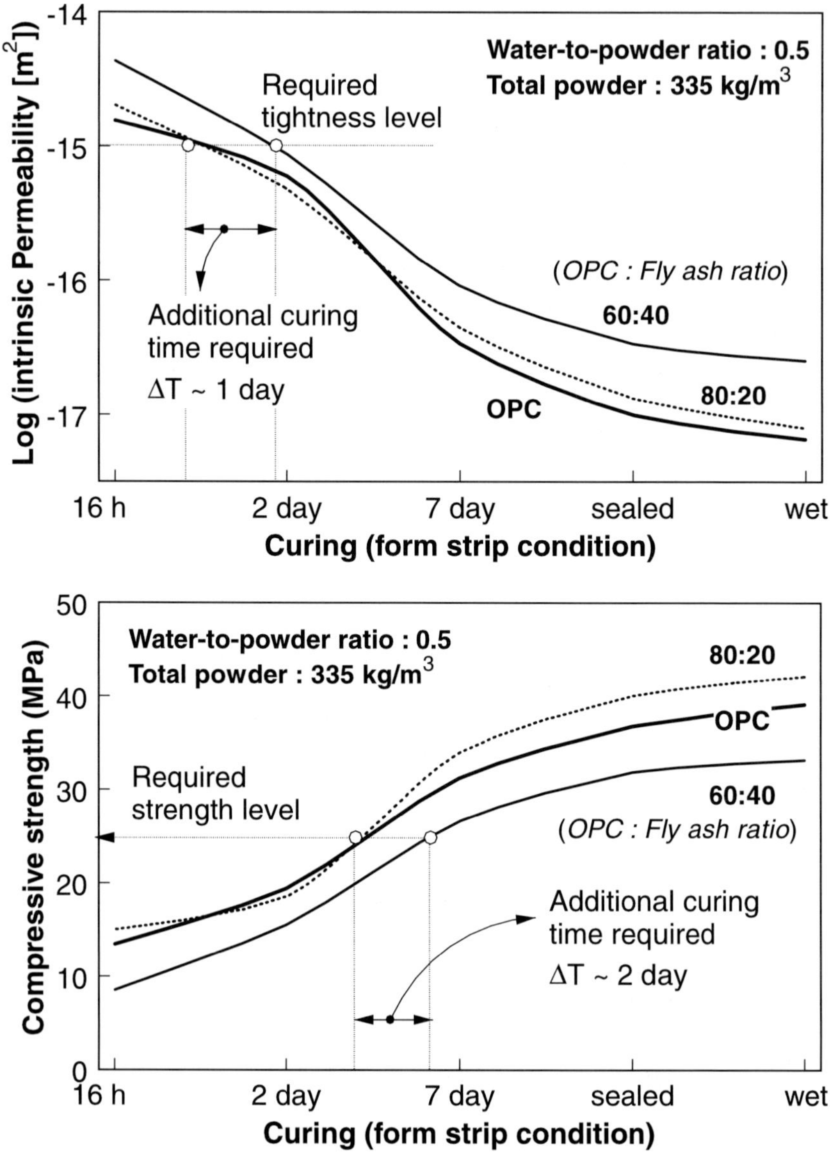

Figure 6.12 Influence of (sealed) curing period on the quality of cover concrete and prediction of curing periods for different mix types to achieve a given performance level.

hydration processes. The modelling of concrete performance can explicitly evaluate the concrete quality achieved inside concrete structures. This is the direct way of using performance assessment. At the same time, the modelling can be implicitly used at the trial-and-error stage of designing for deciding some detailed items (cover size of skin concrete, period of curing, curing procedure applied) which are deemed to satisfy the required performances.

6.4.7 Influence of fly ash on structure and strength development of concrete

In this section, the effect of arbitrary ratios of OPC and fly ash on a structure and its strength development will be discussed. It is envisaged that, through computational simulation, the effect of arbitrary ratios of OPC and fly ash powder on the microstructural properties and strength of concrete will be studied. To this end, dynamic changes of microporosity distribution with time for different OPC to fly ash powder mix ratios are computed using the coupled structure development, hydration and moisture transport model. A comparison of strength and porosity development is also made for different powder material ratios.

In the computations, the OPC to fly ash ratio was varied from pure OPC powder mix to 50% OPC and 50% fly ash powder mix. Five different powders' mix ratios, 100 : 0, 90 : 10, 80 : 20, 70 : 30 and 50 : 50, were analysed. The W/P ratio was kept constant at 0.3 by weight for all the cases. The test specimen was a two-dimensional infinite bar 4 cm × 4 cm in cross-section (Figure 6.13). After one day of sealed curing, a fully moist (RH = 1.0) boundary condition was used up to the end of the analysis period. This ensured that enough moisture would be available for hydration at all times and any secondary effects arising due to the coupled nature of the hydration-moisture content would not influence the results. The extrinsic model

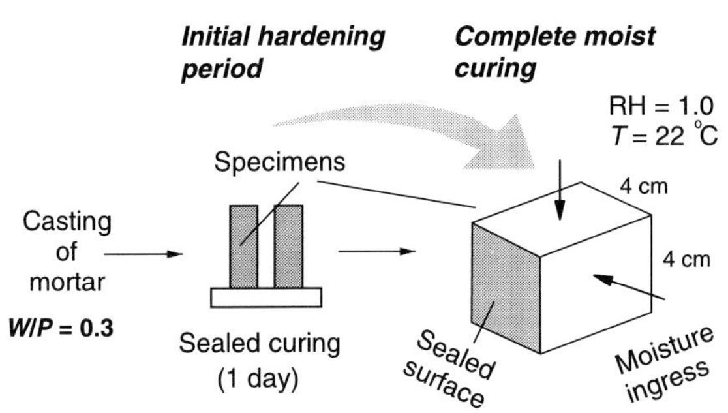

Figure 6.13 Schematic representation of the analysis.

parameters used in the model are W/P ratio, composition of cement, aggregate content, weight percentages of cement, slag and fly ash. (The average powder particle radius and Blaine fineness index parameters are considered but they are kept constant in all the analyses.)

The microstructural growth in the simulations is influenced by the gel crystal length parameters as well as the specific surface area of the gel as defined in Chapter 3. It will be observed from equations (3.14b) and (3.8) that the slag and fly ash contributions to these parameters are significantly different from those of normal Portland cement. The chemical interaction between the OPC and fly ash is considered by the multi-component hydration model. All the parameters related to these phenomena are computed using the material models described earlier.

The sample geometry analysed is an infinite prismatic bar (Figure 6.13). All four surfaces are fully conducting of moisture and heat. The bar is exposed to an environment of unity RH and a temperature of 22 °C after one day of sealed curing. The boundary conditions are kept constant thereafter. The results of the computed microstructure as well as the strength development for different ratios of OPC : lime mixes were obtained at different intervals of time. The results are shown in Figure 6.14 and Figure 6.15.

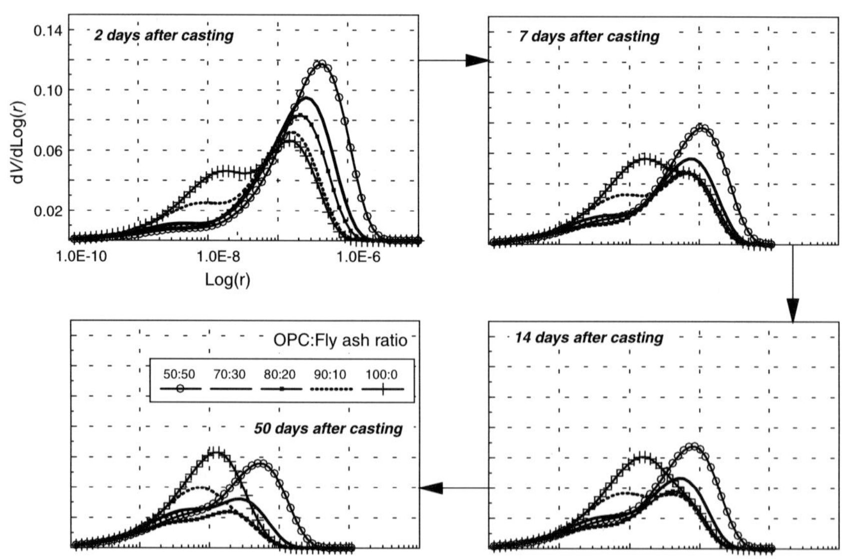

Figure 6.14 Simulated porosity distributions of an OPC : fly ash mix at different times after casting. Note that the same axis is used for all graphs.

It can be noticed from these figures that fly ash mixed powder mixes shows a higher peak at an early age of hydration. This can be primarily attributed to the net hydration rate retarding effect of fly ash during early ages. The average degree of hydration also significantly decreases with the increase of the proportion of fly ash in the powder mix due to the relative abundance of Ca^{++} ions and lack of enough alkalinity. During later stages of hydration, owing to the higher specific surface area of precipitates formed by the dissolution of fly ash grains' glass network products and the availability of activators such as free $Ca(OH)_2$ and probably sulphates, fly ash hydration contributes to a significantly finer pore structure and higher strength matrix. At larger fly ash content, however, there is not enough $Ca(OH)_2$ liberated from the hydration of Portland cement to aid in fly ash hydration. Also, precipitation of early CSH products on fly ash grain surfaces further slows the dissolution of fly ash grains. As a result, a low strength results and coarser hydration products are formed. It is interesting that though total porosity might decrease with the addition of fly ash, the microstructure eventually tends to be coarser above an optimum mix of OPC and fly ash. In this analysis, the optimum point turns out to be 80 : 20 weight ratio of OPC and fly ash. Both the strength and microstructural properties deteriorate after that. Development of strength with time and comparison of average compressive strengths for five different cases are shown in Figures 6.15(a,b).

It must be recalled that, in this study, we have assumed that the products of hydration being formed in nature are constant throughout the process of hydration. This is incorporated into the model by assuming a constant specific surface area of the interlayer mass and average gel particle (equations (3.14b) and (3.8)). However, we know that in the early stages

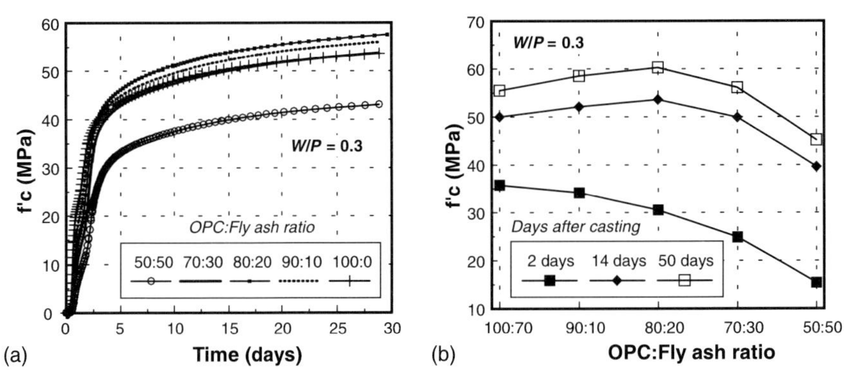

Figure 6.15 Simulated development of strength with time and optimal mix design for different OPC : fly ash mix mortars: (a) compressive strength development; (b) *fc′* versus powder composition.

of hydration there is hardly any hydration of slag or fly ash, therefore the specific surface area of the hydration products will be comparable to the hydration products of normal Portland cement. In a more realistic model, the hydration level of fly ash and slag should be incorporated into the specific surface area parameters of equations (3.14b) and (3.8) to take into account the dissolution of glass networks of fly ash. Probably incorporation of such an effect into the model would show more prominent differences between the microstructures of hydration products for different cases of arbitrary mixes of powder materials.

6.5 Summary and conclusions

Using simple physical models of hydration, moisture transport and microstructure development, early age development processes and various parameters relevant for long-term durability can be quantitatively obtained. A finite-element based computational model has been developed by combining the pore-structure development, heat of hydration and moisture transport models. With this computational model, the interdependency of hydration, structure development and moisture transport mechanisms can be simulated in a rational way. The highlights of the various models incorporated into the proposed durability evaluation system are summarized below.

The pore structure development is dependent on the state of maturity of the hydrating matrix. Also, the pores in the developing microstructure are primarily subdivided into interlayer, gel and capillary pores. The moisture transport and retention characteristics of hardening concrete have a direct relation to the developing microstructure and the same relationships are obtained directly from the computed microstructures. Regarding hydration, the cement clinkers are classified into four minerals with which five patterns of hydration are linked. The hydration rate is directly coupled to the free water and temperature, which represent the thermodynamic environment of cement in concrete.

Overall the combined system has very few empirical parameters and as such can be applied with confidence to study the cases of different mix proportions under various curing conditions in early age development and related problems. Preliminary verifications have shown reasonable agreement with experiments for various aspects of early age development. As for future development, a combination of the coupled mass transport, hydration and pore structure formation theory with the structural mechanics model of reinforced concrete is required for examining total performance of reinforced concrete from birth to death.

References

[1] Zeinkiewicz, O.C. and Taylor, R.L., *The Finite Element Method*, Vol. 2, Solid and Fluid Mechanics, Dynamics and Non-linearity, McGraw Hill, New York, 1989.

[2] Suzuki, Y., Harada, Y., Maekawa, K. and Tsuji, Y., Quantification of heat of hydration generation process of cement in concrete, *Concrete Library of JSCE* 1990, No. 16, 111–124.

[3] van Breugel, K., Simulation of hydration and formation of structure in hardening cement-based materials, Ph.D thesis submitted to Delft Technological Institute, Netherlands, 1991.

[4] Smith, J.M., *Chemical Engineering Kinetics*, McGraw Hill, New York, 1981.

[5] Tanaka, T., Experiments on the drying shrinkage of HPC mortars, Bachelor Thesis submitted to the University of Tokyo, 1992.

[6] Shimomura, T. and Maekawa, K., Analysis of the drying shrinkage behaviour of concrete using a micromechanical model based on the micropore structure of concrete, *Magazine of Concrete Research*, 1997, **49**(181), 303–322.

[7] Shimomura, T. and Uno, Y., Study on properties of hardened high performance concrete stripped at early age, *Proceedings of JSCE*, 1995, **26**(508), 15–22.

Chapter 7

Multicomponent model for the heat of hydration of Portland cement

- Multiple component system of clinkers for arbitrary cement mixtures
- Hydration rate control by evaluating temperature dependency, hydration level and free water consumption
- Application for thermal stress analysis of mass concrete

7.1 Introduction

7.1.1 *Thermal cracking of massive concrete structures*

The hydration of cement in concrete is accompanied by the generation of heat with the development of micropore structures and associated mechanical strength. The thermal energy liberated by the cement hydration is transferred in concrete with elevated temperature and finally released to the environment through the surfaces of the structure. When the structure has a relatively thin cross-section and larger surface area, the temperature of the structure does not rise greatly. In the case of massive concrete structures, generated heat which leads to the temperature rise inside the structure is hardly released at the surface due to the comparatively low conductivity of heat in concrete and the longer path through which the heat flux flows. Here, the volumetric change of structure associated with thermal expansion and contraction rooted in temperature variation give rise to selfequilibrated thermal stresses. The thermal stress itself does not bring any serious change of properties of concrete material in normal situations, but it may cause structural cracking and damage when the tensile strength is exceeded. The thermal stress-induced cracking generally deteriorates the structural serviceability and durable performance very seriously, because it often penetrates whole sections of structures and the cracks have relatively wider and more irreversible openings. The occurrence of thermal cracking is one of the clear limit states to be assessed in performance-based design.

There are two types of mechanism producing thermally induced stresses, i.e. internal and external confinement. Let us consider a new concrete placement cast on a previously constructed one or on a rock foundation as shown in Figure 7.1 [1]. The first mechanism is that the gradient of temperature over the section of the structure gives rise to different thermal expansions. Due to the continuity of structure, a selfequilibrated stress will be introduced. The sectional force (integral to the local stress developing over a section) should be zero but the local stress is not zero even if there is no external confinement. In general, tension arises around the surface of a massive concrete structure and compression is observed inside the structure by this mechanism. On the other hand, if the newly cast structure has external confinement through a construction joint or a basement interface, externally induced compression and tension forces will arise according to the change of temperature and solidification. Once thermal cracking occurs during the stage of temperature decay in structures, the cracks never close but continue to open with time. This cracking tends to pass through whole sections of the structure.

The engineering challenge of thermal crack problems arises from the irreversibility of the mechanical properties, since concrete grows according to the progress of the hydration of cement. If the mechanical properties of a material are perfectly reversible and path-independent in terms of temperature and strain, the stress states must return to the initial conditions when the temperature is initialized. On the contrary, it is well known that in concrete under external confinement that the compressive stress experienced at the early stage of hydration is relatively small, and the tensile stress increment produced by volumetric contraction is largely induced when the temperature starts to fall. The stress induced at early age is relatively small due to the premature stiffness of green concrete while, at the phase of temperature drop, the hydration has almost finished generating the full performance of the mechanical properties, such as stiffness. Thus, a larger

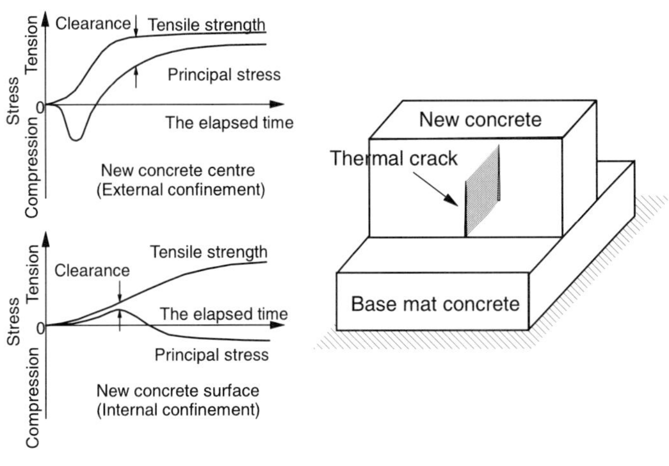

Figure 7.1 Thermal stress and thermal crack in a mass concrete structure [1].

stress increment toward tension is produced. This irreversible change of stiffness finally creates absolute tensile stresses, even if a reversible temperature path is followed in concrete structures.

It is clear that the histories of temperature, stresses and the varying mechanical properties that occur in concrete structures must be simulated in elapsed time to evaluate the risk of thermal crack occurrence. At any time, the heat generation rate is different in each location of structure since the temperature and the degree of hydration are not uniform due to the conduction of heat and the temperature dependency of cement hydration. Further, each location has a nonunique strength development in tension. In performance assessment at early age, the risk of thermal cracking must be evaluated quantitatively on the basis of thermal conduction analysis with the heat generation of cement, prediction of tensile strength of concrete with time and thermally induced stresses. Here, it should be noted that the exothermic process of cement hydration depends so much on the updated temperature and its history. The accuracy of predicting heat generation rate determines the reliability of the thermal stress analysis as well as the constitutive laws for young concrete.

7.1.2 Exothermic hydration process of cement [2]

Cement clinker is manufactured with limestone, clay, quartzite and small amount of additives as raw materials by burning at 1400 °C (1673 K) and Portland cement is made by pulverizing cement, burning it, and intergrinding the clinker with gypsum. Cement clinker consists of four major minerals, i.e. alite (mainly C_3S), belite (mainly C_2S), aluminate phase (mainly C_3A) and ferrite phase (mainly C_4AF) with some impurities. Alite and belite are calcium silicate and occupy around 85% of the cement clinker, while aluminate and ferrite occur as the interstitial phase which fills the space around the calcium silicates when cement clinker is burned. The fraction of each mineral in the cement clinker varies according to the type of Portland cement.

Portland cement reacts with water and hydration products, which form a hardened structure with porosity, are precipitated. The hydration process of cement is not simple since the rate of hydration and sorts of hydrates vary greatly among clinker minerals. Alite and belite produce both calcium silicate hydrates called CSH and calcium hydroxide ($Ca(OH)_2$), while aluminate and ferrite phases produce ettringite or monosulphate with gypsum. After the gypsum is consumed, calcium aluminate hydrates are produced. All reactions that simultaneously occur in hardening cement paste are accompanied by some amount of heat generation. It was reported that the theoretical heat of hydration of clinker minerals (pure hydration) at complete hydration are C_3S: 120, belite C_2S: 62, C_3A: 207 and C_4AF : 100 (cal/g), respectively. The reactivity of clinker minerals is not similar. The degree of hydration of each clinker mineral at a constant temperature in terms of the elapsed time measured by an X-ray diffraction method in ordinary Portland cement are shown in Figure 7.2 [2].

To identify the hydration of cement, a conduction calorimeter that can measure the heat liberation rate of cement at a specified constant temperature is often used. One example of the phases of the exothermic hydration process observed in typical ordinary Portland cement [3] is shown in Figure 7.3. The hydration mechanisms of alite and belite are quite similar; however, the rate of hydration of belite is very much slower than that of alite. Then, it can be considered that the alite hydration dominates the overall behaviour of the exothermic hydration process when the mineral composition is similar to ordinary Portland cement in which the fraction of alite reaches near 50% in cement. In this division of the hydration process, it is classified into five steps as follows.

Step 1: The rapid reactions which show the larger rate of heat generation (the first peak) occur for a short time just after the cement comes into contact with water. This heat generation results from the wetting of cement particles and the dissolution of ions in water. Some semistable phase of CSH is formed and the formation of ettringite is mainly included in this phase.

Step 2: The rate of heat generation is too small and the hydration seems to be stagnant. This phase is called the *dormant period*. This phase is brought about by the formation of a protective layer on the surface of the cement particles or delaying the nucleation of hydrates. Though the concentration of ions in the solution becomes gradually higher during this phase according to the solution of solid phase, the hydrates made of the main compounds C_3S and C_2S are not crystallized yet.

Step 3: The hydration proceeds actively where the rate of heat generation increases. This phase follows the termination of the dormant period that is induced by the increase of permeability of the protection layer and the beginning of the crystallization of the CSH.

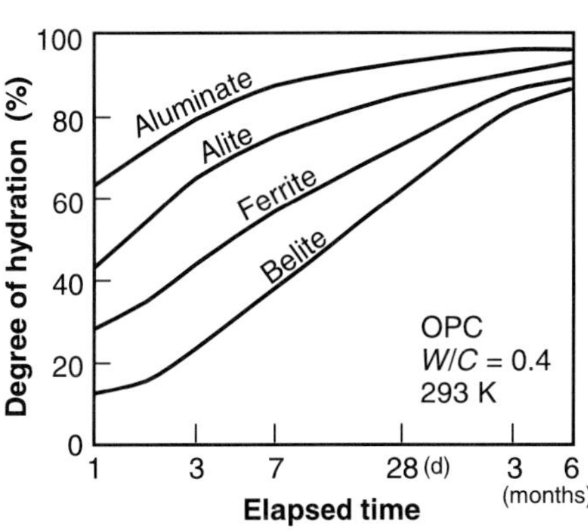

Figure 7.2 Rate of hydration of clinker minerals in OPC [2].

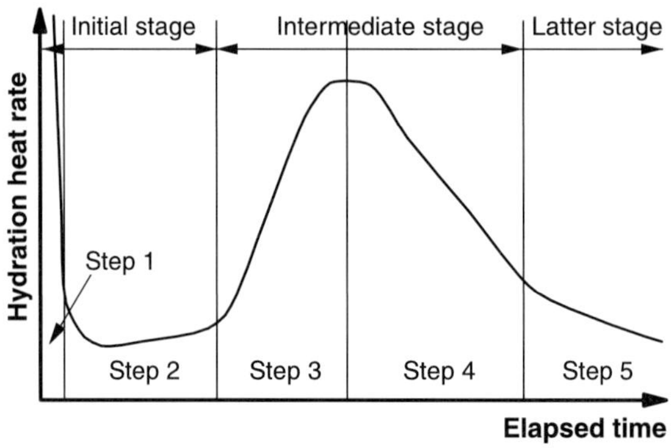

Figure 7.3 General stages in the exothermic hydration process in OPC [3].

Step 4: The rate of heat generation becomes gradually slower. In this phase, the thickness of the hydrate layer which covers unhydrated particles increases and the surface area of the unhydrated parts are reduced according to the progress of hydration. The layer of cement hydrates plays a role as the diffusion area, which governs the permeability of the water and dissolved ions.

Step 5: This phase follows Step 4 but the rate of hydration is remarkably reduced by the thicker layer of hydrates around particles. The space originally filled by liquid water is almost occupied by cement hydrates. Then, it becomes difficult for hydrates to be precipitated. Steps 4 and 5 are called the *diffusion control* phases.

7.1.3 Adiabatic temperature rise and conventional thermal analysis

The caloric value of concrete measured by the conduction calorimeter cannot be applied to the thermal analysis of massive concrete without additional procedures since it is measured at constant temperature. In contrast, the adiabatic temperature rise test is often conducted to investigate the caloric property of concrete for massive structures and its measured values are applicable to thermal analysis. Here, it must be noted that the reliability of test results has much to do with how perfectly the thermal isolation is realized. Suzuki *et al.* [4] developed a testing apparatus with high accuracy. It was verified through comparison with actual temperature rises in massive concrete blocks. The type of cement, unit cement content and initial temperature at casting are primary factors which affect the adiabatic temperature rises. Series tests results with systematically arranged parameters of concern are shown in Figures 7.19–7.21 [5, 6] with the DuCOM analytical results discussed later. We will see the unique adiabatic temperature rises in terms of unit cement content and the casting temperature as well as type of cement. Further, the nonlinearity among them due to the temperature dependency of the reaction and the differences of mix proportion is also recognized.

The adiabatic temperature rise curve is usually expressed by approximate formulae, and several types of formulae are proposed as follows [4].

$$
\begin{aligned}
Q(t) &= Q_\infty(1 - \exp(-rt)) \\
Q(t) &= Q_\infty(1 - \exp(-rt^s)) \\
Q(t) &= Q_\infty(1 - \exp(-r(t - t_0))) \\
Q(t) &= Q_\infty(1 - \exp(-r(t - t_0)^s))
\end{aligned}
\tag{7.1}
$$

where $Q(t)$ is the adiabatic temperature rise t days after casting, Q_∞ is the ultimate temperature rise, and r, s, t_0 are experimental constants. The first formula has been well used to represent the adiabatic temperature rise due to the convenience of computation [7]. But its applicability is not sufficient for expressing the dormant period when the casting temperature is quite low or a delaying type of chemical admixture is mixed with a superplasticizer.

The adiabatic temperature rise curve that is measured under the thermally isolated situation is directly proportional to the heat released on a cumulated basis with time, because the heat capacity can be assumed constant due to the large volume occupied by aggregates having thermal stability. In the conventional analysis of temperature prediction, this heat generation rate, directly derived from the adiabatic temperature rise, is given as the thermal property of cement hydration at arbitrary locations in the structure. We should note that the adiabatic temperature rise and the related heat of hydration rate are not general properties of cement hydration, because the rate of hydration strongly depends on temperature which greatly varies in space. The adiabatic temperature rise represents the particular heat generation under the condition of being thermally isolated and with the particular concrete mix and casting temperature.

If we wish to compute the temperature field assuming that the rate of heat generation varies with time alone, the same cement hydration could be assumed regardless of location. But the temperature hysteresis that the concrete undergoes actually changes with its position in a structure, and the heat generation rate concerned is affected. In the case of relatively small structures, there exist places near the surface of structure where the exothermic hydration process is very different from that in the adiabatic state, owing to considerable cooling by ambient atmosphere. If we assume the same heat generation rate at all positions to be that of the adiabatic temperature rise tests, the computed temperature most likely exceeds the actual temperature [5, 8].

7.1.4 Quantification of the exothermic hydration process of cement in concrete (Suzuki model) [5]

There are attempts to predict the hydration exothermic process by establishing a hydration model of cement [9]. The approach based upon a hydration reaction model of cement can be an excellent method, because it enables one to predict the heat generation rate of hydration for any temperature history in a generalized way. It is essential that the heat of hydration model must be applicable to cement hydration existing in concrete under an arbitrary temperature history. Suzuki *et al.* [5] successfully developed a quantification technique of the exothermic hydration process of cement in concrete dependent on the temperature, and provided a general approach for deriving a heat of hydration model capable of assessing any temperature hysteresis.

The feature of this model is that the exothermic hydration properties of cement in concrete are specified by two material functions, i.e. the thermal activity and the intrinsic heat rate. Suzuki *et al.* offered a practical way to identify inversely the cement properties from concrete-based measurements. The accumulated heat generation is specified as a main parameter for the rate of heat generation as well as the temperature in order to represent the past hydration process. Arrhenius' law of chemical reaction is successfully introduced to express the temperature dependency of cement hydration [9]. The generalized governing equation of heat of hydration generation of cement yields

$$H = H(T, Q) = H(T_S, Q)\exp\left[-\frac{E(Q)}{R}\left(\frac{1}{T} - \frac{1}{T_S}\right)\right] \tag{7.2}$$

$$H(T_S, Q) = H_\infty(Q)\exp\left(-\frac{E(Q)}{RT_S}\right) \tag{7.3}$$

where H is the heat generation rate per unit weight of cement, $H(T_S, Q)$ is the reference heat generation rate of cement at constant temperature T_S (also a function of the accumulated heat Q), $H_\infty(Q)$ is the ultimate reference heat generation rate of cement at infinite temperature, which is an imaginary situation, $E(Q)$ is the activation energy of cement, R is the gas constant, and T is the absolute temperature of concrete.

The activation energy and intrinsic reference heat rate characterize the process of heat generation of cement in concrete. If zero energy of the igniting reaction is considered as an extreme case, we will have a temperature-independent reaction rate. In other words, we can obtain the activation energy determining the thermal activity of a reaction by measuring the temperature effect on the rate of the chemical reaction. In using adiabatic temperature rise curves for three different initial temperatures, we can derive those material functions (characteristics) in terms of the accumulated heat per unit weight of cement as shown in Figure 7.4 [5]. The various heat of hydration rates can be obtained with different temperatures from the adiabatic temperature rises of concrete cast at different temperatures. Arrhenius' plots, which show the correlation between the inverse of absolute temperature T and the logarithm of heat of hydration generation rate H, can be drawn. Here, a unique correlation could be found with respect to each accumulated heat that can be the indicator of how fast the hydration proceeds. The typical activation energy (thermal activity) and intrinsic reference heat rate confirmed

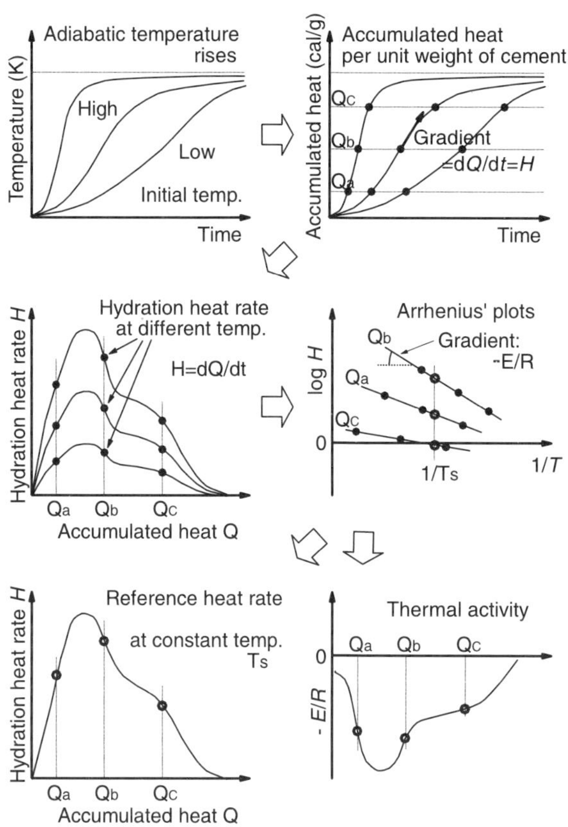

Figure 7.4 Procedures to derive thermal properties of cement hydration from adiabatic temperature rise tests [5].

experimentally are shown in Figure 7.5. Both thermal properties are apparently nonlinear and strongly dependent on the degree of hydration, since cement consists of several chemical compounds with different hydration processes with some interactions.

The activation energy is regarded as the energy barrier to the chemical reaction [1]. The necessary energy for the forward reaction is generally supplied by the molecular kinetic energy. When collision between molecules

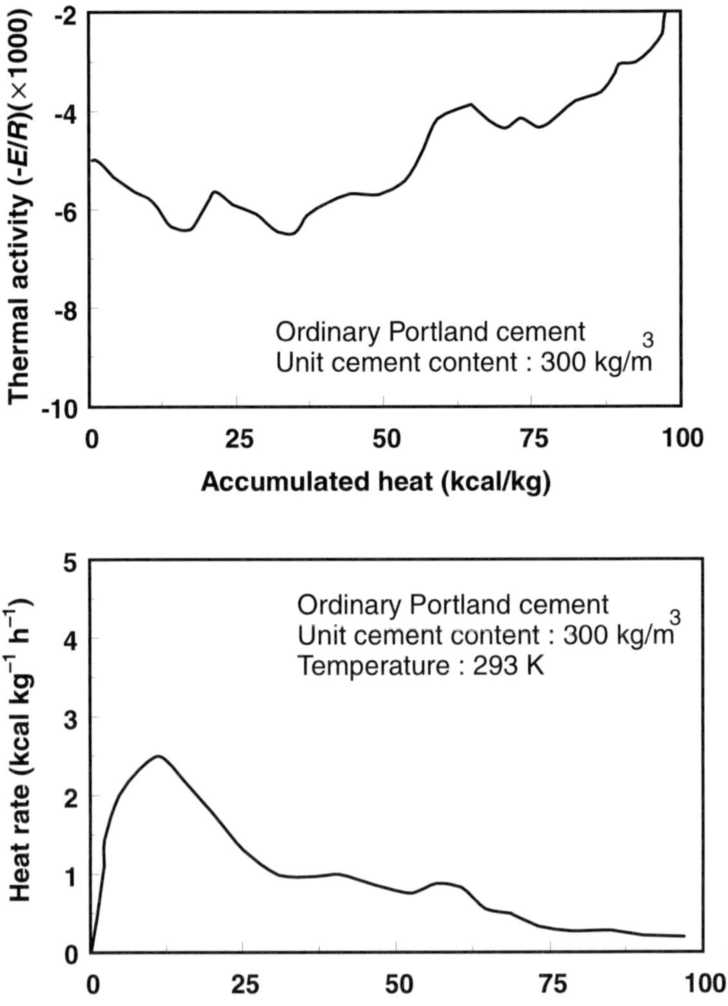

Figure 7.5 Thermal activity and reference heat rise with accumulated heat confirmed through adiabatic temperature rise tests [5].

of cement and water takes place, molecules that have a higher kinetic energy than the activation energy can react. The temperature dependence of the chemical reaction rate, which is expressed as an exponential function of temperature, represents the probability of the kinetic energy of the molecules being greater than the activation energy. Therefore, the activation energy is associated with the sensitivity of the reaction to the temperature (temperature dependence). If the activation energy is zero, the heat of hydration rate is independent of the temperature and the reaction will definitely take place when molecules collide. The temperature can be regarded thermodynamically as the sum of the kinetic energy of the molecules. Under an elevated temperature, large numbers of molecules whose energy exceed the activation energy are thought to exist in the system and the reaction can proceed smoothly. Lower temperature implies that fewer molecules have kinetic energy greater than the activation energy and the reaction progresses relatively slowly.

In a chemical reaction, the reference heat rate is understood to represent the occurrence of particle collision [1]. If we deal with an idealized gas and a bimolecular reaction, the rate of reaction is proportional to the number of remaining molecules per unit space (concentration) and reduced according to the degree of reaction. In reality, the cement exhibits complexity in the reference heat rate $H(T_S, Q)$ at constant temperature T_S in terms of the accumulated heat Q which represents the degree of hydration of cement in this model. The hydration process of cement can be considered to be multistage. At the very beginning of the reaction, the protection layer of CSH and the hard coating of ettringite are produced on the surface of the cement particles, and the occurrence of collisions with water will be less. After some progress of the reaction, the precipitation of hydrates is accelerated and the rate of heat generation shows the second peak in the exothermic hydration process. The penetration of water from outside into unhydrated particles will be followed through the cover of the hydrated layer and the rate of hydration is governed by the diffusivity of water and ions. When the process is close to termination, the surface area of unhydrated cement particles, which implies the probability of contact between water and active molecule of cement, becomes small due to the thick layer of hydrate products, then the reaction slows down. Further, it is also recognized that cement consists of several chemical compounds with different hydration processes, which affects the complexity of the reference heat rate.

Here, let us assume a reaction in which the activation energy is constant though the actual activation energy of cement is a function of the accumulated heat. It was found from the definition that the accumulated heat, i.e. the degree of hydration of cement, can be expressed as a function of the maturity, which is derived by integrating the temperature term with respect to time. That is, it can be said that the constant temperature dependence of the reaction (constant activation energy) is tacitly assumed in the maturity model. In other words, the maturity model corresponds to the special situation where constant thermal activity is assumed in the hydration of cement. Further, if we assume that the activation energy is zero in the reaction, the accumulated heat can be expressed simply as the function of the elapsed time. This assumption corresponds to the conventional model of the heat of hydration of cement, where the adiabatic temperature rise curve directly corresponds to the rate of hydration. It is recognized that the conventional temperature analysis is a fictitious one without the consideration of temperature dependence of cement hydration, and the adiabatic temperature rise must be given as a particular solution of a generalized heat generation governing equation [1, 5]. The heat of hydration model based on the chemical reaction rate theory is regarded as the more generalized one, which includes the aforementioned various special models proposed in the past.

7.1.5 Scheme of the multicomponent heat of hydration model

The multicomponent heat of hydration model which is described in this book was originally established for the temperature analysis of massive concrete structures without conducting adiabatic temperature rise tests [10, 11]. In thermal stress analysis, a heat of hydration model of cement applicable to any given condition is required. This section presents a concrete performance model that describes the hydration of cement in terms of the reactions of individual mineral components and expresses the differences between various types of Portland cement as differences in mineral composition. Referring to the model proposed by Suzuki *et al.* [5] for the entire hydration of cement, the rate of heat of hydration generation for each mineral component is expressed by using two material functions, an intrinsic reference heat generation rate at specified temperature and the thermal activity. Taking into account the temperature dependence of the individual mineral component's reactions, the exothermic behaviour of the cement as a whole during hydration is quantified for any given temperature history. The proposed model has been verified through analysis of adiabatic and semi-adiabatic temperature rises.

In conventional thermal analysis, adiabatic temperature rises obtained from experiments are often used to represent the exothermic processes of cement in massive concrete during hydration. An adiabatic temperature rise, however, is not a general physico-chemical characteristic of the material, but rather a thermal characteristic only found under the specific conditions of adiabatic temperature change. Since concrete near the surface of a structure experiences a temperature history that deviates from adiabatic conditions due to the influence of thermal loss, there is a need to take into account the temperature dependence of hydration in carrying out a thermal analysis. As the authors stated in section 7.1.3, Suzuki *et al.* [5] successfully quantified the process of hydration-heat generation by cement in concrete for any temperature history by carrying out two or more adiabatic temperature rise experiments at different casting temperatures. Harada *et al.* [8] then proposed a nonlinear temperature analysis method that couples a temperature-dependent exothermic hydration process with thermal conduction. These studies made it possible to carry out temperature analysis of concrete structures taking into account the temperature dependence of the cement's hydration reaction. As mentioned, though, this approach requires two or more sets of highly precise adiabatic temperature rise results so as to generalize the heat of hydration characteristics of the cement in concrete. The model is therefore suitable for high-precision analysis in situations where the materials to be used and their mix proportions have already been determined, but if used as a performance evaluation method for use in selecting materials and their mix proportions at the design stage, it would require repeated experiments for each possible material mix proportion. Thus, the model is clearly inappropriate for such design-stage consideration of various mix proportions.

This has important ramifications for the durability inspection [1] of selfcompacting high-performance concrete designed to offer high durability regardless of the compaction performance (high-performance concrete). High-performance concrete contains a higher powder content to maintain selfcompactability, so it is necessary to use a certain minimum amount of cement over and above the standard content. This limits freedom to reduce the unit cement or powder content, although reduced unit cement content is considered as an effective way of avoiding thermal cracking. The remaining options are then to change the binding materials, by using low-heat cement, for example, or to add pozzolans and other mineral powders. In a performance-based design, material design is also an iterative process that continues until the required performance is obtained. If the requirements cannot be met in terms of avoiding thermal cracking, it is necessary to return to the first step of materials selection and then to repeat the test to identify the thermal properties again. In actual material design, past experience enables the number of suitable candidates to be narrowed down.

However, many requirements apart from absence of thermal cracking have to be satisfied, and mix proportions that satisfactorily avoid thermal cracking are not necessarily capable of meeting the other requirements. If one mix proportion fails to fulfil all demands, another mix must be chosen and again subjected to thermal crack checking. The more requirements there are, the greater the demand for easier identification of material characteristics whenever this process returns to the initial selection of materials and their mix proportion. In other words, if a mix proportion has to be evaluated for its ability to meet several performance requirements in each experiment, the design scheme covering a great number of check items is virtually impossible to develop. To solve this problem, a versatile model of concrete performance is needed that can evaluate the various characteristics of concrete by inputting the types of materials, their mix proportion, and the environmental conditions.

The heat of hydration model presented is designed so that various types of cement are represented by their mineral compositions. Based on the studies carried out by Suzuki *et al.* [5], the hydration of each mineral component was modelled separately, and then a multicomponent model was developed in which the rate of heat of hydration generation of the cement is obtained as the sum of the heat of hydration rates of each component according to their proportions [12]. To accommodate interactions among the reactions of mineral compounds in the common reaction environment, the effects of the shared environmental temperature and mixing water as well as changes in the heat generation rate of each mineral compound depending on its proportion are also included.

To expand the proposed model to be applicable to blended cement with blast furnace slag or fly ash, the amount of calcium hydroxide ($Ca(OH)_2$) in the system should be focused on, because the reactions of slag or ash are strongly dependent on the supply of $Ca(OH)_2$ as an activator of the reaction. That is, it is necessary for rational evaluation of the exothermic process of blended cement that the production from cement hydration and the consumption by slag and fly ash are simultaneously simulated in terms of calcium hydroxide. Further, the delaying effects delivered by the chemical admixture, and fly ash to cement and slag reaction are also taken into account. These factors are mutual interactions which take place between the exothermic hydration process of Portland cement and admixtures, and it becomes possible to deal with them rationally by using the multicomponent system adopted here [13].

7.2 Modelling of the exothermic hydration process

7.2.1 *Basic concept of the multicomponent heat of hydration model*

The major minerals making up cement are cement clinker compounds and gypsum, which is added after the baking of the clinker. Clinker consists of alite (mainly C_3S), belite (mainly C_2S), an aluminate phase (mainly C_3A), and a ferrite phase (mainly C_4AF). The proportion of each mineral component varies according to the type of cement. Thus, if a heat of hydration model is to cope with any given type of cement, it must properly describe the exothermic process of that cement under hydration in accordance with its mineral composition. Further, heat generation is generally restrained not only by reducing the heating efficiency of the

Portland cement itself, but also by replacing the cement by various types of pozzolans. Thus, the model will have to cope with situations involving pozzolans. In developing a model for exothermic behaviour, it is possible to treat the cement as a single material. The mineral composition of Portland cement tends to be fairly characteristic for a particular type of cement; for example, C_3S and C_3A are often found in greater proportions in types of cement that emphasize earlier strength development, such as ordinary Portland cement or early-strength Portland cement. Other types of cement characterized by reduced heat of hydration contain a relatively high proportion of C_2S and C_4AF. Consequently, selecting a suitable index to represent the overall exothermic characteristics of the cement might make it possible to treat Portland cement as a single material and to model the varying heat output of different types of cement. An example of this is the work of Suzuki *et al.*, who introduced, as thermal properties necessary to quantify the overall exothermic process of cement, the reference heat rate at constant temperature and the thermal activity in terms of the accumulated heat (representing the degree of hydration) [5]. It may also be possible to develop such a model from a macroscopic point of view, such that the thermal properties of the entire cement reflect changes in mineral composition. If the cement can be treated as a single entity in this way, the interdependence among the various reactions occurring in the cement will be included in the modelled thermal properties.

In a simple reaction system, the relationship between reaction and heat release is usually one to one. But this is, of course, not so when multiple reactions are observed, as in cement. If the temperature dependence of each mineral reaction is different, the reaction rate of each mineral cannot be given as a single value related to the exothermic output of the entire cement. This is because, with the different temperature dependence of each reaction, the rate at which each mineral component reacts varies according to the temperature history of the cement. If a model treats the cement as a single phase, however, it must be assumed that the hydration process is determined by a single value according to the accumulated heat regardless of the temperature history. In other words, if it is technically reasonable to assume that the generated heat and the reactions of the mineral components correlate well, an approach in which cement is treated as a single system may be deemed appropriate. In order to expand the applicability of a model to cover any temperature history, it is necessary to express the heat generation rate and thermal activity as a function of temperature history. Consequently, a single system model cannot always be regarded as a generalized case.

A further consideration is that, in order for a model to be applicable to the temperature analysis of massive concrete structures in practice, it will be necessary to expand it to cover systems containing pozzolans. In a blended cement containing various powdered materials which further complicate the interdependence among reactions, it is not possible to guarantee the reliability of the assumption that hydration effects can be predicted solely from the accumulated heat of the entire system, especially when the temperature history changes in a variety of ways. When a model of the heat of hydration of Portland cement is combined with a model for another blending powder, the evaluation of their interdependence is the key to a successful approach. Today, a considerable variety of powder materials, including blast furnace slag, fly ash, limestone powder, silica fume and expansion agents, are used as admixtures. With the increasing number of reactions resulting from these additives, the interdependence of reactions becomes more complicated. Thus, a complex model would appear necessary to treat rationally cement as a single entity.

The alternative approach employed in this book entails dividing the simultaneously occurring reactions into appropriate reaction units, and then attempting to describe them in as rational a manner as possible. That is, the reactions taking place during cement hydration are resolved into mineral units, and the exothermic hydration process of each component is described separately. When the reactions taking place in a system are described individually in this way, their interactions must also be taken into consideration. The result of this type of representation is that the model is expandable and more general. Further, by describing reactions in separate reaction units, this approach leaves open the possibility of introducing other components, corresponding to blending powders, and considering the resultant interactions.

The hydration process for each mineral compound is, in this model, basically described in the same manner as introduced by Suzuki *et al.* [5]. That is, the reaction process is expressed by using two material functions: the reference heat rate, which gives the heat rate at a specified constant temperature, and the thermal activity, which describes the temperature dependence of the reaction. The minerals present in the cements covered by the model are alite (C_3S), belite (C_2S), an aluminate phase (C_3A), a ferrite phase (C_4AF), and gypsum (CS2H), and the exothermic reaction for each of these minerals is individually described. Gypsum is treated as the dihydrate (gypsum dihydrate). For the blended cement, blast furnace slag and fly ash are regarded as single units of reaction and incorporated into the model as individual components. Slag and fly ash are not classified like Portland cement because the part reacting is only glass phase at normal temperature, which can be regarded as a homogeneous material. However, at this stage, the reaction of slag is assumed to be unaffected by the existence of gypsum [14]. Therefore, the heat rate of the entire cement, H_C, including blending powders, is given as the sum of the heat rate of all reactions as follows.

$$\begin{aligned} H_C &= \sum p_i H_i \\ &= p_{C_3A}(H_{C_3AET} + H_{C_3A}) + p_{C_4AF}(H_{C_4AFET} + H_{C_4AF}) \\ &\quad + p_{C_3S}H_{C_3S} + p_{C_2S}H_{C_2S} + p_{SG}H_{SG} + p_{FA}H_{FA} \end{aligned} \tag{7.4}$$

where H_i is the heat generation rate of mineral i per unit weight, p_i is the weight composition ratio, and H_{C_3AET} and H_{C_4AFET} are both heat generation rates in the formation of ettringite. The ettringite formation model is included in the proposed system because ettringite is first formed from C_3A and C_4AF prior to hydration when gypsum is present. The generation of heat of hydration by C_3A and C_4AF, expressed by $H_{C_3}AET$ and $H_{C_4}AFET$, is assumed to begin once the ettringite formation reaction has completed due to the disappearance of unreacted gypsum.

Uchida and Sakakibara [9] confirmed in their studies with cement paste and Suzuki *et al.* [5] with concrete that Arrhenius' law is applicable to cement hydration, which is the reaction of a composite containing multiple mineral compounds, by regulating the thermal activity in terms of the accumulated heat of the cement as a whole. The authors stretched this applicability and, in the proposed heat of hydration model, assumed that the temperature-dependent heat generation rate of each mineral was expressed by the equation below. Since mineral-based reactions in the cement are described individually, the coefficients expressing the interdependence of the reactions were introduced into the model. In this study, the following factors are taken into account as mutual interactions: free water consumed are shared by components; chemical admixture and fly ash delay reactions of Portland cement and blast furnace slag; the reactions of slag and fly ash are dependent on the supply of calcium hydroxide produced by cement hydration.

$$H_i = \gamma\beta_i\lambda\mu s_i H_{i,T_0}(Q_i)\exp\left\{-\frac{E_i}{R}\left(\frac{1}{T}-\frac{1}{T_0}\right)\right\} \tag{7.5}$$

$$Q_i \equiv \int H_i \, dt \tag{7.6}$$

where E_i is the activation energy of component i, R is the gas constant, H_{i,T_0} is the reference heat generation rate of component i at constant temperature T_0 (and is also a function of the accumulated heat Q_i), γ is a coefficient expressing the delaying effect of chemical admixture and fly ash in the initial hydration exothermic process, β_i is a coefficient expressing the reduction in heat generation rate due to the reduced availability of free water (precipitation space), λ is a coefficient expressing the change of the heat generation rate of blast furnace slag and fly ash due to the lack of calcium hydroxide in the liquid phase, and μ is a coefficient expressing changes in the heat generation rate in terms of the difference of mineral composition of Portland cement. The coefficient s_i changes the reference heat rate according to the fineness of powders. The coefficients γ, β_i, λ and μ are assumed to give a changing ratio when there are no effects, respectively, by other factors. Then the minimum coefficient is adopted as a reducing ratio when several coefficients give less value than 1. Here $-E_i/R$ is defined as the *thermal activity*.

7.2.2 Reference heat generation rate of components

Exothermic hydration process

The exothermic characteristics of the hydration process of cement and synthesized clinker minerals have been studied in detail using conduction microcalorimeters. Uchikawa, aware of the many approaches to dividing the hydration process time-wise, adopted the time-domain division shown in Figure 7.6 [14]. In this scheme, the exothermic process comprises the first stage (Stage 1) in which the exothermic peak appears as a result of initial ettringite formation and then a dormant phase appears. The second stage (Stage 2) is that in which active hydration of alite occurs and ettringite converts to monosulphate hydrate, and the third stage (Stage 3) in which subsequent diffusion-dominated hydration proceeds gradually. This basic framework forms the basis for discussion in this study also. After the peak, where rapid heat liberation results from the formation of ettringite, the cement reaction moves into a dormant period during which only slight exothermic reactions continue. The heat generation rate during this dormant period is extremely low, but the heat which is generated may be attributed mainly to the elution of ions from mineral particles and continued ettringite-forming reactions. Ion elution from the minerals proceeds continuously at this stage, increasing the ion concentration in the solution to the level where precipitation of hydrates takes place. Once the dormant period ends, cement reactions once again become active, and the second exothermic peak, accompanied by hydrate precipitation, is seen. It is generally accepted that

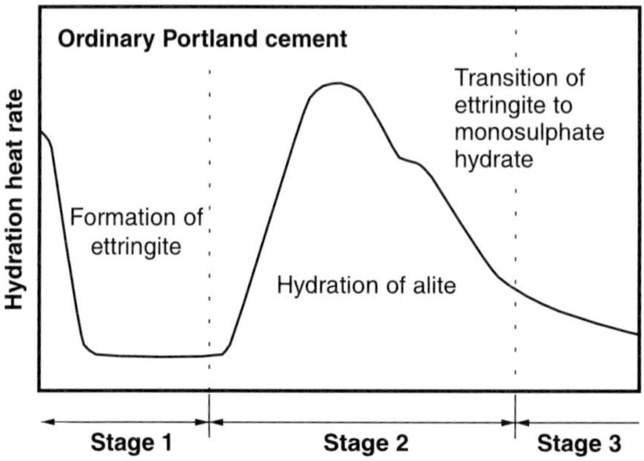

Figure 7.6 Stages of the exothermic hydration process in OPC [14].

the dormant period comes to an end because hydrate nuclei form or the protective layer on the particle surfaces is destroyed [2]. At this time, the $Ca(OH)_2$ concentration in the liquid phase oversaturates. After the second heat-generation peak, the reactions move on to the diffusion control stage, where long-term heat generation continues as hydration progresses.

The aim of this proposed model is to predict the exothermic process for each mineral component after isolating the individual hydration processes in cement chemistry. The ettringite-formation reaction, however, is not handled as part of the model for the heat of hydration of C_3A and C_4AF, but was modelled separately.

Modelling of the reference heat generation rate

The reference heat generation rate, H_{i,T_0}, at the reference temperature, T_0, which are set as material functions for each individual mineral reaction, are shown in Figure 7.7. These rates are generally set as values for the mineral composition of ordinary Portland cement. The reference temperature T_0 was set at 293 K (20 °C). Heat generation resulting from the wetting of each mineral was not considered. The process of the exothermic hydration reaction in each mineral was divided into stages in terms of the cumulative heat generation. Stage 1 was defined as the period until 1% of the total heat output was reached; Stage 2 was up to 25% for C_3S and up to 30% for C_2S; Stage 3 was above these values, respectively. In the exothermic process of C_3A and C_4AF, the border is not assumed between Stage 2 and Stage 3. These ranges were set by analysis using our model. That is, Stage 1 was assumed to represent the temperature increment corresponding to the initial dormant period, and Stage 2 was assumed to represent the characteristic changes in heat generation corresponding to the second exothermic peak.

The exothermic peak in Stage 2 of the C_3S and C_2S reactions is thought to be attributable to two sources of heat: the heat generated during precipitation of products from ions that were already eluted and oversaturated during the dormant period, and the heat generated by new elution and precipitation occurring in Stage 2. In setting the heat generation rate of Stage 2, it was assumed that the former was more dominant, so Stage 2 was treated as a reaction control process in which products are formed from the ionic phase. The heat output of C_3S and C_2S during Stage 2 was assumed so as to fit the initial characteristic temperature increment corresponding to the second exothermic peak in an adiabatic temperature rise test. However, since the division of the exothermic process for each mineral was fixed, the heat generated by each mineral in Stage 2 was made constant. Where a large amount of superplasticizer was added, both the heat generation and its rate in Stage 2 are known to be affected. Thus, such a situation falls outside the applicable scope of our proposed model.

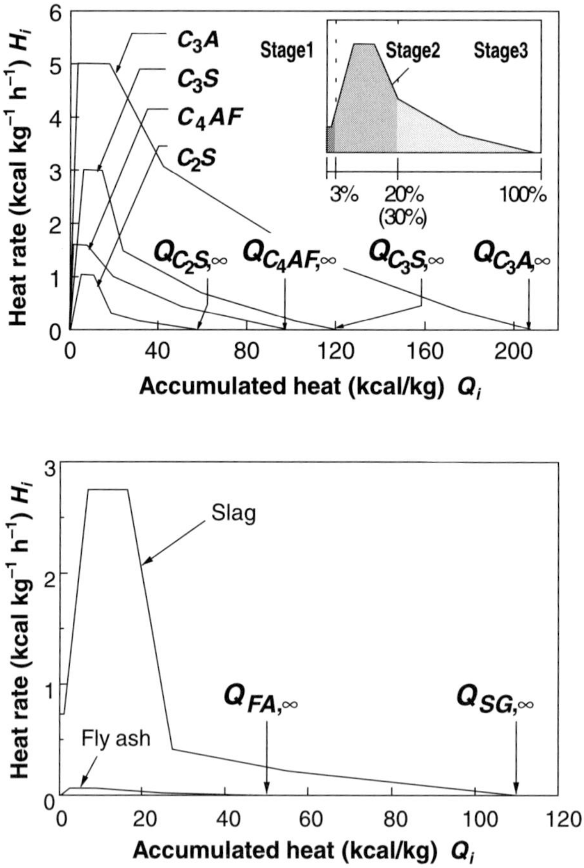

Figure 7.7 Reference heat rate set for each reaction.

Stage 3 is a diffusion control process in which the heat generation rate is considerably slower than that in Stage 2. The reference heat generation curve for the entire cement developed by Suzuki *et al.* [5] was used as a reference in determining how much the reference heat rate should be reduced to correspond to the increase in diffusion resistance. This diffusion control stage, in which precipitated hydrates cover unhydrated portions, is characterized by the elution of ions from unhydrated surfaces. These then diffuse through the internally formed layer and external formation layer, finally reaching some appropriate location at which precipitation occurs. For this reason, it is assumed that the reaction rate is highly influenced not only by the diffusion area itself, but also by the characteristics of elution from unhydrated parts. The reaction rate was set by considering the differences in reactivity of the minerals. To be more specific, the reaction rates were arranged in the order $C_3A > C_3S > C_4AF > C_2S$. The accumulated heat at which the reference heat rate of a particular mineral is certainly zero corresponds to the final heat output achieved by 100% hydration ($Q_{i,\infty}$). This value was determined from the theoretical heat generation of each mineral [2].

The reference heat generation rates of blast furnace slag and fly ash are assumed to be under conditions where water and calcium hydroxide are sufficiently supplied to their reactions. The reaction rate of fly ash, which is the reference heat rate divided by total heat generation, is set much smaller than the slag reaction in the model because a reaction between the glass phase in fly ash and Ca^{2+} in the solution is relatively slow and is accelerated after 1 to 3 days from mixing at the normal temperature [14].

In general, it is difficult to derive the reaction heat generation rate of blending admixtures individually from experiment, since it cannot continue to react without an activator. Besides, it is well known that heat rate measured under the addition of a reagent is not similar in blended cement, since the type of reagent affects the reactivity of admixtures [14]. Thus, the reference heat generation rates of slag and fly ash are set by comparing the analytical results to the experimental data as in Suzuki [6]. The total heat generation per unit weight set in the model, which corresponds to 100% reaction ($Q_{i,\infty}$), can also be derived from qualitative observations of reactions besides analytical trials. In the case of blast furnace slag, it is well known that replacement of slag up to 50% hardly changes the ultimate temperature rise of ordinary Portland cement in adiabatic tests and gradually raises that of moderate heat Portland cement [15]. Therefore, the total heat generation of slag was assumed to be approximately the same as ordinary cement and slightly greater than moderate heat cement. Then, 110 kcal/kg was assumed for the total heat generation of unit weight blast furnace slag in this study. On the other hand, quantitative information for fly ash reactions is scarce and it is not clear that the reaction of fly ash terminates within the testing period of adiabatic temperature rise. The total heat generation of fly ash in the reference heat generation rate was set as 50 kcal/kg in this book.

The proposed model should also be applicable to changes of powder fineness. In general, reaction rate per unit mass proceeds faster with finer particles. Powder fineness is enhanced for early hardening Portland cement with changes of mineral composition rather than ordinary cement for early strength development. In the case of blast furnace slag, wide variations in fineness are also produced. Thus, the material functions assumed in the model can be regarded as values corresponding to certain fineness of powders which are used in experiments and adopted for comparison. Here, it is assumed that the fineness of powder affects only the probability of contact between particles and the surrounding water which is the reacting object. In this study, only the reference heat generation rate is changed according to powder fineness and coefficient s_i represents the change of reference heat rate expressed by means of the Blaine value as an indicator of fineness,

$$s_i = S_i / S_{i0} \tag{7.7}$$

where S_i is the Blaine value of component i (cm^2/g), and S_{i0} is the reference Blaine value of component i. The reference Blaine values of blast furnace slag and fly ash are values used in experiments which were adopted for comparison [5, 6]. These are assumed to be 3380, 4330 and 3280 (cm^2/g) from material specifications of experiments. In Stage 2 of C_3S, C_2S and slag, however, the effect of fineness is not taken into account. At the current stage $s_i = 1$.

7.2.3 Temperature dependence of mineral reactions

Temperature dependence of minerals

The relationship between thermal activity and the accumulated heat output for the cement as a whole, as determined in the study by Suzuki *et al.* [5], is shown in Figure 7.8. This verifies that the thermal activity does in fact change with accumulated heat. The temperature dependence is thought to be different for each mineral and also for each control process in the exothermic hydration process, and this is why the thermal activity of cement as a whole changes nonlinearly with the accumulated heat.

In the overall hydration process, it is mainly minerals of high reactivity that undergo rapid reactions in the initial period; later, minerals of lesser reactivity take over. If this is the case, the heat generation characteristics of cement as a whole should reflect the characteristics of the minerals that play a major role in each reaction. If the temperature dependence of each mineral

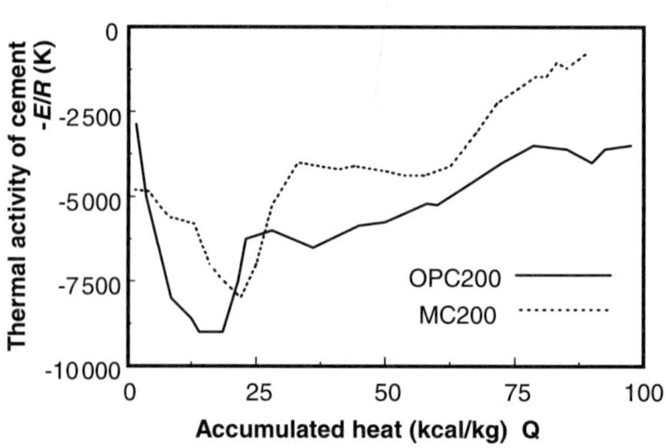

Figure 7.8 Thermal activity of cement from adiabatic temperature rise tests [5].

is different, it is reasonable to expect the thermal activity of the cement as a whole to change as the accumulated amount of heat increases, since the minerals take turns to be the major players in the reaction depending on how far the hydration process has progressed. Further, the hydration process of each mineral may be associated with different levels of thermal activity depending on the reaction control factors and other factors affecting the rate change. The expressions for the temperature dependence of minerals used in this study are those developed by Suzuki *et al.* [5], except that thermal activity was set differently for each mineral. That is, we take the position that the superficially complex thermal activity of cement as a whole can be described as an accumulation of reactions for each component, each with an individual thermal activity.

Modelling of temperature dependence

The thermal activity of each mineral included in this model is shown in Figure 7.9. The authors know of no studies that quantitatively determine the temperature dependence of the minerals making up Portland cement. Consequently, we decided to use Suzuki's values for the thermal activity of cement as a reference and determine the thermal activity of individual mineral reactions for the purpose of this study by considering the reactivity of each mineral. Suzuki's values of thermal activity for cement [5] tend to fall from around −6500 to −2500 (K) when the accumulated heat exceeds 25 kcal/kg. This is the point at which the process is considered to become one of diffusion control (Figure 7.8). As already hinted, we inferred that changes in the thermal activity of cement as a whole as the reaction progresses arise because minerals of different thermal activity take turns as the major player. In other words, the variation in thermal activity with

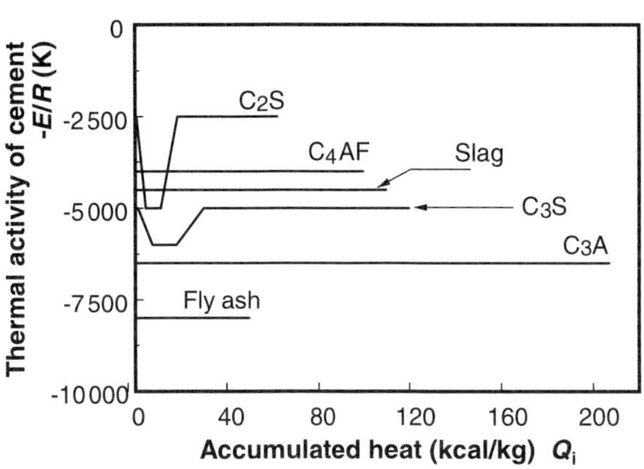

Figure 7.9 Thermal activity for reaction of each component.

increasing accumulated heat reflects the thermal activity of the individual minerals taking part in reactions at each stage.

Taking $C_3A > C_3S > C_4AF > C_2S$ to be the order of mineral activity and assuming a constant thermal activity for all minerals, the overall activity values ranging from −6500 to −2500 (K) were mapped to the thermal activity of each mineral while ensuring that the changes in each minerals reactivity corresponded to the overall thermal activity of the cement. It was also found possible to determine the thermal activity of cement as a whole using Suzuki's technique [5], but substituting the experimental results of the adiabatic temperature rise history by those calculated using the proposed heat of hydration model. The results of this reverse analysis, which will be discussed later, were used as reference values for setting the thermal activity of the minerals.

The thermal activities of C_3A and C_4AF were assumed to be constant throughout the hydration process. This is because of our understanding that in reactions involving C_3A and C_4AF there is a simultaneous elution of ions from unhydrated surfaces and precipitation of hydrated products throughout the reaction process. It is also our understanding that the temperature dependence exhibited by the heat generation rate is governed by the temperature dependence of elution from unhydrated mineral surfaces.

Different thermal activities were set for Stage 2 and Stage 3 of the hydration processes of C_3S and C_2S. At Stage 3, the activity was assumed to be the activity related to ions eluding from unhydrated mineral surfaces. On the contrary, in Stage 2, it was set according to the assumption that thermal activity related to the precipitation of oversaturated ions as hydrates has a prevailing influence. Therefore, using as a reference the overall thermal activity of cement up to an accumulated heat of 25 kcal/kg, a value higher than that set for Stage 3 was set for the thermal activities of C_3S and C_2S in Stage 2.

The thermal activity of blast furnace slag was given a slightly smaller value, −4500 K than the one given to Stage 3 of C_3S which is a major mineral compound in cement because the thermal activity of blended cement with 40% slag replacement showed a slightly smaller value than the entire ordinary Portland cement in a previous study conducted by Suzuki *et al.* [6]. For analysis, the same value was applied to Stage 2 and Stage 3 of the slag reaction process. For fly ash, it is difficult to deduce the thermal activity from the measured value of fly ash blended cement since the heat generation of the fly ash reaction is much smaller than cement. An appropriate value for thermal activity was assumed for the fly ash reaction through an analytical approach. In the analytical reproduction of adiabatic temperature rise of fly ash blended cement, it is difficult to ignore the heat generation of fly ash even if the fly ash reaction is not active at the normal temperature. Thus, it is supposed that the fly ash reaction is greatly accelerated at high curing temperatures. That is why a relatively large thermal activity, −8000 K, was provisionally assumed at the current stage.

7.2.4 Ettringite formation model by reaction of aluminate and ferrite phase with gypsum

Ettringite formation process

The so-called interstitial materials C_3A and C_4AF are compounds that undergo hydration when wetted. In Portland cement, however, where gypsum is present, they react energetically with gypsum prior to hydration to produce ettringite ($C_3A \cdot 3C\overline{S} \cdot H_{32}$) as shown in Figure 7.10 [2]. In the early stages of the reaction, ettringite formation is very rapid, but the reaction slows down as the remaining C_3A and C_4AF becomes covered in the ettringite reaction product. In this study, we assume that no hydration of C_3A and C_4AF takes place as long as there is unreacted gypsum in the liquid phase, so ettringite formation continues. Thus, if a considerable excess of gypsum is added as compared with the amount of C_3A and C_4AF, ettringite will continue to be formed for an extended time and hydration of C_3A or C_4AF will not occur. Gypsum is generally added to curb the hydration of C_3A, which would otherwise cause rapid setting of the cement, and thus ensure a certain period of workability under fresh conditions. It is usually consumed within 24 hours of hydration beginning in a standard Portland cement.

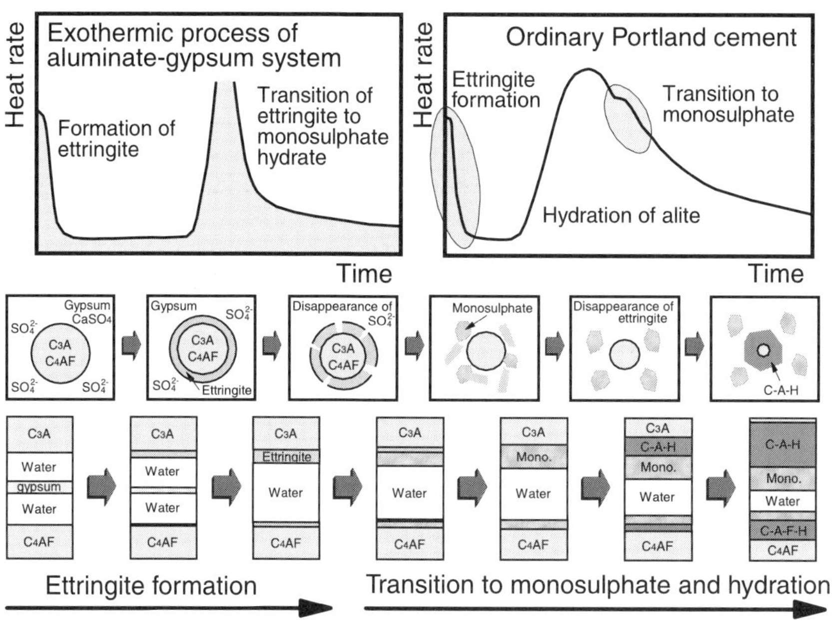

Figure 7.10 Reactions of aluminate and ferrite phase with gypsum.

It is known that ettringite formation comes to a halt when there is no more SO_4^{2-} in the liquid phase because the gypsum has been consumed. This disappearance of SO_4^{2-} from the liquid phase undermines the stability of the ettringite, and the ettringite layers covering unreacted parts crumble. The ettringite then reacts with unreacted C_3A or C_4AF and easily converts into monosulphate ($C_3A \cdot C\overline{S} \cdot H_{32}$) [2]. Here, we assume that conversion from ettringite to monosulphate continues for as long as ettringite is present, and that after full conversion of the ettringite, hydrates from the unreacted C_3A and C_4AF start to precipitate. Consequently, the heat generated by the formation of ettringite and the timing of both the ettringite–monosulphate conversion and the hydration of unreacted interstitial materials will depend on the amount of C_3A and C_4AF and the amount of gypsum added.

Ettringite formation model and representation of the monosulphate conversion reaction

It has been reported by Suzuki *et al.* [4] that the ultimate temperature rise in the adiabatic test is somewhat affected by the initial temperature at casting and the higher temperature rise was observed generally when the casting temperature was lower. One possible reason for the dependence on initial temperature for ultimate temperature rise is the microstructure formation of the hydrate which plays a role as diffusion resistance changes according to temperature history [16]. It is also possible that the heat generation accounted for the temperature rise at the initial stage after casting changes because the heat release during the time from mixing to the beginning of measurement is not constant. In the thermal history of adiabatic temperature rise measured by Suzuki [5], there is a period in which no temperature rise takes place just after casting at an initial temperature of 30 °C, while at an initial temperature of 10 °C, a small temperature rise comes first and subsequent stagnation follows. The temperature history changes slightly according to the initial state, because the heat generation before casting changes as a result of the temperature dependence of the reaction. It is possible that when the initial temperature is higher, most of the heat generation corresponding to the first peak of the exothermic process just after mixing raises the initial temperature itself, since it almost finishes before casting, while in the case of the lower peak, a part appears as a slight temperature rise just after the start of temperature measurement.

To explain how a higher initial temperature brings a lower ultimate temperature rise, it was considered that heat generation at the initial stage counted as a part of the adiabatic temperature rise changes which depended on the initial temperature and the change of the micropore structure under different temperature histories. Thus, it is necessary to accurately model the ettringite formation which brings rapid heat generation just after mixing in order to reproduce an adequate temperature rise according to the initial temperature, and the formation of ettringite is treated separately. If an admixture, such as blast furnace slag or an expansive agent, is added, the reaction processes of C_3A and C_4AF can be expected to be influenced since they produce ettringite by reaction with gypsum, as C_3A or C_3AF. However, the reactions of blended powders with gypsum are still not taken into account in the current model.

The reference heat generation rates for ettringite-forming reactions involving C_3A, C_4AF and gypsum are shown in Figure 7.11. The thermal activity values of these reactions were assumed to be the same as those set for the hydration reactions of C_3A and C_4AF (Figure 7.9). The model ettringite-forming reaction is initiated concurrently with the start of the calculation, and the first exothermic peak in the exothermic cement-hydration process as a whole, which occurs when the water is added, is expressed by this model. Judgment of whether ettringite formation has ended or not depends on a calculation of the amount of remaining gypsum. Gypsum consumption can be calculated from the degree of reaction ($Q_{iET}/Q_{iET,\infty}$) and the rate of combination of C_3A and C_4AF, and then the amount of unreacted gypsum can be obtained by deducting this gypsum consumption from the total gypsum content of the cement. The bonding ratios of C_3A, C_4AF and gypsum dihydrate in ettringite formation are obtained in this model by using the following equation to describe the ettringite formation reaction [2].

$$
\begin{aligned}
&C_3A + 3C\overline{S}H_2 + 26H \rightarrow C_3A \cdot 3C\overline{S} \cdot H_{32} \\
&C_4AF + 3C\overline{S}H_2 + 27H \rightarrow C_3(AF) \cdot 3C\overline{S} \cdot H_{32} + CH
\end{aligned}
\tag{7.8}
$$

where

$$C \equiv CaO,\ A \equiv Al_2O_3,\ F \equiv Fe_2O_3,\ H \equiv H_2O,\ CH \equiv Ca(OH)_2,$$

and $\overline{S} \equiv SO_3$.

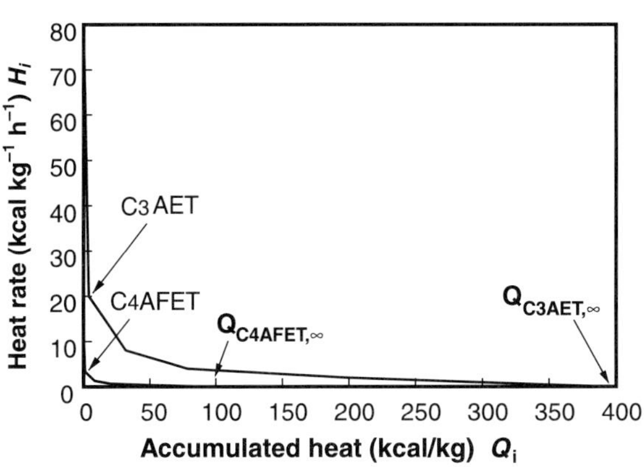

Figure 7.11 Reference heat rate set for the ettringite formation reaction.

It is assumed that the casting time from mixing to beginning of measurement is 0.015 days (around 20 min) and the heat generation during this period is excluded from the analytical result of the temperature rise. This results from the fact that a part of the heat generation after mixing is already included in the initial temperature which is the starting point of computation in this study. Therefore, in all analyses in this study, heat generation after the start of computation up to 0.015 days is not counted in the temperature rise. This means the starting point of the temperature rise is at 0.015 days after hydration starts. This model of the ettringite formation reaction continues to operate until all the gypsum is consumed. Then, the heat of hydration generation of interstitial phases expressed by H_{C3A}, H_{C4AF} starts immediately after ettringite formation terminates due to the disappearance of unreacted gypsum.

Our proposal for this model represents conversion to monosulphate after the formation of ettringite and the hydration of unreacted C_3A and C_4AF using the heat generation curves for hydration shown in Figure 7.7. That is, our interpretation is that after the collapse of ettringite layers due to an absence of SO_4^{2-} in the liquid, elution from unreacted C_3A and C_4AF takes place continuously, and the eluted elements are then used in reactions which convert ettringite to monosulphate as long as ettringite remains. Upon completion of these reactions, the pattern of reactions taking place among the eluted elements changes to hydration. The timing of the change from monosulphate reactions to hydration reactions is consequently determined by the amount of SO_4^{2-} in the ettringite, or the amount of gypsum added. For the sake of convenience, in modelling the generation of heat of hydration generated by C_3A and C_4AF, as with the ettringite formation model, gypsum consumption and unconverted gypsum are calculated and we assume that conversion to monosulphate stops when there remains no unconverted gypsum. The following equation is used to represent the conversion to monosulphate from ettringite and to give the unreacted amounts of C_3A and C_4AF as

$$\begin{aligned}&2C_3A + C_3A \cdot 3C\overline{S} \cdot H_{32} + 4H \rightarrow 3[C_3A \cdot C\overline{S} \cdot H_{12}]\\&2C_4AF + C_3(AF) \cdot 3C\overline{S} \cdot H_{32} + 6H \rightarrow 3[C_3(AF) \cdot C\overline{S} \cdot H_{12}] + 2CH\end{aligned} \tag{7.9}$$

The heat of reaction of these compounds can be calculated from the formation enthalpy of each, provided that there is an identified chemical equation to use. Here, the heat generated in the conversion to monosulphate is determined by equation (7.7), which shows that 3 mols of monosulphate are formed from 2 mols of unreacted C_3A. The standard formation enthalpy for each of these compounds has been given by Osbaeck [17], and, using this data, one obtains a figure of about 210 calories of heat generated per gramme of unreacted C_3A, an amount almost equal to the heat of hydration of C_3A. In this proposed model, 1 mol of C_3A already converted to ettringite is treated as unreacted C_3A in calculating the heat of hydration, and the heat generated in the conversion to monosulphate is expressed by multiplying the reference heat rate associated with hydration by $\frac{2}{3}$.

7.2.5 Evaluation of interdependence among component reactions

Since the modelling of cement in this book includes the reactions of separate minerals individually, it is necessary to account for any interdependence they exhibit (see Figure 7.12). One possible factor that might cause mineral reactions to interact with one another in Portland cement is the amount of water they share for hydration. Czernin concludes that the amount of water required for cement to fully hydrate is about 40% of its weight, and that some cement remains unchanged if the water-to-cement ratio is less than 40% [18]. In a hydration reaction model proposed by Tomosawa, the residual concentration of water is used as a parameter indicating a rate decline toward the reaction end point, thus expressing the possible effect of water content on the reaction [19]. This led us to consider the possibility that the reaction rate decreases in a mix with low water-to-cement ratio. Another effect taken into consideration after repeated analysis-based review was that a difference in mineral composition causes the heat generation rate of each mineral to change during the diffusion control process. It was further found

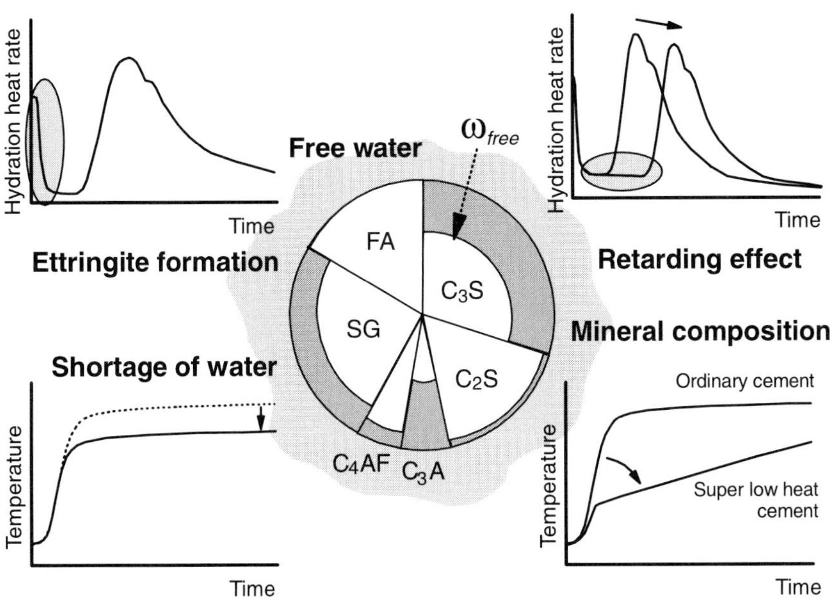

Figure 7.12 Interdependency of mineral reactions.

that the temperature of the system as a whole, which changes with the heat generation of the reactions, is a factor that affects interdependence as a feature of the common environment governing the individual exothermic hydration processes.

For treatment of blended cement reactions, the mutual interaction among Portland cement, blast furnace slag and fly ash should be taken into account in the system. Both slag and fly ash are unable to continue their reactions without $Ca(OH)_2$ as an activator which is produced by C_3S and C_2S hydration in cement and they are quite different in their reactivity. That is, reactions of both slag and fly ash are dependent on the hydration of cement and have a correlation in which $Ca(OH)_2$ is shared in ternary blended cement. Therefore, the interdependence related to $Ca(OH)_2$ should be suitably evaluated to apply the model to blended cement. Further, the delaying effects of chemical admixture and fly ash to cement and slag reactions should also be included.

Fall in heat of hydration generation by reducing free water

Since minerals react with the mixing water, it follows that free water is shared by the processes of reaction and hydration. Hence, there is an interdependence between reaction and hydration. Free water is not only required for continued hydration, it also provides the space in which hydrates precipitate. As already mentioned, there is a general acceptance that the amount of water necessary for complete hydration of ordinary Portland cement is about 40% of the total cement weight [2]. It is assumed that free water may run short toward the end of the reaction in high-performance concrete where the water-to-cement ratio is around 30% or less.

Given the need for a model that is generally applicable, it is necessary to express the decline in hydration rate if the supply of free water is inadequate. The reference heat rate set in the proposed model is based on the assumption that hydration proceeds with an ample supply of free water. However, the actual hydration rate is reduced against the reference heat rate at a low water-to-cement ratio since it is impossible to assume that sufficient free water exists around the powder particles. In a system with a low water-to-cement ratio, it is highly likely that the reaction will stagnate due to a lack of free water, leaving unhydrated material once the reaction terminates. It has been demonstrated that mix proportions with lower water-to-binder ratios exhibit a greater drop in adiabatic temperature increment per unit powder weight, and this is probably because of the lack of free water. In the proposed model, it is assumed that the reduction of heat rate due to a shortage of free water results from the reduction of the probability of contact between the reacting surface of particles and free water. The authors' heat of hydration model makes use of the following equation to express the degree of heat generation rate decline (Figure 7.13) as

$$\beta_i = 1 - \exp\left\{ -r\left[\left(\frac{\omega_{free}}{100 \cdot \eta_i}\right) s_i^{1/2}\right]^s \right\} \tag{7.10}$$

where r and s are material constants common for all minerals. From comparison between the experimental and the analytic results, it was determined that $r = 5.0$ and $s = 2.4$. The coefficient β_i represents the reduction of heat rate and is simply formulated in terms of both amount of free water and the thickness of internal hydrates layer. The function adopted varies from 0 to 1 and hardly gives the reduction when sufficient free water exists and the hydrates layer is thin. On the other hand, it sharply reduces the heat rate when the amount of free water is reduced and unhydrated particles are covered by thick clusters of hydrates. The coefficient β_i is also a function of coefficient s_i which represents the change of reference heat generation rate due to the fineness of powder. This is because the surface area of unit weight particles, which can come in contact with free water, varies according to fineness as well as reference heat rate changes. Here ω_{free} is the free water ratio and η_i is the thickness of internal reaction layer of component i. These are defined by the following equations.

$$\omega_{free} = \frac{W_{total} - \sum W_i}{C} \tag{7.11}$$

$$\eta_i = 1 - \left(1 - \frac{Q_i}{Q_{i,\infty}}\right)^{1/3} \tag{7.12}$$

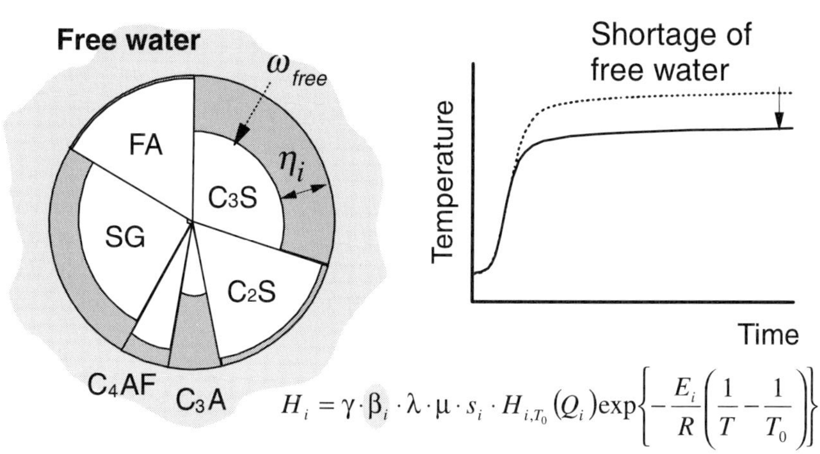

Figure 7.13 Reduction of heat of hydration rate by insufficient free water.

where W_{total} is unit water content, W_i is the water consumed and fixed by the reaction of constituents, C is the unit cement content, Q_i is the accumulated heat of component i, and $Q_{i,\infty}$ is final heat generation (see Figure 7.7). The water used for hydration or the elution of ions is dispersed throughout the hydration formation layer, so the diffusion resistance increases as the formed layer grows. It is assumed that the resistance to diffusion can be represented by the thickness of the internal formation layer, η_i, which has microtexture structure.

The amount of water bound by each reaction can be arrived at by multiplying the bound water ratio, as obtained from the reaction equation of each mineral, by the degree of reaction computed by the model ($Q_i/Q_{i,\infty}$). The reaction equations used in this model are shown as below.

$$\begin{aligned}
&C_3A + 6H \rightarrow C_3AH_6 \\
&C_4AF + 2CH + 10H \rightarrow C_3AH_6 - C_3FH_6 \\
&2C_3S + 6H \rightarrow C_3S_2H_3 + 3CH \\
&2C_2S + 4H \rightarrow C_3S_2H_3 + CH
\end{aligned} \tag{7.13}$$

where $S \equiv S_iO_2$. It must be noted at this point that it is not only chemically bound water that is consumed in the various reactions. This is water that is physically trapped by the formed texture. Although such constrained water is generally assumed to be dependent on the area and condition of the hydrate surface, it is known to be typically about 15% of the total cement weight [18]. In this model, this figure for the constrained water ratio was assumed for all minerals undergoing reaction. Consequently, the water consumption, W_i, needed for equation (7.11) will be the calculated bound water volume plus another 15%.

The reactions of clinker minerals can be assumed to terminate with the arrival of a degree of hydration of 100% or the reduction of hydration rate caused by a shortage of free water when only Portland cement is used in the model. In the blended cement case, however, the reaction rates of blast furnace slag and fly ash are reduced because of an insufficient supply of $Ca(OH)_2$. Nevertheless, the amount of free water is sufficient. If the shortage of $Ca(OH)_2$ brings the reduction of admixture reactions, free water left in the system can be assumed to fulfil its function in hydration of Portland cement as the reacting object or the space where the hydrates can be precipitated. When the reaction of admixtures are stagnant due to a shortage of Ca(OH), the effect of free water reduction on Portland cement is expressed by modifying equation (7.11) as follows (Figure 7.14).

$$\omega_{free} = \frac{W_{total} - \sum W_i}{C \cdot (p_{PC} + m_{SG} \cdot p_{SG} + m_{FA} \cdot p_{FA})} \tag{7.14}$$

$$m_i = \lambda/\beta_i \tag{7.15}$$

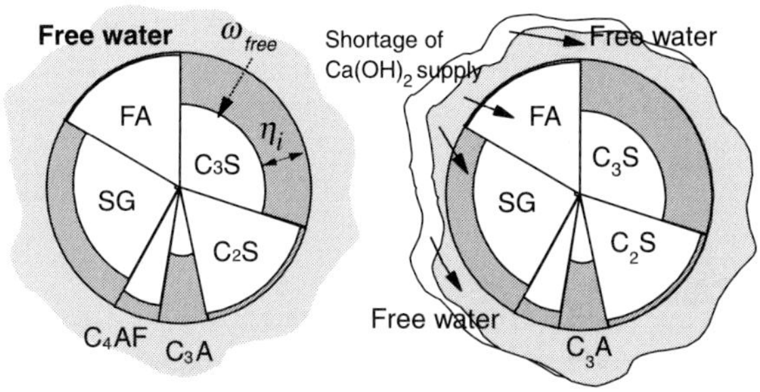

Figure 7.14 Free water distribution under stagnant reactions of admixtures.

The water consumed in ettringite formation is not a concern, because the reaction ends in the initial period and the water bound in the ettringite changes to monosulphate-bound water in the conversion. Mixing water is also assumed to be consumed only by hydration and physical constraint as a result of hydration. Therefore, the heat of hydration model assumes a virtually sealed condition. When the cement dries, it should be solved by combining a heat of hydration model with a model of water transport, in which the amount of free water can be given as a state variable for the heat of hydration model.

Change of heat rate by difference of mineral composition

The adiabatic temperature rise history of ordinary Portland cement or moderate-heat Portland cement indicates that the rapid rise in temperature continues even when the hydration process proceeds to Stage 3 because of the temperature dependence of the reaction. On the other hand, the adiabatic temperature rise of low-heat Portland cement, which is increasingly being used in massive concrete structures, is known to exhibit a long slow rise long after the rapid rise that corresponds to the initial period of Stage 2 [20]. This difference is more apparent in cements with greater C_2S content and with less C_3S content than in other types of cement.

A review of the proposed heat of hydration model was carried out by performing a repeated temperature analysis of low-heat Portland cement for various reference heat generation rates on the assumption that the reference heat generation of each mineral remained constant regardless of the mineral composition. This revealed that the reference heat rates set for ordinary Portland cement poorly predict the lessened temperature rise rate at and beyond Stage 3 of the adiabatic temperature rise of low-heat Portland cement. This is probably because the reference heat generation rate for each reaction changes depending on the type of Portland cement, i.e., the difference in constitutive mineral composition. In this study, therefore, we concluded that the heat generation rate of each mineral changes with its proportioning in the Portland cement in question, since mineral reactions depend greatly on the composition (Figure 7.15).

It is not yet clearly understood what causes the heat rate of minerals to change according to constitutive mineral composition. One possible factor is changes of pH as a result of change of $Ca(OH)_2$ concentration in the liquid phase or changes of other ions in the vicinity of the particles. $Ca(OH)_2$ usually precipitates in great quantities upon hydration of alite whereas almost no precipitation is observed from belite [21]. If one focuses on the liquid phase in the vicinity of particles, once the mineral composition of Portland cement changes, the existence probability of C_3S and C_2S adjacent to some cement particles undergoes a relative change and the reaction environment, as represented by pH changes or various ions' concentration, changes.

The issue of reaction dependence among clinker minerals still leaves many facts unexplained, and to date it is still difficult to make quantitative evaluations of possible causes and their degree of influence. It was therefore decided, for the time being, that the effect of mineral composition on the heat rate would be simply expressed by using a C_3S/C_2S ratio that is considered to represent the mineral composition of Portland cement and heat generation characteristics. A function was adopted which gives larger reduction to the heat rate when the constitutive fraction of C_2S is higher and inversely increases the heat rate when the C_3S/C_2S ratio is greater than in the composition of ordinary Portland cement [12]. In this study it is assumed that this effect is the interdependence between C_3S and C_2S, and the coefficient μ is not applied to C_3A and C_4AF. In order to cover early hardening and special low heat Portland cement, the coefficient μ of the reference heat rate is expressed as

$$\mu = 1.4 \cdot \left\{ 1 - \exp\left[-0.48 \cdot \left(\frac{p_{C_3S}}{p_{C_2S}} \right)^{1.4} \right] \right\} + 0.1 \tag{7.16}$$

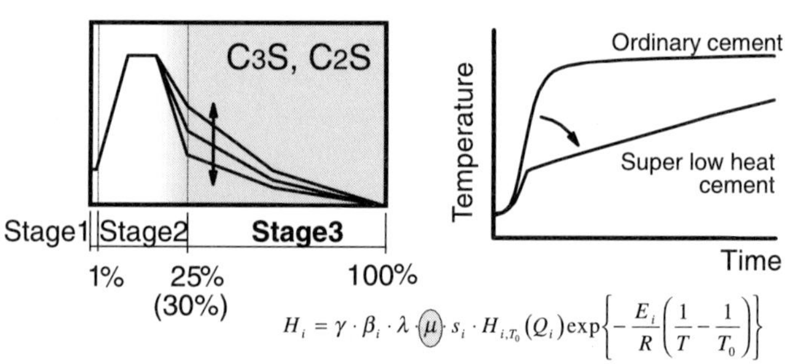

Figure 7.15 Effect of mineral composition on heat of hydration rate.

The reference heat rates of C_3S and C_2S in stages other than Stage 3 were assumed to be constant regardless of any change of cement mineral compositions ($\mu = 1$). This assumption was made because it is possible to reproduce the initial adiabatic temperature history quite precisely within the range of interest in this study without varying the heating rate in Stage 2 according to the composition. The effect of the liquid phase pH on the heat generation rate in Stage 2 was not considered, since it was assumed that the dominant influence in Stage 2 is the precipitation of ions eluted in the dormant period. However, when $Ca(OH)_2$ is added to the cement, the reaction is known to accelerate in both Stage 1 and Stage 2. Thus, if the pH of the liquid phase changes considerably, the effect of pH on the heat of hydration generation rate should be considered even in Stages 1 and 2. As this discussion makes clear, interdependence among mineral reactions, including the effects of liquid phase pH, are treated very simply in this model as given in equation (7.16), but the authors understand the need to further discuss this matter in future.

The evaluation of reaction rate of admixtures dependent on calcium hydroxide produced by Portland cement hydration

Santhikumar studied the interdependence between Portland and blast furnace slag where the heat of hydration rates of blended cement at various replacements were measured by conduction calorimeter as shown in Figure 7.16 [22]. The heat generation rate of blast furnace slag was calculated by extracting the heat of hydration rate of Portland cement according to its constitutive ratio from that of blended cement on the assumption that the reaction of Portland cement was not affected by slag reaction. Through his study, it was found that the slag reaction is stable at any replacement of slag in the former term of reaction process where it is considered that $Ca(OH)_2$ was sufficiently supplied and the heat generation of slag was reduced in accordance with the reduction of Portland cement content which indicates the shortage of $Ca(OH)_2$ supply. It is assumed that a similar mechanism to the slag reaction exists in the fly ash reaction though quantitative studies of fly ash reactions having less reactivity are difficult. Therefore, the evaluation of interdependence in which the amount of $Ca(OH)_2$ is focused on is indispensable in order to evaluate the exothermic process of blast furnace slag and fly ash in blended cement [23].

In ternary blended cement, $Ca(OH)_2$ is simultaneously consumed by blast furnace slag and fly ash. The consumption of $Ca(OH)_2$ should be computed step by step with the production from Portland hydration in the analysis. Here, it is assumed that the reducing ratio of the reaction rate of admixtures can be expressed by the following equation by means of a quantitative ratio between the amount of $Ca(OH)_2$ left in the system and what is necessary for active reaction of slag and fly ash at that time and which is

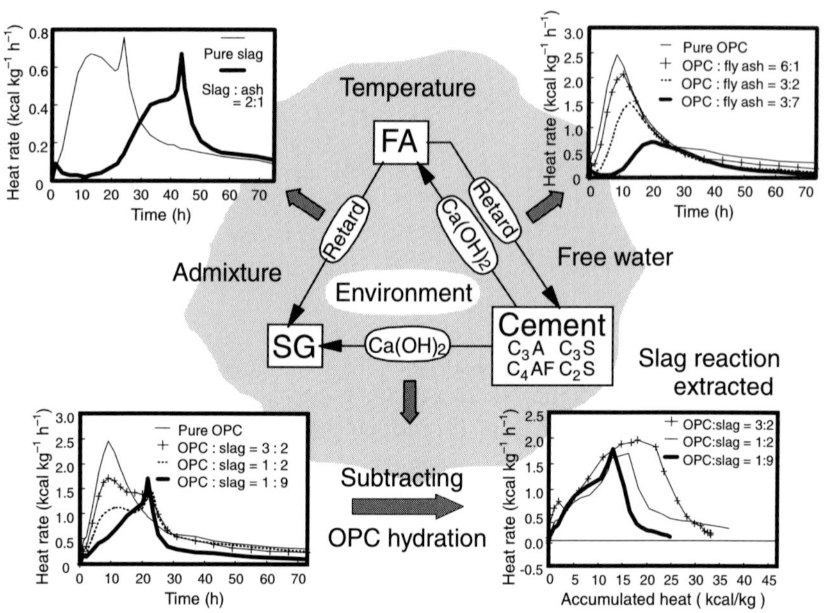

Figure 7.16 Connection between heat rates of blast furnace slag and fly ash with OPC.

simply expressed by

$$\lambda = 1 - \exp\left[-2.0\left(\frac{F_{CH}}{R_{SGCH} + R_{FACH}}\right)^{5.0}\right] \tag{7.17}$$

where F_{CH} is the amount of $Ca(OH)_2$ which is produced by hydration of C_3S and C_2S, and not yet consumed by C_4AF reaction, and R_{SGCH} and R_{FACH} are the amount of $Ca(OH)_2$ necessary for reactions of slag and fly ash when $Ca(OH)_2$ is in sufficient supply. Here, F_{CH} can be computed from the reaction degree of each component and the ratios of production and consumption of $Ca(OH)_2$ which is given by chemical equations adopted in this study, and R_{SGCH} and R_{FACH} are computed by the consumption ratio of $Ca(OH)_2$ by blast furnace slag and fly ash, and their reaction rates when $Ca(OH)_2$ is sufficient.

It is not clear how much $Ca(OH)_2$ is consumed by unit weight of reacted slag and fly ash and also the Ca/Si ratio of products changes according to the amount of $Ca(OH)_2$ supplied and the character of admixtures [14]. So, it is simply assumed that the consumption ratio of $Ca(OH)_2$ is constant through the whole reaction process irrespective of the character of the admixture. The values are determined through the analysis by qualitative observation. The reaction of blast furnace slag is restrained just after contact with water because of the formation of a low penetrable layer on the surface [14]. The supply of OH^- ions from the ambient solution is essential for the continuity of active hydration of slag besides what is eluted from the slag itself. In general, it is assumed that a sufficient supply of $Ca(OH)_2$ can apply to slag reactions of up to 60–70% replacement because the adiabatic temperature rise barely changes within these replacement ranges. On the other hand, the reaction of fly ash does not proceed without the active stimulation of OH^- ions from outside since the condensation degree of silicate ion is high in the glass [14]. Therefore, $Ca(OH)_2$ consumption when a unit weight of fly ash reacts seems to be much higher than in the slag case. In fact, around 60% or 70% replacement ratio to cement is often adopted in the slag case but fly ash replacement is 30% at most. Thus, the consumption ratios of slag and fly ash reactions are assumed to be 22% and 100% of reacted mass, respectively, for the analysis. In ternary blended cement, it is assumed that $Ca(OH)_2$ in the liquid phase is almost completely consumed by the slag reaction since it is faster than fly ash.

Delaying effect of chemical admixtures and fly ash on the exothermic hydration process

It is common knowledge that the addition of a superplasticizer delays initial setting and active heat generation. In the case of high-fluidity concretes using three-component cements or high belite cement, delays exceeding 12 hours can be experienced depending on effects related to the type of cement when a superplasticizer is added [24]. In a heat of hydration generation model of high-performance concrete, it is necessary to account for such effects because a superplasticizer is an essential component of such concretes. Uomoto *et al.* [25] developed an explanation for the delay mechanism of superplasticizers based on past studies. When functional groups of the added superplasticizer, which is designed to contribute to particle dispersion, react with Ca^{2+} formed by cement hydration, a calcium salt is formed. This consumption of Ca^{2+} delays the formation of crystal cores of $Ca(OH)_2$, which are the triggers for active heat of hydration generation, or at least slows the formation rate of the crystals. According to Uomoto's study, some of superplasticizer gets incorporated into the hydrates of interstitial materials as a result of rapid hydration of C_3A or other interstitial materials, and so is consumed regardless of any delay. According to a report by Uchikawa *et al.* [26], the delaying effect of an organic admixture varies depending on how it is added; the effect is apparently greater with post-addition than with simultaneous addition. This corresponds with the observation that the water-reducing effect varies depending on the addition method. The reason is that consumption of Ca^{2+}

formed by cement hydration in the solution, which results from reaction with organic admixtures, delays the formation of a crystal core of $Ca(OH)_2$ that triggers the start of active heat of hydration generation or delays the formation rate of crystals [25].

It is also known that the addition of fly ash to Portland cement delays hydration [14]. Santhikumar [22] measured the heat of hydration rate of binary blended cement in various replacements of fly ash and then observed that the exothermic peak corresponding to active hydration of cement is retarded according to increases of fly ash replacement (Figure 7.16). It was also found that fly ash has a delaying effect in the slag reaction as well as in cement through the study of the reaction heat rate of a binary composite of slag and fly ash with a reagent [22]. If fly ash exists, Ca^{2+} in the liquid phase is adsorbed on the surface of fly ash particles instead of Al^{3+} eluted just after it comes in contact with water. That is why fly ash delays the initial reaction of cement [14]. Thus, it can be assumed that the mechanism of the delay effect for fly ash is similar to that of an organic admixture. Hence, the delay effect of fly ash in reactions of Portland cement and blast furnace slag is dealt with in the same way as chemical admixtures.

The proposed heat of hydration model treats the delaying effect of chemical admixtures and fly ash as an extended dormant period (Figure 7.17). For the sake of convenience, the model incorporates this effect by reducing the heat generation rate in Stage 1, which corresponds to the dormant period, according to the delaying effect of them. Within the normal dosage range, the delaying effect of chemical admixtures has mostly disappeared by the start of Stage 2, so it was temporarily concluded that their effect on exothermic characteristics in Stage 2 and beyond does not need to be considered. However, it must also be noted that the combination of a higher amount of chemical admixture and fly ash delays severely the hydration of cement as an extreme case. In the multicomponent heat of hydration model, the delay due to chemical admixture and fly ash was simply modelled as a first approximation.

Uomoto *et al.* [25] also reported that the amount of superplasticizer consumed depends on the type of superplasticizer, regardless of the delay. Uchikawa *et al.* [26] reported that the delaying effect depends on the type of chemical admixture. Given these observations, our model must take account of the degree of the delaying effect imposed by different chemical admixtures. In this study, we use the term *admixture delaying effect* to describe this delaying effect which depends on the admixture's Ca^{2+} consumption. The coefficient χ_{sp}, which represents the delaying effect per unit weight of admixture, is used to describe characteristics of admixtures. By multiplying this coefficient χ_{sp} by the admixture dosage as a ratio of addition to the cement ($C \times \%$), ϑ_{total}, the total delaying effect of the admixture can be calculated. A part of the admixture, however, is restrained by C_3A or C_4AF, and loses its delaying effect as soon as the reaction starts.

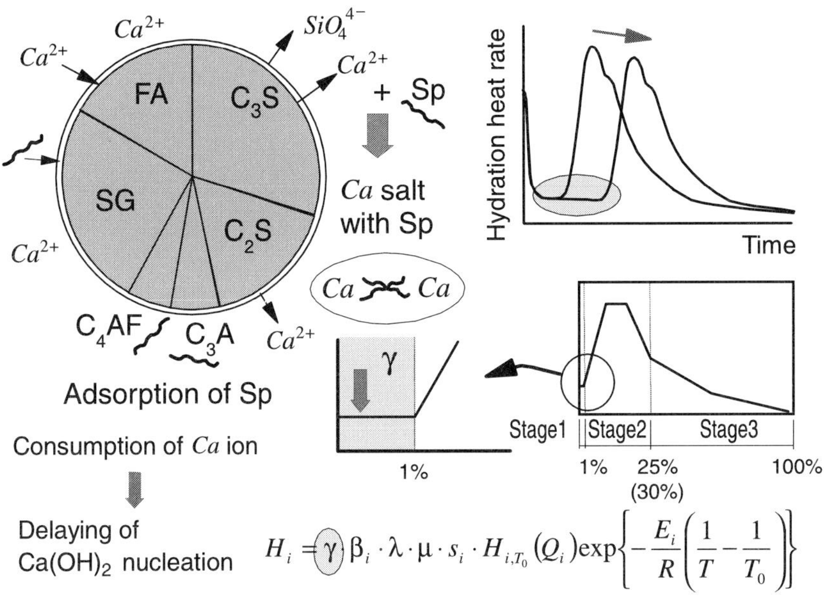

Figure 7.17 Modelling of delaying effect of chemical admixture and fly ash.

Then the value obtained by subtracting the invalid value corresponding to the amount of admixture restrained from the total delaying effect is now defined as the *effective delaying effect* of admixture in the exothermic process of cement hydration, ϑ_{ef}. It is assumed here that the aforementioned chemical admixture corresponds to that adsorbed on the surface of particles immediately after mixing since the delay effect is markedly different between simultaneous addition of water and post-addition [24]. Thus, assuming that a part of the organic admixture is adsorbed on the surface of C_3A, C_4AF, blast furnace slag and fly ash which have an Al composite, the amount of organic admixture consumed regardless of delaying effect is expressed as

$$\vartheta_{Waste} = \frac{1}{200}\left(16p_{C_3A}s_{C_3A} + 4p_{C_4AF}s_{C_4AF} + p_{SG}s_{SG} + 5p_{FA}s_{FA}\right) \tag{7.18}$$

where ϑ_{Waste} represents a composite of chemical admixtures which do not exert a delay effect due to adsorption on the surface of powder particles. Thus, the effective delaying capability, ϑ_{ef}, may be expressed by the following equation as

$$\vartheta_{SPef} = p_{SP} \cdot \chi_{SP} - \vartheta_{Waste} \tag{7.19}$$

where p_{SP} is a dosage of organic admixture expressed as additive ratio to binder ($C \times \%$), and χ_{SP} is a coefficient representing the delay effect per unit weight of organic admixture. Thus, $p_{SP}\chi_{SP}$ represents the total delay effect brought about by an added organic admixture. When the adsorption capacity of the components exceeds the total capacity of the delay effect by an organic admixture, ϑ_{SP} is assumed to be zero. Assuming that the delay effect brought about by fly ash is in proportion to its replacement ratio, we have

$$\vartheta_{FAef} = 0.02 p_{FA} s_{FA} \tag{7.20}$$

In contrast with the delaying effects of chemical admixtures and fly ash, cement minerals are providers of Ca^{2+} ions. If the cement provides a copious supply of Ca^{2+}, the delaying effect wanes rapidly. On the other hand, if the supply of Ca^{2+} is slow, the delaying effect may be long-lived. Elution of Ca^{2+} must vary from mineral to mineral. Given the content of Ca and its reactivity, it follows that the supply from C_2S is smaller than from C_3S. It may therefore be deduced that low-heat Portland cement, with its relatively poorer elution of Ca^{2+}, should be more prone to delaying effects than ordinary Portland cement, even when the same amount of superplasticizer is added.

We decided to express this delaying effect of chemical admixtures and fly ash by incorporating the delaying effect by them and counteracting it with the rate of Ca^{2+} supply from the minerals. In this case, C_3S and C_2S were treated as minerals supplying Ca^{2+}, so the delaying effect of chemical admixtures and fly ash on the heat generation can be expressed by the following equation.

$$\gamma = \exp\left[-\frac{1000(\vartheta_{SPef} + \vartheta_{FAef})}{10 p_{C_3S} s_{C_3S} + 5 p_{C_2S} s_{C_2S} + 2.5 p_{SG} s_{SG}} \right] \tag{7.21}$$

where γ is the coefficient of heat generation reduction in Stage 1, where heat generation is reduced by the delaying effect due to chemical admixture and fly ash. The delaying effects are regarded as an extension of the dormant period and then the delay of reactions are expressed by reducing the heat rate of components concerned in Stage 1 which correspond to the dormant period by means of γ in this model.

Admixtures other than superplasticizers have similar mechanisms as regards delaying effect. These admixtures are handled by adding a suitable coefficient, χ_{sp}, to express the difference in delaying effect depending on the type of admixture. The values applied to a superplasticizer consisting mainly of β-naphthalene sulphonate, another superplasticizer comprising mainly polycarbonate, and a retarding air-entraining (AE) water-reducing agent are 1.2, 1.2 and 5.0, respectively. At this moment, precise quantification of the delaying effect of each type of admixture is very difficult, and the values given here are rough ones derived from analysis. These various values, however, are considered acceptable in qualitative terms since it has been shown by Uchikawa *et al.* [26] that the delaying effect is greater for polycarbonate-based superplasticizers than for naphthalene sulphonate-based ones at standard proportions and also because the retarding AE agent is designed primarily to have a delaying effect. It should be noted that this model is specifically applicable to normal ranges of organic admixture dosage; this is because the method of expressing the delay in the reaction, by individually reducing the heat generation rate of each mineral in Stage 1, results in different Stage 2 starting times for some minerals as the dosage of the admixture increases. Since the amount of heat generated by reactions between the admixture and Ca^{2+} is very small, it was judged appropriate to assume that the heat in Stage 1 does not change with the amount of admixture. This proposed model is predicated on the assumption that the delaying effect of an admixture only causes an extension of the dormant period, but if a larger dose of superplasticizer is added, its effect on heat generation and generation rate in Stage 2 and beyond needs to be evaluated. This is a task for the future.

7.2.6 Constitution of the multiple heat of hydration model

Figure 7.18 shows the structure of a temperature analysis model in which the exothermic hydration process, with its multicomponent model for heat of hydration, is coupled with thermal conduction [8]. The information input into the model includes the mineral composition of the cement used, the amount of gypsum added, the unit cement volume, the unit water content, the amount of organic admixture, coefficients expressing the types of blending powders, the casting (initial) temperature, the ambient temperature, and thermal constants.

The mineral composition of the cement is determined by Bougue's equation [2] based on the chemical element analysis table. It should be noted that Bogue's equation fails to exactly reflect the mineral composition of cement since it makes various suppositions; for instance, it ignores the presence of trace components in the cement. We judged that this equation was appropriate for two major reasons: (1) simplicity of mineral composition was a priority; and (2) its use was not expected to have any technically negative effects in terms of precision. The gypsum content is calculated based on the SO_3 content by assuming that it is all in the form of gypsum dihydrate. The organic admixture dosage is treated as an addition rate as a proportion of cement content. There is a delaying effect of the admixture at the dormant period in this model, but it is not taken into account that the heat rate significantly drops after Stage 2 when the addition of organic admixture is remarkably large. For the time being, differences in delaying effect between admixtures, including AE water-reducing agents, retarding agents, and superplasticizers, are all represented by the coefficient χ_{SP}.

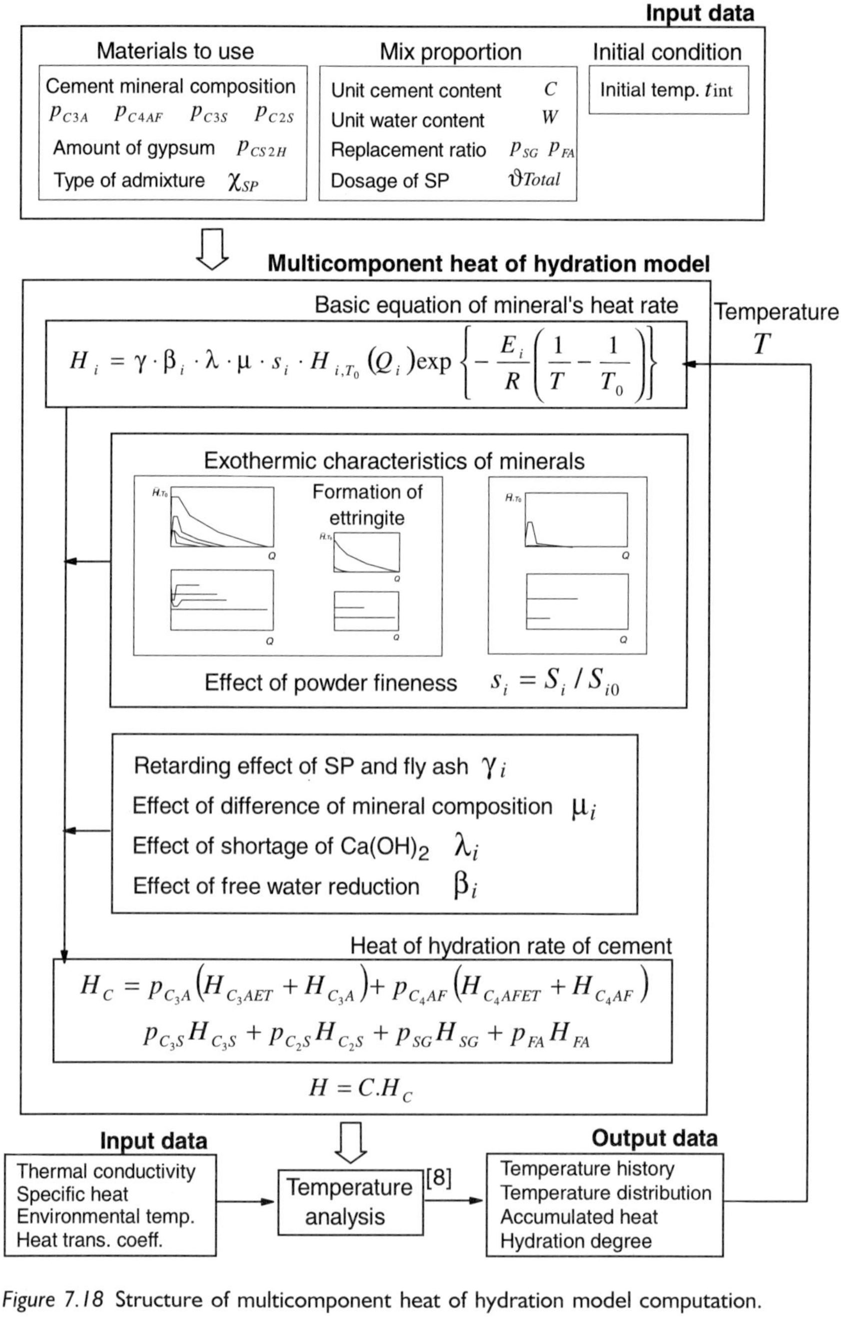

Figure 7.18 Structure of multicomponent heat of hydration model computation.

In the multicomponent heat of hydration model, there is initially a heat of hydration model for C_3S and C_2S, with an ettringite formation. The ettringite formation reactions terminate when there is no gypsum dihydrate left, and then a heat of hydration model takes over for C_3A and C_4AF. The heat generation rate for each mineral can be calculated from just two material functions, the reference heat generation rate and the thermal activity, after evaluating the effects of interdependence among the minerals and the effect of the organic admixture. The overall temperature of the system is then calculated using this temperature analysis by obtaining the heat generation rates of individual component minerals taking into account the temperature dependence.

The temperature analysis was implemented by incorporating the proposed heat of hydration model into a method of nonlinear finite-element analysis developed by Harada *et al.* [8] that allows for the coupling of the exothermic hydration process with thermal conduction. Of the thermal constants needed for the temperature analysis, the specific heat and heat capacity are calculated from specific heat values of the individual materials used and their proportioning. Other thermal constants are appropriate values established by considering the mix proportion and the curing environment. Once the constants are set, the temperature history of each part of the structure in question, as well as the cumulative heat output of each mineral and the amount of free water, are calculated. This calculated value of cumulative heat is used as an index to describe the degree of hydration in determining the heat generation rate and thermal activity. By dividing the cumulative heat output by the heat generation rate at the time hydration reaches completion (the theoretical heat), the reactive activity of each mineral can be defined.

7.3 Review of adiabatic temperature rise history

7.3.1 *Several types of Portland cement*

Suzuki *et al.* [5, 6] carried out a series of systematic adiabatic temperature rise tests with several types of Portland cement (early hardening cement, ordinary cement and moderate-heat cement), and we compared the results with analysis by the proposed method. Since a retarding type of AE water-reducing agent was used in these tests, the heat generation appears to have started later than would have been the case with an ordinary water-reducing agent.

In implementing our analysis, the delaying effect per unit weight of retarding AE water-reducing agent was set as $\chi_{SP} = 5.0$. Time-step intervals in all analyses of this study are 0.01 day from the beginning of hydration till 0.3 day and, after that, 0.05 day was adopted for the time interval. The mineral compositions of the cements are shown in Table 7.1, while the

Table 7.1 Cement mineral composition in adiabatic temperature rise test [5]

	C_3A	C_4AF	C_3S	C_2S	CS_2H	Blaine (cm^2/g)
EPC	9	8	63	12	6.45	4210
OPC	10.4	9.4	47.2	27.0	3.9	3380
MC	3.7	12.5	44.4	33.7	3.9	3040

concrete mix proportions are shown in Table 7.2. Both experimental results and analytical values are shown in Figures 7.19, 7.20 and 7.21, respectively.

Three casting temperatures (10 °C, 20 °C and 30 °C) and three unit cement weights (200 kg, 300 kg and 400 kg) were examined for early hardening Portland cement (EPC), ordinary Portland cement (OPC) and moderate-heat Portland cement (MC). Certain problems remain unsolved in terms of the analytical precision when the water-to-cement ratio of the moderate-heat Portland cement is low, but the measured adiabatic temperature rise history is almost perfectly matched by the analysis. The results reflect the temperature dependence of the reaction at each casting temperature.

As already described, the effects of a low water-to-cement ratio on heat of hydration generation were considered in this study. For unit cement weight of 400 kg, the water-to-cement ratio is around 40%. It is well known that approximately 40% cement weight of water is required for the complete hydration of ordinary Portland cement at least. The reduction in heat generation rate and in the adiabatic temperature rise at the completion of the reaction when the water-to-cement ratio is low was correctly modelled.

Table 7.2 Concrete mix proportions in adiabatic temperature rise test [5]

	W/C (%)	W (kg/m^3)	C (kg/m^3)	S (kg/m^3)	G (kg/m^3)	AE water reducing agent (C %)
EPC400	38.0	152	400	730	1058	0.25
EPC300	47.7	143	300	813	1092	0.25
EPC250	57.2	143	250	869	1071	0.25
OPC400	39.2	157	400	658	1129	0.25[a]
OPC300	49.3	148	300	765	1129	0.25[a]
OPC200	78.5	157	200	862	1089	0.25[a]
MC400	39.2	157	400	663	1129	0.25[a]
MC300	49.3	148	300	770	1129	0.25[a]
MC200	78.5	157	200	865	1089	0.25[a]

[a]Delaying type agent

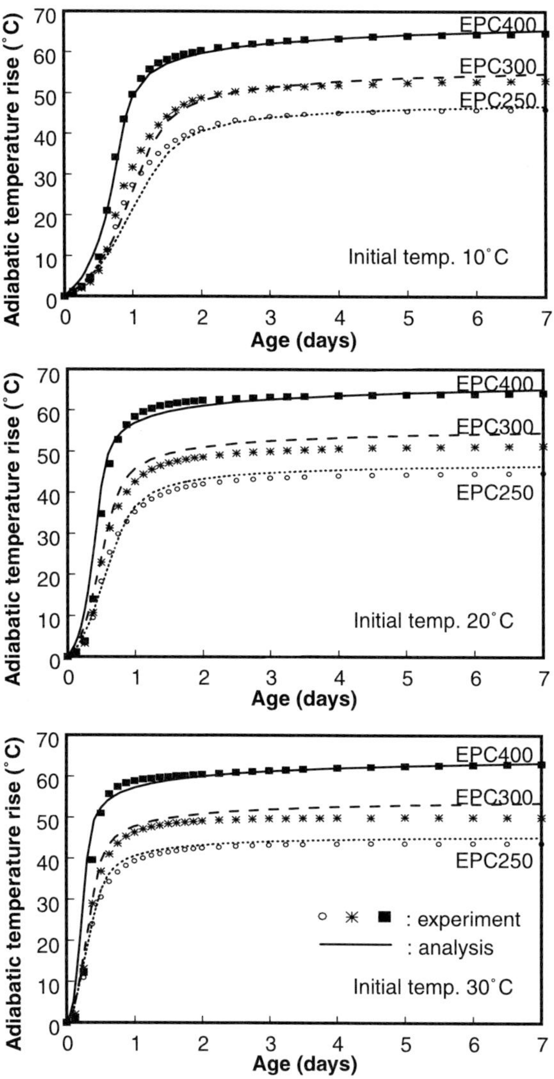

Figure 7.19 Analytical result for adiabatic temperature rise history (early hardening OPC).

7.3.2 Inverse analysis of thermal activity by using the calculated results of the heat of hydration model

Suzuki *et al.* [5] computed the hydration exothermic characteristics of ordinary Portland cement and moderate-heat Portland cement from experimental data of the adiabatic temperature rise tests. On the other

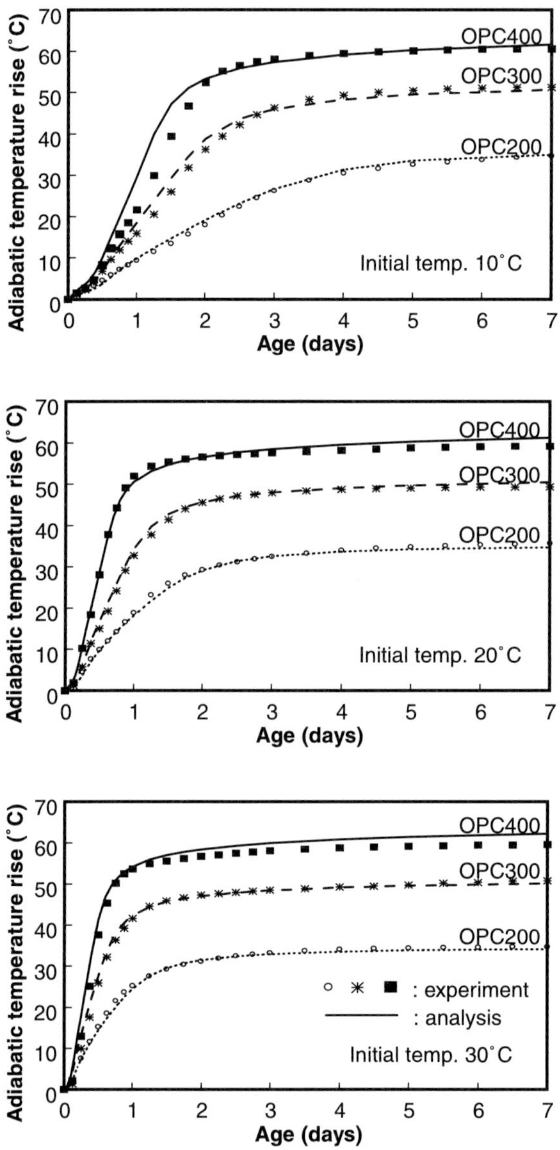

Figure 7.20 Analytical result for adiabatic temperature rise history (OPC).

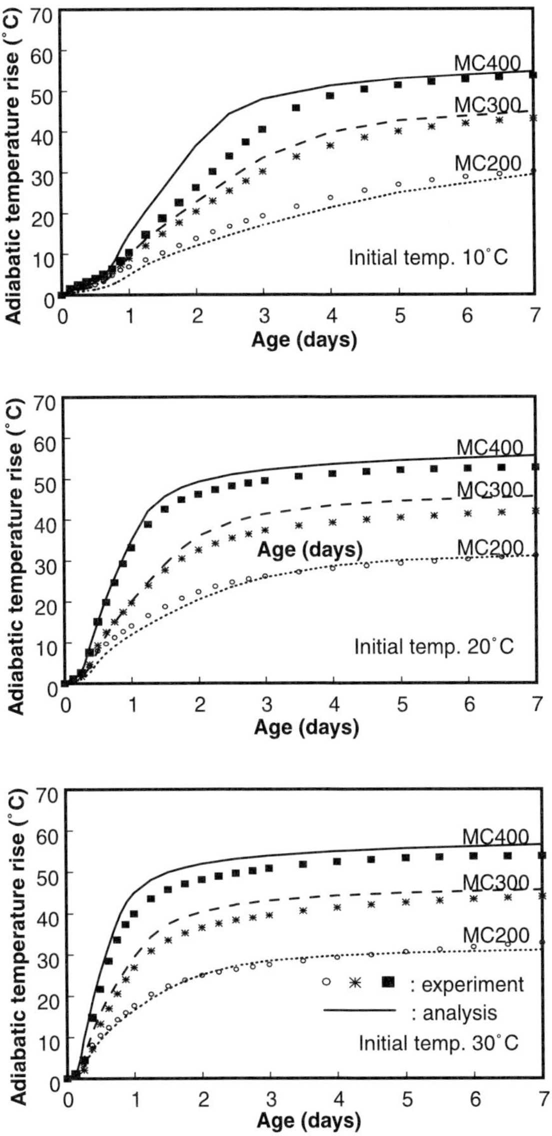

Figure 7.21 Analytical result for adiabatic temperature rise history (moderate-heat Portland cement).

hand, the adiabatic temperature rise histories of several types of Portland cement for three casting temperatures were successfully modelled as described above.

Adopting the method used by Suzuki *et al.* [5] to quantify the exothermic process of hydration, the analytical results obtained for three different casting temperatures were used instead of actual measurements in calculating the averaged thermal activity of the cement as a whole. The thermal activity values obtained in this inverse analysis are shown together with those determined by Suzuki *et al.* [5] in Figure 7.22. Our values for the cement as a whole are similar throughout the hydration process, although they do not fully match those obtained by Suzuki in the range equivalent to Stage 2. Also, the average thermal activity values obtained in reverse analysis from our multicomponent heat of hydration model, in which constant thermal activity is established for each mineral, change nonlinearly with total cumulative heat output. This is because the reaction rate of each mineral component of the cement varies; the minerals that generate heat mainly in the initial, middle, and terminal periods of the reaction are different. This supports the validity of the assumptions made in setting up the proposed model.

7.3.3 Binary blended cement with blast furnace slag or fly ash

The adiabatic temperature rise tests of binary blended cement including blast furnace slag or fly ash and ordinary Portland cement as a basis were carried out systematically by Suzuki [6]. Three casting temperatures, 10 °C, 20 °C and 30 °C, and three unit cement contents, 200 kg/m^3, 300 kg/m^3 and 400 kg/m^3, were examined for both slag blended cement and fly ash in the experiment (SG200, SG300, SG400, FA200, FA300, FA400). Mineral

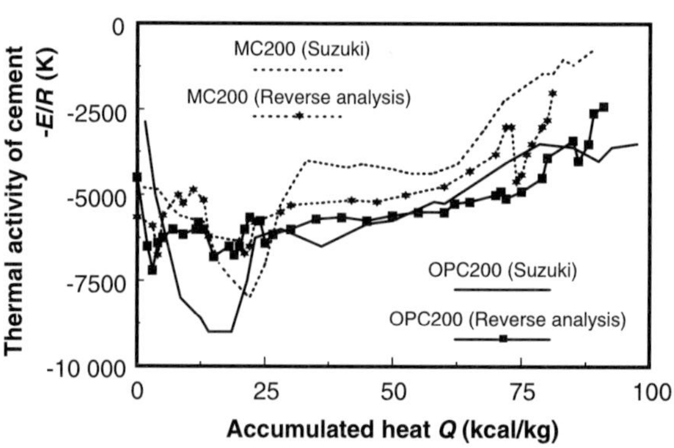

Figure 7.22 Comparison of thermal activities derived from experimental and analytical adiabatic temperature rises.

compositions of the base cement and mix proportions of concrete are shown in Table 7.1 and Table 7.3. The delaying capability per unit weight of the delaying type AE water reducing agent $\chi_{SP} = 5.0$ was used in the analyses due to modification of the delaying effect model of admixtures. Experimental and analytical results are shown in Figures 7.23 and 7.24.

The reference heat rate and the thermal activity were assumed so as to reproduce experimental results. The adiabatic temperature rise history was quite accurately reproduced, reflecting temperature dependence of the reactions at each casting temperature. But, in the case of fly ash blended cement when the initial temperature is 10 °C, the accuracy of the proposed model is not sufficient, since it could not closely reproduce the delaying of heat generation at the latter part of Stage 2. It can be assumed that this disagreement is not caused by the data setting of fly ash characteristic, since heat generation from Portland cement hydration delivers the temperature rise of this part in the analysis. Although delaying effects by organic admixture and fly ash at other stages except Stage 1 and combining effects are possible reasons for the disagreement, the temperature dependence of the delaying effect should also be investigated.

Further, the characteristic of the AE water reducing agent used in the experiment needs to be properly modelled. The water consumption ratios of slag and fly ash were assumed to be 30% and 10%, respectively, through the analytical discussion. For a unit cement content of 400 kg, reductions in the heat rate and in the adiabatic temperature rise in the terminal period of the reaction was closely reproduced since the effect of a low water-to-cement ratio of 40% on the heat of hydration rate was considered. The ultimate temperature rises at different initial temperatures was almost the same, because the loss of heat generation mainly derived from ettringite formation from mixing to casting was taken into consideration, although the experimental results showing a higher initial temperature bringing about a lower ultimate temperature rise was not accurately reproduced.

Table 7.3 Concrete mix proportions in adiabatic temperature rise [6]

	W/C (%)	*W (kg/m^3)*	*C (kg/m^3)*	*S (kg/m^3)*	*G (kg/m^3)*	*AE water reducing agent (C%)*
SG400	39.2	157	400	645	1129	0.25
SG300	49.3	148	300	757	1129	0.25
SG200	78.5	157	200	854	1089	0.25
FA400	39.2	157	400	639	1129	0.25
FA300	49.3	148	300	749	1129	0.25
FA200	78.5	157	200	852	1089	0.25

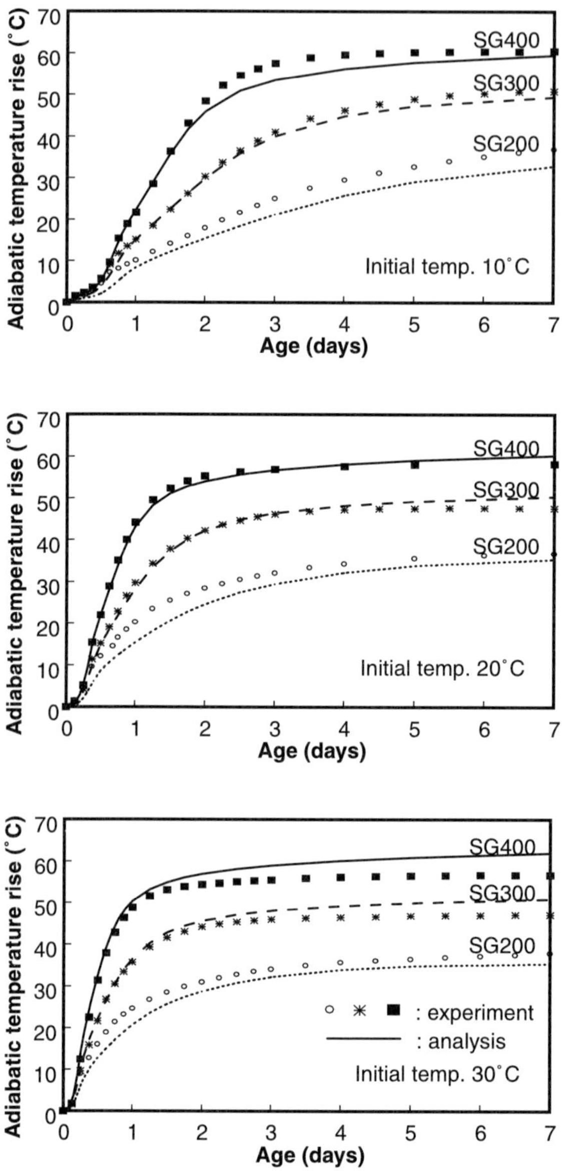

Figure 7.23 Analytical results for adiabatic temperature rise history (blast furnace slag blended cement).

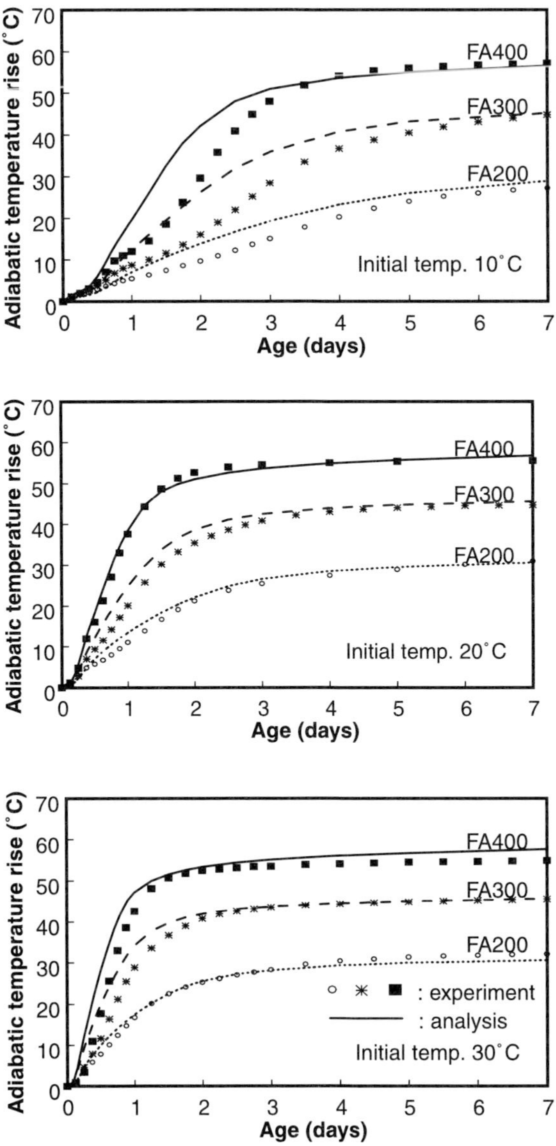

Figure 7.24 Analytical results for adiabatic temperature rise history (fly ash blended cement).

7.4 Verification by quasi-adiabatic temperature tests

7.4.1 Experimental outline

To verify the applicability of the proposed heat of hydration model and the material functions established for the model, temperature measurements at centre points were carried out on concrete blocks completely covered in styrofoam to a depth of 8 cm. In order to focus on model verification, water migration was prevented around the measurement locations. These experiments were conducted in an experimental room where no wind blows and the temperature was kept at almost 20 °C (293 K). The specimen is outlined in Figure 7.25.

7.4.2 Temperature analysis by the heat of hydration model

Several types of Portland cement

Let us pick five concrete specimens: selfcompacting high-performance concrete containing early hardening Portland cement, ordinary Portland cement, moderate-heat Portland cement, belite rich Portland cement and super belite-rich Portland cement, respectively. The mineral compositions of cements used and the mix proportions of concretes are shown in Table 7.4 and Table 7.5 respectively. The thermal constants used in the thermal conductivity analysis are shown in Table 7.6. These values are popular in practice and were again checked through analytical trials and applied for all computations.

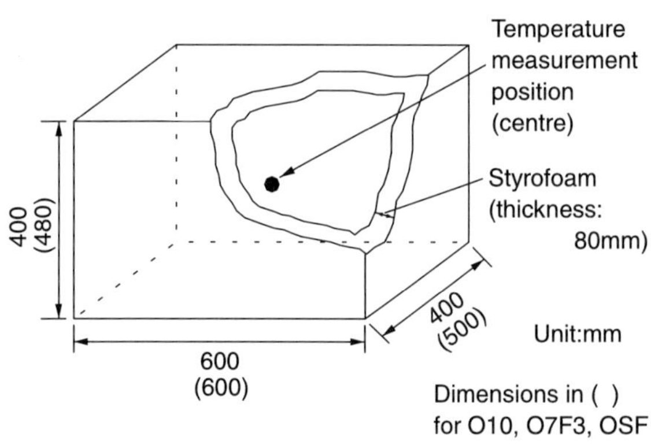

Figure 7.25 Quasi-adiabatic temperature rise test specimen.

Table 7.4 Cement mineral proportion used in quasi-adiabatic temperature rise tests

	C_3A	C_4AF	C_3S	C_2S	*CS2H*	*Blaine* (cm^2/g)
EPC	9.1	7.9	61.5	13.0	6.5	4210[a]
OPC	10.4	7.9	49.9	24.7	4.5	3380[a]
MC	2.9	12.5	49.5	28.7	4.1	3240
BC	2.5	13.1	26.3	52.8	4.5	3040
SBC	2	6	16	71	4.3	3480

EPC: Early hardening Portland cement; OPC: Ordinary Portland cement; MC: Moderate-heat Portland cement; BC: Low-heat Portland cement; SBC: Super low-heat Portland cement; CS2H: Gypsum dihydrate
[a]Assumed in analysis

Table 7.5 Concrete mix proportions used in quasi-adiabatic temperature rise tests

	W/C (%)	*W* (kg/m^3)	*C* (kg/m^3)	*S* (kg/m^3)	*G* (kg/m^3)	*Superplasticizer* (*C* %)	*Initial temperature* (°*C*)
EPC	31.9	174	545	857	827	1.5[a]	23.5
OPC	31.6	174	550	857	827	1.5[a]	23.5
MC	31.5	182	577	828	827	1.0[a]	19.5
BC	24.3	185	761	722	798	0.9[b]	17.5
SBC	25.0	187	749	722	798	0.9[b]	13.1

[a]β-naphthalene sulphonate, [b]Polycarbonate

Table 7.6 Thermal constants used in analyses

Thermal conductivity	41 kcal m^{-1} day^{-1} K^{-1}
Heat transfer coefficient	18 kcal m^{-1} day^{-1} K^{-1}

The superplasticizers used are those based on β-naphthalene sulphonate for early hardening, ordinary and moderate-heat cements and on polycarbonate for low heat belite rich and special low heat cements. The value of χ_{SP} was taken to be 1.2 for those admixtures to reproduce the starting time of rapid temperature rise shown in the experiments. The results of experiment and analysis are shown in Figure 7.26. The experimental results were judged to be roughly reproduced by the heat of hydration model according to various cement types from early hardening cement to super

Table 7.7 Cement mineral proportion used in quasi-adiabatic temperature rise tests

	C_3A	C_4AF	C_3S	C_2S	CS2H	Blaine (cm^2/g)
O5S5	10.4	7.9	49.9	24.7	4.5	3380[a]
M6S4	2.9	12.5	49.5	28.7	4.1	3240
O10	10.3	8.2	53.7	19.8	4.3	3380[a]
O7F3	10.3	8.2	53.7	19.8	4.3	3380[a]
OSF	10.3	8.2	53.7	19.8	4.3	3380[a]

[a] Assumed in analysis

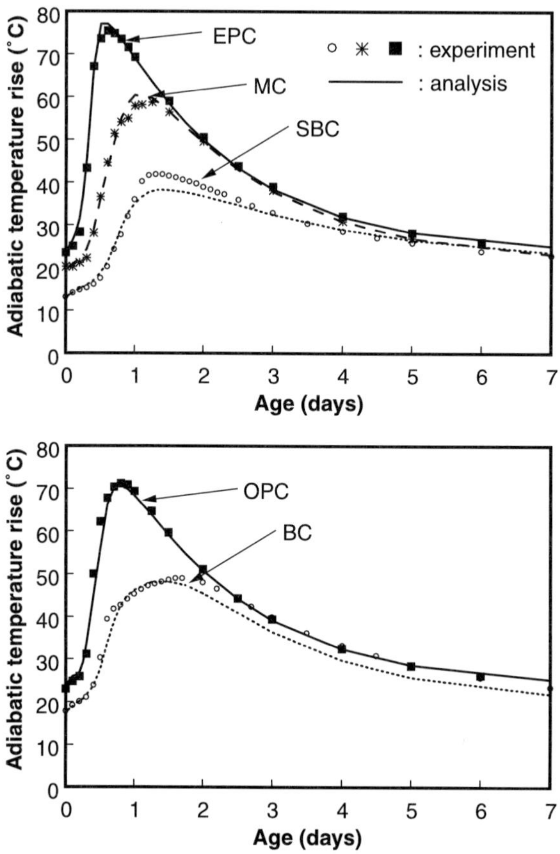

Figure 7.26 Analytical results of quasi-adiabatic temperature rise tests (various types of Portland cement).

belite rich cement, but the soft temperature drop after the peak temperature in the low-heat type cements was not sufficiently reproduced.

These problems need to be solved in the future. The slight decline of precision in the temperature drop after the peak may be attributable to the face that the minuteness of the hardened texture which controls diffusion resistance is dependent on the precipitation rate of the hydrates. In general, when the temperature history has high values, the progress of hydration is said to stagnate in the terminal period of the reaction. So, the authors consider that it is necessary to incorporate into the model the effect of change of microporous texture depending on the difference of reaction rate. Further, it is also important for accurate prediction of the temperature history at the period corresponding to Stage 3 that the effect of mineral composition change on the exothermic characteristic is suitably evaluated in the model.

Binary blended cement

To verify the applicability of the proposed model to binary blended cement, quasi-adiabatic temperature rise tests were also conducted. Two mixtures were adopted: blast furnace slag blended cement with 50% replacement of ordinary Portland cement (O5S5) and 40% replacement of moderate heat Portland cement (M6S4) by blast furnace slag. Regarding the fly ash blended cement, a mixture of 30% replacement of ordinary Portland cement (O7F3) by fly ash was used. A 100% pure ordinary Portland cement mixture (O10), which is the same as the base cement of the fly ash blended one, was also checked for comparison. The test specimen is the same as that shown in Figure 7.25, but the sizes for O7F3 and O10 are slightly larger than the others as indicated by parentheses in Figure 7.25. Mineral compositions of base cements used and mix proportions of cements are shown in Table 7.7 and Table 7.8, respectively. The results of experiments and analyses are shown in Figure 7.27.

Table 7.8 Concrete mix proportions used in quasi-adiabatic temperature rise tests

	W/C (%)	*W (kg/m³)*	*C (kg/m³)*	*S (kg/m³)*	*G (kg/m³)*	*SP (C%)*	*Initial temperature (°C)*
O5S5	32.9	182	553	828	827	1.1[a]	20.0
M6S4	31.3	180	576	803	827	1.5[b]	16.5
O10	36.0	180	500	757	902	0.25[c]	25.5
O7F3	36.0	180	500	731	873	0.25[c]	25.5
OSF	30.1	160	531	765	879	1.2[c]	25.0

O5S5: OPC 50% + slag 50%, M6S4: MC 60% + slag 40%, O10: OPC 100%, O7F3: OPC 70% + fly ash 30%, OSF: OPC 30% + slag 28% + fly ash 42%

[a] β-naphthalene sulphonate, [b] Polycarbonate [c] AE water reducing agent

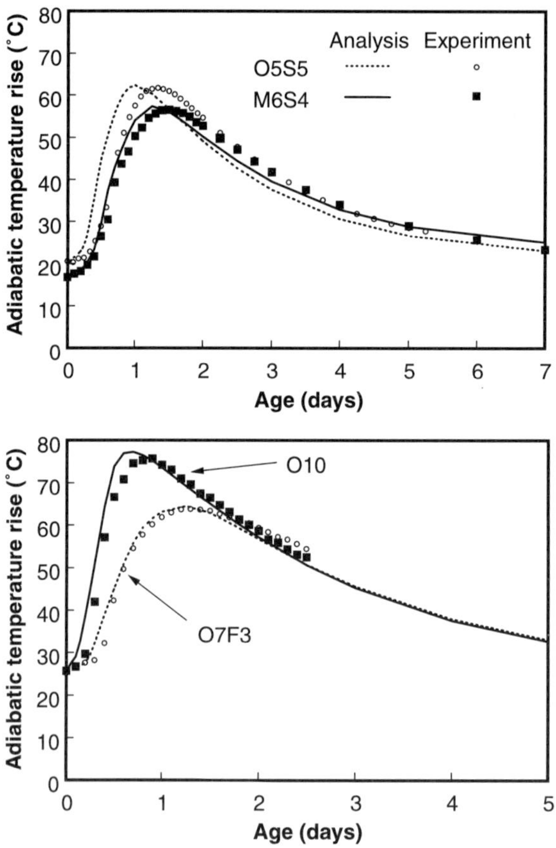

Figure 7.27 Analytical results of quasi-adiabatic temperature rise tests (two composite blended cement).

The temperature histories including maximum temperature rise are satisfactorily reproduced in both mixtures of O5S5 and M6S4 of blast furnace slag blended cement. For the O5S5 case, however, the computed temperature rise is achieved approximately 0.25 days earlier than the reality. But this discrepancy is not influential on the thermal cracking. A possible explanation is that the effect of the post addition of an organic admixture is not taken into account in the proposed model although a part of it was mixed by post addition in the experiment.

For fly ash blended cement in O7F3, analytical and experimental results approximately coincide with each other with 100% base cement mixture O10. The descending temperature after the peak is predicted a little earlier than the experimental measurement. The temperature histories, including the maximum temperature rise, are suitably reproduced in both mixtures of O5S5 and M6S4 of blast furnace slag blended cement. For the O5S5 case, however, about 0.25 day shift of time on temperature rise was computed. The possible reason is that the effect of the post addition of an organic admixture is not taken into account in the proposed model although a part of the superplasticizer was mixed in the way of post addition. For fly ash blended cement in O7F3, the analytical and experimental results coincide well with each other with 100% base cement mixture O10. The decline after the peak temperature was predicted to be a little earlier than the measured data as in the case of slag blended cement.

Ternary blended cement

The quasi-adiabatic temperature rise test of concrete using ternary blended cement with blast furnace slag and fly ash was conducted. The size of the specimen is noted in parentheses of Figure 7.25. The mineral composition of base cement and mix proportions are shown in Table 7.7 and Table 7.8. The results of experiment and analysis are shown in Figure 7.28. The beginning of the temperature rise is satisfactorily reproduced because the delaying effects of organic admixture and fly ash are taken into account in the model. The temperature history after rising is not closely simulated although this does not matter in practice. The computed reactions of slag and fly ash are greatly restrained by a shortage of calcium hydroxide where the base cement contains less than 30%. On the other hand, coefficient β_i was moderated for clinker minerals in the computation, since reactions of slag and fly ash are

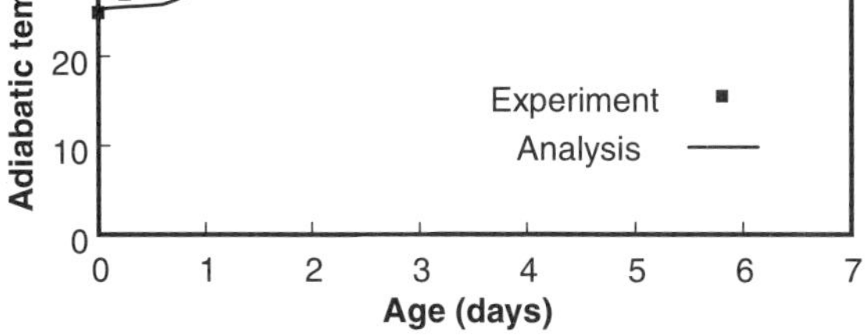

Figure 7.28 Analytical results of quasi-adiabatic temperature rise tests (three-composite blended cement).

restrained and free water is assigned mainly to clinkers in spite of the low water-to-cement ratio.

7.5 Verification by thermogravimetric analysis of combined water

One of the primary aims of the multicomponent heat of hydration model of cement is to predict the early age temperature rise accurately, to facilitate the evaluation of the risk of thermal cracks. It can be observed from the discussions in the previous section that the model indeed can be applied with confidence for the prediction of early age temperature rise for a wide range of powder materials. The integrated nature of the model also allows us to verify its accuracy by checking several other properties of the hardening cementitious matrix. To estimate accurately the impact of curing conditions in the early age of hardening, it is usually desired that the amount of water essential for obtaining a stipulated performance could also be predicted in the same framework. This multiple check of modelling concrete performance is vital since we deal with a highly nonlinear system with a complexity of interactions. Two erroneous assumptions could bring about a solution close to reality. Experimental verification with systematically arranged parameters is the basis of reliability.

To check the versatility and applicability of the current analytical model in predicting the amount of chemically bound water during hydration reactions, several thermogravimetric analyses (TGA) of various kinds of cementitious pastes were performed. The amount of combined water for several pastes obtained experimentally by TGA was compared with the analytical results of the proposed model, which indeed shows reasonable agreement. This gives further reasons for confidence in the core assumptions of the model. A brief discussion of the experimental methodology and the analytical results is given next.

7.5.1 Experimental outline

As a major aim of these experiments was to evaluate the water consumption, only pastes were prepared for various powder material type, i.e. without any addition of aggregates. This was also necessitated in part due to the requirements of the TGA apparatus to decrease the relative sensitivity of the experimental results to the specimen sample size. In the experiments, the samples were sealed cured so that no exchange of moisture could take place with the environment. This was essential to create a stable, reliable and reproducable experimental setup. At certain periods after the casting of specimens followed by sealed curing, the amount of combined water was measured by performing differential TGA up to a maximum of 100 days of sealed curing. In the series of experiments discussed here, the water-to-powder ratio by weight was kept at 0.3 which is typical for powder-rich high-performance concrete being used nowadays at site. Using a low water-to-powder ratio also meant enhancing the sensitivity of hydration reactions to the amount of available water. Initial mixing and curing temperature in all the cases was maintained at about 20 °C. The mix-proportion for the various mixes is shown in Table 7.9

The specimens were broken at certain intervals of time after casting and samples of about 15 mg to 20 mg were taken for TGA. The temperature was increased in steps of 10 °C up to 100 °C where it was kept constant for 5 min. Subsequently temperature was increased at a rate of 10 °C/s until 1200 °C. It was however observed that there was no significant decrease in the samples weight after 800 °C. The loss of weight in the sample from 105 °C to 800 °C was taken as the loss of unevaporable combined water. Suitable corrections were also made to take into account the weight loss due to oxidation of unhydrated cement mineral compounds. This loss was however small compared to the weight loss observed for hydrated cementitious paste. Figures 7.29–7.31 shows a comparison of the experimentally measured combined water at certain intervals after casting for several specimens with the analytical models that are discussed next.

7.5.2 Analysis by coupled simulation model

The multicomponent heat of hydration model considers the availability of free water for hydration in a dynamic way by keeping the chemical mass balance of free water as defined by stoichiometric equations of hydration reactions (equations (7.9) and (7.13)). Of course, the actual rate of consumption will be dependent on the rate of hydration, which is itself dependent on the amount of free water available besides several other parameters. Therefore, prediction of the combined water with time not only involves consideration of reasonable representative stoichiometric chemical

Table 7.9 Mix-proportions for various pastes used in thermogravimetric analysis

	W/P (%)	*Water (W) (kg/m³)*	*Cement (C) (kg/m³)*	*Fly ash (FA) (kg/m³)*	*Slag (SG) (kg/m³)*
OPC	0.3	485	1620	–	–
High belite	0.3	492	1641	–	–
FA 40%	0.3	449	899	599	–
FA 90%	0.3	411	137	1233	–
SG 40%	0.3	477	954	–	636
FA 20% + SG 20%	0.3	463	925	308	308
FA 40% + SG 40%	0.3	442	295	589	589

Specific gravity: OPC, 3.15; High belite cement, 3.23; Slag, 2.89; Fly ash, 2.26

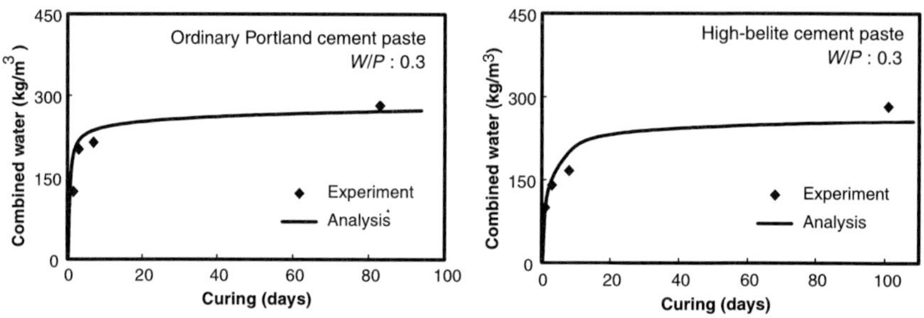

Figure 7.29 Analytical results of combined water due to hydration with time for OPC and high-belite cement pastes.

balance equations of hydration in the multicomponent heat of hydration model but also, a proper sensitivity of the hydration rate upon the amount of free water available. In the computational analysis, the amount of combined water due to hydration reactions can be computed in an incremental manner for given initial and boundary conditions. This solution enables us to compare directly the computed and measured chemically combined water due to hydration as a further check on the integrated multicomponent heat of hydration model. Plain cement pastes as well as binary and ternary blends of powder pastes were used for the comparison.

Several types of cements

Figure 7.29 shows the comparison of computed and measured bound water with time for ordinary Portland cement paste and high-belite (C_2S) type cement pastes. The chemically combined water obtained by the analytical method is essentially the bound water as given by the set of stoichiometric equations (7.9) and (7.13). It can be observed that there is a slower initial rate of hydration and therefore a lower rate of water binding for high-belite cement can be reasonably predicted. This is primarily due to the way the reference heat rate model for mineral reactions is formulated (see section 7.2.2 for detailed explanations). It can be seen that a substantial increase in the amount of bound water occurs even after several weeks in the case of high-belite cement pastes whereas, for OPC paste, the reactions essentially stop after about four weeks in sealed conditions. Of course, an external supply of moisture would accelerate the third stage of mineral hydration for OPC pastes, too (section 7.2).

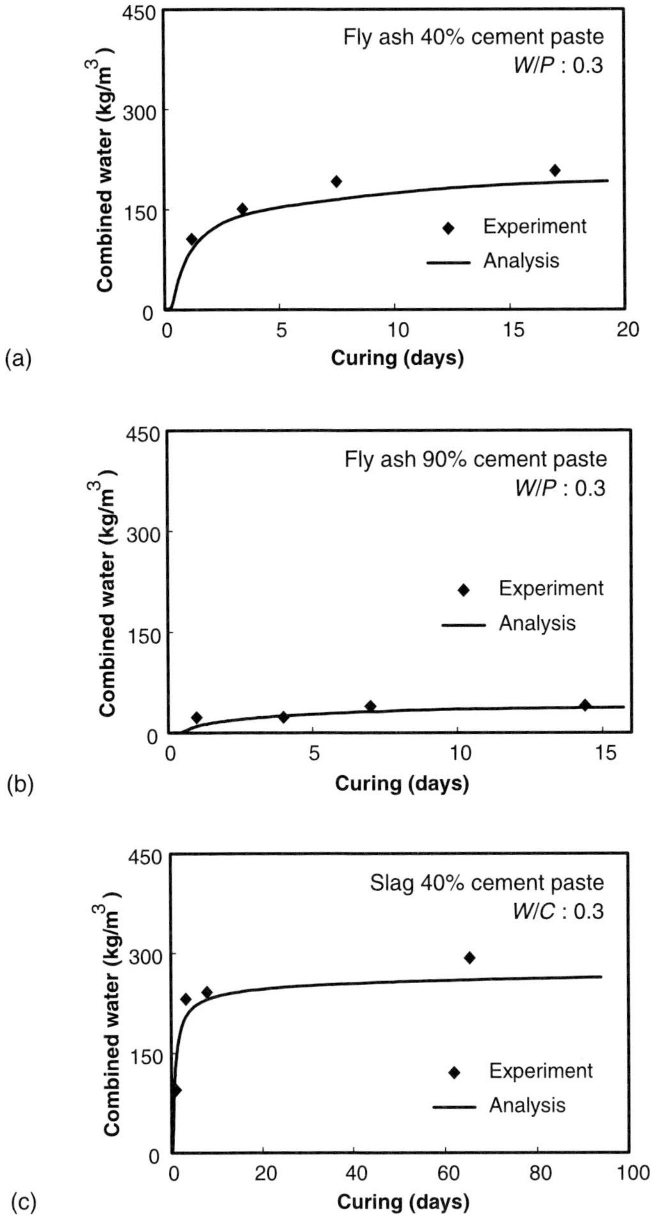

Figure 7.30 (a, b) Analytical results of combined water due to hydration for binary mixes containing fly ash and OPC. (c) Analytical results of combined water due to hydration for binary mixes containing blast furnace slag and OPC.

Binary blended cements

To test the validity of assumptions regarding the interdependence of mineral reactions as well as the heat rate models for fly ash and blast furnace slag, cementitious pastes of binary blends of powder materials were used. A certain weight fraction of the ordinary Portland cement was replaced by fly ash and blast furnace slag. Figures 7.30(a) and 7.30(b) show a comparison of the analytical and experimental combined water for pastes having 40% and 90% fly ash replacement of ordinary cement by weight. The heat of hydration rate of fly ash is almost negligible compared to the other mineral components. Moreover, practically no reaction of fly ash occurs in the absence of Ca^{2+} ions, liberated by the reactions C_2S and C_3S. It might be more appropriate to consider fly ash as a filler material as regards the amount of chemically combined water. The reduction in the amount of bound water is almost in direct proportion to the weight fraction of the fly ash. Indeed, for a paste containing 90% fly ash, the rates of reaction are extremely slow and the amount of bound water corresponds roughly to the 10% of the cement paste.

A reasonable agreement can be observed in the magnitude as well as the rate of binding of combined water between the analysis and the experimental data for both cases (Figure 7.30(a, b)). Slag reactions, on the other hand, are significantly affected by the amount of free water available for hydration, once sufficient $Ca(OH)_2$ is available in the pore solutions. The rate of heat of hydration for slag is also assumed to be comparable to that of C_3S. The proposed model seems to give reasonable estimates of combined water for 40% replacements (Figure 7.30b).

For extremely large slag replacements (about 90% by weight of powder materials), some discrepancies have been observed between the experimental data and the analysis. It is thought that a complex sensitivity of slag component reactivity on $Ca(OH)_2$, especially under initial stages of hydration, needs to be investigated carefully. In fact, the interlinked hydration pattern of slag with Portland cement may be influenced by the phase concentration of pore solutions and pH level. For most of the practical cases of interest, however, i.e. when $Ca(OH)_2$ does not becomes a critical factor in the hydration reactions, the proposed model can give reasonable estimates.

Ternary blended cements

Figure 7.31 shows a comparison of the combined water for pastes having a ternary blend of powder materials (slag, fly ash and OPC). The ternary blends were used to check the interdependence among blast furnace slag and fly ash with Portland cement (Figure 7.16). It can be observed that a satisfactory agreement exists between experimental and measured results for a replacement even as high as 80% of cement by slag and fly ash. These results show the viability of the proposed model of heat of hydration and mutual interactions among mineral components.

7.6 Thermal crack control design of mass concrete

7.6.1 *Scheme of thermal crack control design*

The proposed scheme of thermal crack risk assessment is shown in Figure 7.32. As discussed in reference [27], a particular HPC mix proportion which meets the requirement of superfluidity without segregation derives from the qualitative features of aggregates, cement and admixture used. At this stage, it cannot be determined whether thermal cracking is avoidable or not. The transient temperature rise of HPC structures is to be computed by idealizing the hydration process of cement in HPC. It is crucial to predict the heat of hydration according to both specified raw materials (quality) and mix proportion (quantity).

At the second stage, the free volumetric expansion owing to the temperature rise and the hardening has to be modelled in the frame of stress analysis. The tensile resistance of young concrete is also formulated in terms of the hydration level and temperature already estimated in the scheme. Finally, the crack index [7] as shown in Figure 7.32 is employed as the safety factor for thermal crack occurrence.

In the 1990s, crack risk evaluation [7] was applied to massive structures of importance. However, the cement hydration process under adiabatic temperature conditions is mistakenly assumed to be common for any location of structures and material age. Therefore, when the concrete mix proportion and/or initial temperature at site was changed, the temperature test was carried out again to maintain the reliability of analysis. It can be

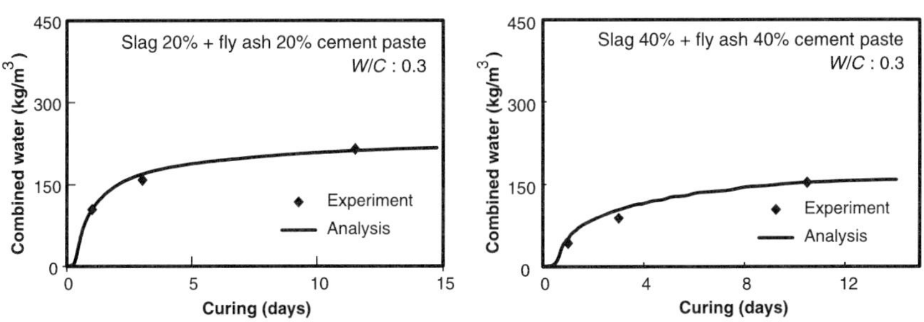

Figure 7.31 Analytical results of combined water due to hydration for different pozzolanic ternary mixes.

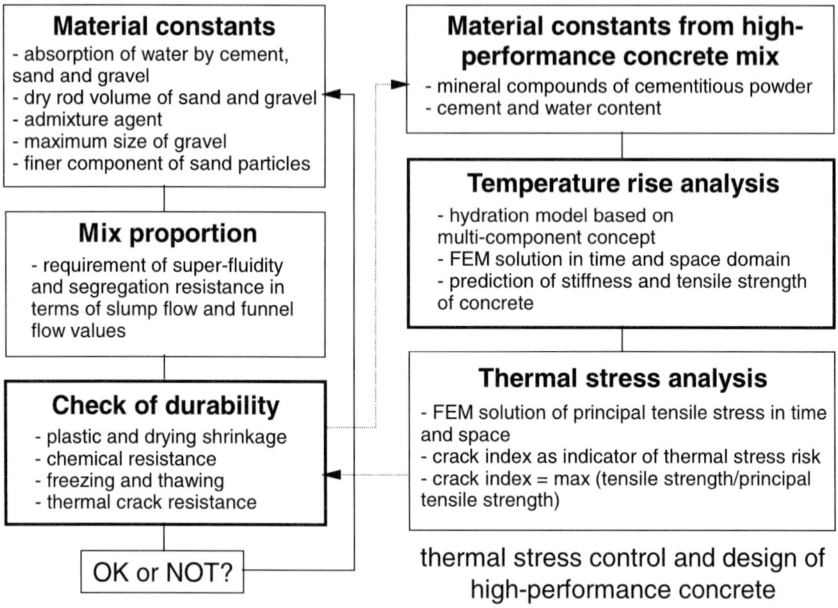

Figure 7.32 Scheme for thermal crack prediction for high-performance concrete.

said that, so far, the thermal crack risk has been checked, but the feedback system to the initial stage of material specification, i.e. control design as shown in Figure 7.32, has been laid aside for normal concrete.

7.6.2 *Modelling of strength development*

Experimental

For making a strength evolution model of young concrete with a temperature rise, compressive loading tests of mortars were conducted under two patterns of temperature histories. Test specimens were cured and tested under the temperature history patterns shown in Figure 7.33.

The specimen was a 5 cm × 10 cm cylinder and cured under sealed conditions. Pattern 1 is supposed to be the analogue of the temperature history of concrete in a massive structure at early ages. For comparison, in Pattern 2, the curing temperature was kept at room temperature for about a week and then the temperature was elevated to accelerate hydration. The beginning of the temperature rise in Pattern 2 is slightly different in each mixture (MC: 7.4 days; SG + MC: 6.8 days; FA + MC: 9.1 days).

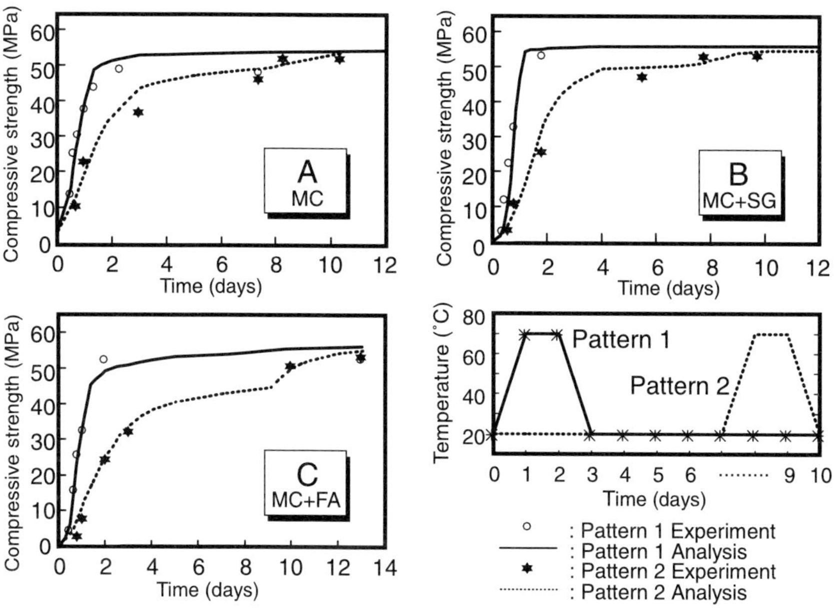

Figure 7.33 Strength development of concrete and induced temperature histories.

Mix proportions are shown in Table 7.10. The degree of hydration of each mineral constituent was computed by the multicomponent model of heat generation.

The evolution of strength at Pattern 1 and Pattern 2 with respect to time are remarkably different, though the strengths at the final stage are almost the same as shown in Figure 7.33. The approach based on the heat of hydration model is essential for modelling the evolution of strength. The degree of hydration of each mineral constituent is not common at the same strength or accumulated heat for the two patterns because of the difference in thermal activity. This indicates that mineral compounds consisting of cement should be formulated with the multicomponent concept in the modelling of the evolution of strength.

Table 7.10 Mix proportion of mortar used in strength development tests

Mix	*Water* *(kg/m^3)*	*Moderate Portland cement* *(kg/m^3)*	*Slag* *(kg/m^3)*	*Fly ash* *(kg/m^3)*	*Sand* *(kg/m^3)*
A	181	761	–	–	720
B	181	360	360	–	720
C	181	478	–	192	720

Modelling

It has been reported that a clear bilinear relation is seen between the maturing strength and the total hydration level of cement which was computed by the multicomponent model and this indicated the possibility of estimating the evolution of strength with the degree of hydration of the cement [11, 23]. But no unique relation between strength and the entire hydration level of the whole cement is obtained. The relation between them should be formulated according to the mix proportion and used powder materials. As the model has to be applicable to any combination of materials, an empirical and simple multicomponent concept based on the mineral constituents was adopted for the strength evolution model in line with the classical water-to-cement ratio law. Four constitutive minerals (C_3S, C_2S, slag, fly ash) are taken as the effective components for the evolution of strength, and C_3A and C_4AF are assumed negligible in the proposed model. Then, the compressive strength is expressed in terms of the total differential equations as

$$f_{c'} = \int df_{c'}, \qquad df_{c'} = 25\ dQ_{3S} + 40\ dQ_{2S} + 27\ dQ_{SG} + 40\ dQ_{FA}$$
$$dQ_i = \frac{p_i}{W_{total}}\, d\phi_i, \qquad \phi_i = Q_i/Q_{i,\infty} \tag{7.22}$$

where $f_{c'}$ is the compressive strength (MPa), and ϕ_i is the degree of hydration indicated by the accumulated heat normalized by final heat generation. The contribution of constitutive minerals are individually formulated and the concept of cement-to-water ratio in the compressive strength is extended to each component in the above equation.

Verification

Figure 7.33 shows the results with respect to the curing time, in which fair agreement can be seen between experimental and analytical results. It is clear that the proposed model can deal with the evolution of strength with respect to time at any temperature history. In general, concrete strength varies in accordance with the curing temperature and is reduced at longer age at elevated temperatures especially experienced at the early stage. Within this study, however, the evolution of strength can be computed only by dealing with the composition of already hydrated minerals as the degree of hydration, though it is not clear how the structural formation of the hydrated products affects the evolution of mechanical properties. The proposed strength model based on chemical components of cement and pozzolans is generally applicable and equivalent to the maturity model.

For a more rational approach, the mechanical performances of hardened concrete should be predicted in terms of micropore structures achieved in the cement-paste matrix as discussed in Chapter 3. One of the proposals is discussed in Appendix A. The linkage of hygrothermal physics with the mechanics of solids and structures is really required for a more versatile system of assessing concrete performance present in structures.

7.6.3 Thermal crack risk and control

The volumetric strain caused by both the temperature rise and the hydration is the chief ingredient of thermal stress induced in massive concrete. The hardening and drying shrinkage evolved by the water transition and transport also activates the free volumetric change. The gradient of the volumetric change in space gives rise to the internal selfequilibrated stress even if the external confinement is not vital. When structural free deformation is restrained at external boundaries, further confinement reaction will be introduced to the structures. Therefore, it is required for the control of cracks to solve the following equations governing the mechanical constitutive law, strain compatibility and equilibrium conditions with time in three-dimensions.

$$
\begin{aligned}
&d\sigma_{ij} = 2G\ de_{ij} + 3K \cdot \delta_{ij}\left(\frac{1}{3}\ \delta_{kl}\ d\varepsilon_{kl} - d\varepsilon_{free}\right) \\
&d\varepsilon_{free} = \alpha\ dT + \gamma\ d\omega_{free} + \chi H \\
&\varepsilon_{ij} = \frac{1}{2}\left(\frac{\partial u_i}{\partial x_j} + \frac{\partial u_j}{\partial x_i}\right), \qquad e_{ij} \equiv \varepsilon_{ij} - \frac{1}{3}\ \varepsilon_{kk}\delta_{ij} \\
&\frac{\partial \sigma_{ji}}{\partial x_j} + \rho g_i = 0
\end{aligned}
\tag{7.23}
$$

where G, K are the shear and volumetric stiffness, α, γ, χ are the thermal expansion rate, drying and hardening shrinkage and ε_{free} is the volumetric change of concrete. If the constitutive model of young concrete could include time dependency of the micropore structural skeleton and pore pressure as a driving force of hygrothermal deformation, the separation of free volumetric change as formulated above will not be necessary in future.

Here, it is crucial to evaluate precisely the stiffness under the transient condition in which the heat of hydration and the mechanical strength evolves. In the scheme proposed, the material stiffness and strength in which creep is implicitly taken into account are evaluated in using the nondimensional hydration level denoted by

$$K = K(\phi) \qquad G = G(\phi), \qquad f_t = f_t(\phi)$$
$$\phi = \sum_i Q_i \Big/ \sum_i Q_{,\infty} \tag{7.24}$$

The so-called maturity model, which has been used for prediction of strength development under different curing conditions from the specified standard, is not apparently used above, but the hydration level is applied. As a matter of fact, the concept of maturity is implicitly incorporated in the system. If we could assume sufficient free water in the mixture, the value of β_i in equation (7.10) approximately becomes unity. Then, we have

$$Z(Q_i) \equiv \int \frac{dQ_i}{F_i(Q_i)} = \int \exp\left[-\frac{E_i}{R}\left(\frac{1}{T} - \frac{1}{T_0}\right)\right] dt \equiv M_i \tag{7.25}$$

Since the right-hand term of equation (7.25) can be defined as the *maturity*, equation (7.25) yields the one-to-one relation of the accumulated heat and the maturity as

$$Q_i = Z^{-1}(M_i) \tag{7.26}$$

This means that the accumulated heat of a single component of clinker is mathematically equivalent to the maturity. But, it must be recognized that the maturity has to be defined for each component of clinker mineral. As the activation energy is not common among the constituents shown in Figure 7.9, it is hardly possible to derive the averaged cement maturity exactly from the governing equation. Therefore, the authors take the averaged entire hydration level as the alternate of maturity. If the activation energies could be equal, the hydration level in equation (7.24) becomes the maturity.

Figure 7.34 is an example analysis of the thermally induced principal stress at the most severe section of the structure and useful for risk evaluation of the HPC massive layer cast on the already hardened concrete with matured stiffness. Moderate-heat Portland cement was used as the base cement. For computing thermal stresses as shown in Figure 7.34, the instantaneous stiffness was used as the conservative way of evaluation since the stress relaxation by creep is not taken into account at all. Through some computational simulation, the thermally induced stress decreases in accordance with the inverse proportion of the creep coefficient. Then, if a creep coefficient of 1.5 is assumed, the induced stress decreases to approximately 70%. If ordinary Portland cement is used as the replacement for moderate-heat Portland cement, the thermal stress level is 40% higher than the computed one in Figure 7.34. It has to be stated that the concrete creep at the elevated temperature and early age of cement hydration is also sensitive to the computed stress and risk evaluation. Regarding this material

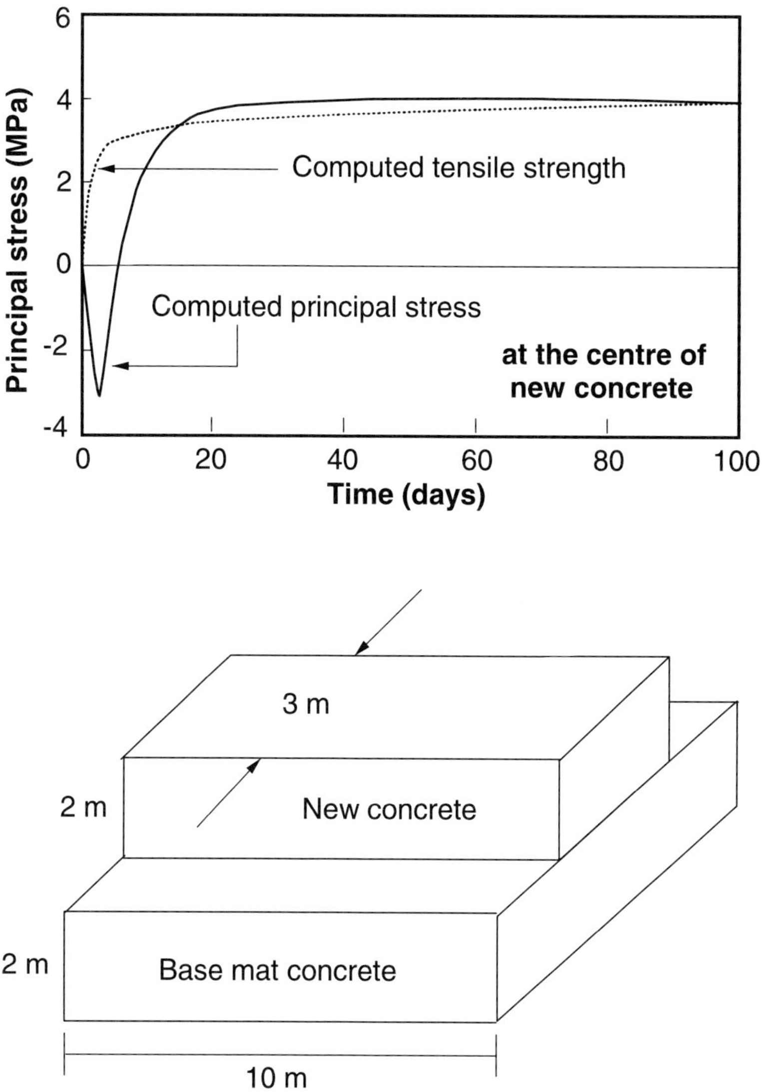

Figure 7.34 Computed thermal stress and tensile strength used for crack risk evaluation of an HPC massive mat slab constructed on the foundation.

property, further investigation is needed so that we can enhance the reliability of the vulnerability assessment in terms of the thermal crack risk.

It is discovered through experience that when the crack index (tensile strength/maximum induced tensile stress) exceeds 1.5, the probability of thermal crack occurrence is about 20% or less except for slag-rich concrete.

In the case of slag-OPC (8 to 2) mixed cement for the purpose of crack resistant cement, the value of χ in equation (7.23) becomes substantial. The prediction of the change of hydrated minerals will be one of the important items in the proposed frame of performance based design for cement-rich selfcompacting concrete based on the mineral compounds and mixture of concrete.

7.7 Summary and conclusions

A unified heat of hydration model of cement in concrete describing the exothermic hydration process for each component is proposed by separating the chemical actions present in Portland cement into minerals, or reaction units, and considering the interdependence among the mineral reactions. Further, reactions of blast furnace slag and fly ash were combined in the system to incorporate mutual interactions. In the model, the amount of free water left in the system and the amount of calcium hydroxide were adopted as control indicators for reactions of mineral compounds. The applicability of the proposed model was verified by analysing the results of existing adiabatic temperature rise tests and temperature measurement of small quasi-adiabatic blocks. The proposed model can describe reactions in the cement by the reaction unit and individually express interdependence of complicated reactions.

The model also permits incorporation of reaction factors that can deal with other blending powders and consideration of the interaction between cement and other chemical units in an application. The tasks remaining include improvement of precision of the model and verification of the parameters assumed in the model. It is also necessary to generalize the model to make it applicable to any type of powder materials and mix proportions. Though plenty of features have to be clarified for crack control design, the structural discussion on the thermal crack risk was made under some simplified conditions.

References

[1] Okamura, H., Maekawa, K. and Ozawa, K., *High Performance Concrete*, Giho-do Press, Tokyo, 1993.

[2] Arai, Y., *Chemistry of Cement Materials*, Dai-nippon Tosho Publishing Co., Tokyo, 1984.

[3] Nagashima, M., Hydration, setting and hardening, *Cement and Concrete*, 1992, No. 544, 36–44.

[4] Suzuki, Y., Harada, S., Maekawa, K. and Tsuji, Y., Evaluation of adiabatic temperature rise of concrete measured with the new testing apparatus, *Concrete Library of JSCE*, 1989, No. 13, 71–83.

[5] Suzuki, Y., Tsuji, Y., Maekawa, K. and Okamura, H., Quantification of

hydration-heat generation process of cement in concrete, *Concrete Library of JSCE*, 1990, No. 16, 111–24.

[6] Suzuki, Y., Tsuji, Y., Maekawa, K. and Okamura, H., Quantification of heat evolution during hydration process of cement in concrete, *Proceedings of JSCE*, 1990, No. 414, 155–164.

[7] Design Specification of Concrete Structures: Structural Design, Vol. 2, Japanese Society of Civil Engineers, Tokyo, JSCE, Tokyo, 1985.

[8] Harada, S., Maekawa, K., Tsuji, Y. and Okamura, H., Non-linear coupling analysis of heat conduction and temperature-dependent hydration of cement, *Concrete Library of JSCE*, 1991, No. 18, 155–169.

[9] Uchida, K. and Sakakibara, H., Formulation of the heat liberation rate of cement and prediction method of temperature rise based on cumulative heat liberation, *Concrete Library of JSCE*, 1987, No. 9, 85–95.

[10] Kishi, T., Ozawa, K. and Maekawa, K., Multi-component model for hydration heat of concrete based on cement mineral compounds, *Proc. of JCI*, 1993, **15**(1), 1211–1216.

[11] Kishi, T., Shimomura, T. and Maekawa, K., Thermal crack control design of high performance concrete, in the Proceedings of Concrete 2000, Dundee, Scotland, E&FN Spon, London, 1993, pp. 447–456.

[12] Kishi, T. and Maekawa, K., Multi-component model for hydration heat of Portland cement, *Journal of Materials, Concrete Structures and Pavements*, 1995, **29**, 97–109.

[13] Kishi, T. and Maekawa K., Multi-component model for hydration heat of blended cement with blast slag and fly ash, *Journal of Materials, Concrete Structures and Pavements*, JSCE, 1996, **33**, 131–143.

[14] Uchikawa, H., Effect of blending component on hydration and structure formation, in the Proceedings of the 8th International Congress on the Chemistry of Cement, Rio de Janeiro, 1986.

[15] Ito, H. and Fujiki, Y., About concrete blended blast furnace slag, in the Proceedings of JSCE Symposium on the Application of Blast Furnace Slag to Concrete, 1987, pp. 37–42, JSCE, Tokyo.

[16] Moriwake, A., Fukute, T. and Horiguchi, K., Study on durability of massive concrete, *Proc. of Japanese Concrete Institute*, 1993, **15**(1), 859–864.

[17] Osbaeck, B., Prediction of cement properties from description of the hydration processes, in the Proceedings of the 9th International Congress on the Chemistry of Cement, New Delhi, India, 1992, Vol. 4, pp. 504–510.

[18] Czernin, W., *Cement Chemistry for Construction Engineers*, translated by Kichiro Tokune, Giho-do Press, 1969.

[19] Tomosawa, F., Cement hydration model, *Annual Report of Japanese Concrete Association*, 1974, Vol. 28, pp. 53–57.

[20] Hanehara, S. and Tobiuchi, K., Low-heat cement, *Journal of Cement and Concrete*, 1991, No. 535, 12–24.

[21] Asaga, K., Daimon M., Konishi K. and Yoshida, K., Effect of curing temperature on hydration of mineral compounds of low-heat cement, *Collection of Theses by Cement and Concrete*, 1991, No. 45, pp. 58–63.

[22] Santhikumar, S., Temperature dependent heat generation model for mixed cement concrete with mutual interactions among constituent minerals, Master thesis submitted to the University of Tokyo, 1993.

[23] Kishi, T. and Maekawa, K., Thermal and mechanical modelling of young concrete based on hydration process of multi-component cement minerals, in the Proceedings of the International RILEM symposium on Thermal Cracking in Concrete at Early Ages, Munich, E&FN Spon, London, 1994, pp. 11–18.

[24] Aoki, S., Miura, N., Takeda, N. and Sogo, S., Strength development of high-strength concrete with belite high-content cement under high temperature rise, *Proc. of JCI*, 1994, **16**(1), 1317–1322.

[25] Uomoto, T. and Ohshita, K., A fundamental study on set-retardation of concrete due to superplasticizer, *Concrete Research and Technology(JCI)*, **5**(1), 119–129.

[26] Uchikawa, H., Sawaki, D. and Hanehara, S., Effect of type of organic admixture and addition method on fluidity of fresh cement paste, Proceedings of Japanese Concrete Institute Symposium on High-fluid Concrete, 1993, pp. 55–62.

[27] Ozawa, K., Maekawa, K. and Okamura, H., Development of high performance concrete, *Journal of Faculty of Engineering*, The University of Tokyo (B), 1992, **XLI**(3), 381–439.

Chapter 8

Conclusions and future development

- Conclusions of this book
- Future development in the scheme of performance based design

Nowadays, concrete structures are designed so that structural serviceability and safety requirements are satisfied with reasonable reliance. Especially, after the Hyogo-ken Nanbu Great Earthquake in Japan in 1995, the concept of safety performance-based design has earned a central position in practice. It requires the transparent assessment of structural seismic performance by simulating dynamic behaviour of structures during and after the specified design actions before construction on site. Although methods of deciding structural detailing, dimensioning and selecting materials so as to satisfy the given requirements can be chosen freely by practitioners, a check of performance must be conducted.

Although durability issues have never been explicitly examined in practical design processes, specifications related to durability performance of concrete and structures have been adopted in implicit ways, such as maximum or minimum amounts of cement content, water-to-cement ratio, minimum cover, allowable concentration of chloride, limitation of slump, possible maximum height of concrete casting, etc. As a matter of fact, deterioration of concrete structures subjected to long-term environmental actions costs hundreds of billions of dollars per year in the way of lost services, maintenance and repair or destruction of damaged concrete. Earthquakes may demolish structures within a couple of seconds, but environmental attack will destroy materials and structures over hundreds of days and years with much less recognition by users.

This is simply because a rational and sound method for assessing durability performance of concrete structures does not exist. There is, however, a huge volume of knowledge and experience that has not yet been systematized. The situation is fairly ironic since, on several research fronts, significant advancements have been made in understanding the various deterioration mechanisms of concrete. However, an integration of these processes is required in a quantitative manner so as to treat the damage process of concrete in the same objective way that the structural response of concrete can be predicted under arbitrary load conditions.

The studies presented in this book are the first step towards the integration of basic phenomena that govern the generation of concrete microstructure, its hardening and its response to external agents, such as liquid water and vapour. In the authors' opinion, the concrete microstructure is one of the most important factors influencing the long-term durability of concrete together with macroscale cracking. Generation of the concrete microstructure is, in turn, dependent on the hydration mechanisms and curing conditions applied. In fact, the foundations of concrete performance related to durability are laid during the first few hours after casting. In this regard, it is important to understand that the underlying mechanisms of these phenomena must be investigated and interconnected in a single unified system to give any meaningful results. In this regard, moisture transport, cement hydration and the associated development of microstructure are the key phenomena that must be investigated and their inter-relationships made clear, quantitatively. It is interesting but difficult to understand that there exists much research on each of these aspects of early age development, yet no attempt has been made to combine the knowledge obtained into a prediction model for the durability properties of concrete.

This book has attempted to quantify the early age development phenomena by quantitatively formulating the underlying mechanisms. These are microstructure development, hydration and moisture transport. During the early life of concrete, the clinker mineral compounds present in the cement and pozzolanic powder dissolve in the water to give what are known as products of hydration. Several of these products formed in the early stages of hydration are short-lived and form stable compounds after combining with several other complex ions present in the pore solution. The chemistry of the reactions involved is quite complex and the models generally proposed are probably a simplified representation of the actual phenomena. The products of hydration that may be granular are generally deposited near the reactant particle surface. The spaces between these grains, as well as the microstructural properties of the grains themselves, define the overall microstructural property of concrete. Transport of liquid and vapour occurs within these micropores that are generally extremely very small, of the order of only a few hundred nanometres. Larger spaces between the reactant particles where hydration products cannot reach, or zones of less particle density, such as near the aggregate surface, give rise to channels of rapid moisture migration. The transport of fluid across the micropores might carry harmful agents, damaging the internal microstructure of concrete. It is indeed amazing that these seemingly complex phenomena can be reasonably explained by simple models that are built upon physical reasoning based on fundamental concepts of thermodynamics and chemico-physics.

Summarizing, the salient points considered or developed in this book are:

- A proposal is made which calls for an integrated approach towards the quantitative evaluation of concrete performance necessary for serviceability and safety evaluation of concrete structures.
- Early age development processes, namely hydration and microstructure formation, heat generation and moisture loss, are identified as the foundation phenomena based upon which any subsequent durability evaluation schemes of materials and structures could be built in both space and time scale.
- A quantitative workable model of microstructure formation of cement is proposed. This model can be extended to arbitrary powder material types and requires only a few fundamental physical quantities to be obtained experimentally that describe the characteristics of building blocks of cementitious composites, e.g. CSH gel.
- The microstructure model is dependent on the degree of hydration at any instant. An existing hydration model is used in this study for this purpose (see Chapter 3).
- The problem of moisture and mass transport in concrete has been reworked considering the microstructural characteristics of concrete as the primary basis.
- The parameters used in the moisture transport formulation are basically nonempirical and have been obtained considering the random state of cementitious microstructures and the basic fundamental laws of thermodynamics and chemical physics. The transport model is generally applicable to all boundary conditions applied as mass flux. A model of the hysteresis effect in moisture isotherms is also newly proposed. However, currently the overall formulation is applicable only to ordinary temperature regimes.
- A finite-element computational code is developed that dynamically couples heat and mass transport in concrete by considering the computational model of hydration, microstructure formation and moisture transport.
- The applicability of the overall computational system is verified by applying it to various test results as well as by studying some objective simulation experiments.

The combined system for concrete performance assessment proposed is versatile in its applications. The same model can be used to analyse the risks of thermal cracks for different kinds of cement. At the same time, the model can be used to estimate the drying shrinkage strains associated with moisture loss. The effect of material properties as well as environmental conditions, such as the effect of different curing conditions, can be studied in the same system of evaluations. The same computational tool can be used to predict the rate of moisture ingress or loss in the concrete microstructure during the hardening stage or a hundred years after casting. This integrated scheme is not possible in the conventional approach. In general, the durability evaluation system proposed here adds new dimensions to the investigation of concrete material engineering from a scientific point of view. It also suggests new directions for the rational use of criteria-based design methods. For a complete life-span simulator for concrete structures, the following issues will have to be addressed:

- Addition of the mass transport and balance of chloride ions, carbon dioxide and oxygen to the coupled moisture and heat transport presented by this book. The state and migration of the corrosion-related ingress used here can be easily overlaid, since the moisture behaviour is thought to be independent of the dissolved chemical substances.
- Addition of steel corrosion (macrocells and microcells) similar to cement hydration, and associated loss of water and oxygen concentration to be also included in each mass conservation. Steel corrosion is an important chemical reaction for reinforced concrete performance. To meet this challenge the electromagnetic field will have to be included for enabling the macrocell type corrosion of steel embedded in concrete.
- Mutual coupling with structural nonlinear mechanics which describes macroscale damaging rooted in cracking, crushing, plasticity and buckling of materials. Here, overall mass transport through both macrodefects and micropore structures of the matured or prematured stage need to be modelled for generic concrete performance.

The concrete committee in the Japanese Society of Civil Engineers (JSCE) has been preparing a standard performance-based scheme of concrete structural design since 1996. Apart from the final volumes due in the next millennium, the seismic performance oriented issue appeared in 1996 and separate design articles for special-purpose selfcompacting concrete and recycle use are avilable at present. The new broader framework for material and structural design of concrete is also sought in ISO and CEN activities. It is expected that creative activity by designers to decide structural/material details and objective tasks to ensure the most likely achieved performances will be clearly separated. The former subjective job should not be automated but just assisted by computerized tools. On the contrary, the latter method of assessing concrete performance should be and, in fact, can be computerized. The authors hope that this book and future development may contribute to the general performance assessment of concrete structures as illustrated in Figure 8.1.

Finally, the authors wish to conclude this book with a poster to represent the projects on concrete engineering (self-compacting high-performance concrete and nonlinear mechanics of reinforced concrete) which have been ongoing since 1980 and still continue at the University of Tokyo as shown in Figure 8.2.

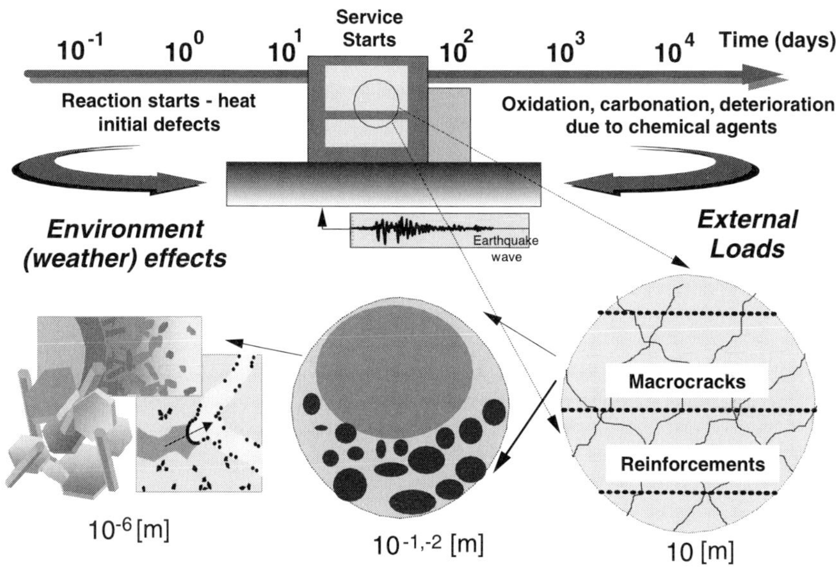

Figure 8.1 Life-span simulation of concrete material and structural performances.

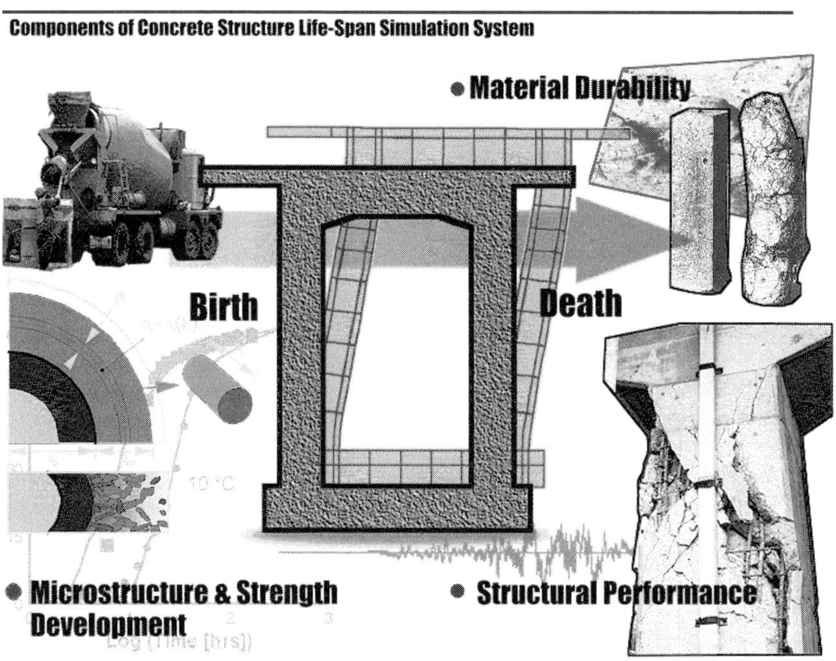

Figure 8.2 Concrete from birth to the end of life.

Appendix A: Computation of autogenous and drying shrinkage strains in mortars

Volume changes in concrete occur even after setting has taken place and may be in the form of shrinkage or swelling. The shrinkage mechanisms are usually related to the moisture movement inside the concrete pores. Even when no movement of moisture is permitted, shrinkage occurs in young concrete. This form of volume change is usually termed *autogenous shrinkage* and occurs primarily due to the internal stresses generated across pore water–air interfaces. These interfaces are formed since a reduction in the internal free water volume due to continuing hydration reactions (self-desiccation) occurs and the total volume of hydration products is smaller than the sum of the volumes of reacted cement and free water. Therefore, to predict the unrestrained drying shrinkage strains as well as the autogenous shrinkage strains, mechanisms that unify the state of moisture distribution inside concrete pore space and shrinkage phenomenon need to be investigated. We will use the term *shrinkage strain* hereafter to denote the unrestrained drying shrinkage strain. Some of the mechanisms that have been suggested in the past to account for a relationship between internal pore humidity of concrete and subsequent shrinkage strains are:

- capillary tension: shrinkage strain induced due to hydrostatic capillary stress (surface tension of liquid);
- disjoining pressure: due to the effect of hindered adsorption;
- surface tension of solids: due to the surface free energy of solids;
- loss of interlayer water: loss of water from the spaces between elementary sheets of CSH;
- stress induced shrinkage.

Actual drying and autogenous shrinkage strains may be a combination of the strains caused due to some or all of the above-mentioned mechanisms. In the past, several investigators have adopted the mechanism of capillary stress to explain the shrinkage behaviour of concrete after initial setting. Under this assumption, the volume change and the deformation of cement paste will occur due to the surface tension force of capillary force caused by a drop of relative humidity in the pore structures. This mechanism will cause the autogenous shrinkage during selfdesiccation as well as the shrinkage caused by drying of concrete. The shrinkage stress in such a case is dependent on the pore water pressure, porosity distribution and water content in the hydrated cement paste. The deformability of the microstructure against shrinkage stress is dependent on the microstructure stiffness that in turn is dependent on the physical properties, such as hydrated cement paste porosity, aggregate content, etc. In this system, where the origin of internal shrinkage stresses is directly related to the pore humidity and microstructure characteristics, autogenous shrinkage need not be distinguished from drying shrinkage caused due to pore water loss. Regarding the physical mechanisms responsible for shrinkage strains, capillary tension theory seems to be adequate enough to explain the behaviour at higher pore relative humidity. It is however recognized that the capillary tension mechanism alone cannot explain the shrinkage behaviour in cementitious materials over the entire range of relative humidity. In any case, it is believed that regardless of the assumptions made about the physical mechanisms responsible for the microshrinkage stresses, an integrated analytical framework that rationally considers the dynamic thermodynamic interactions of pore humidity and cementitious microstructure can explain the drying and autogenous shrinkage behaviour. To illustrate the basic concepts of such an integrated framework, capillary tension theory is adopted here, to explain the shrinkage stresses and subsequent shrinkage behaviour of a hardened cement matrix due to a reduction in concrete pore humidity and selfdesiccation.

A.1 Analytical formulation

A.1.1 *Stress due to capillary tension*

Water present in the capillary pores of a concrete microstructure exists in hydrostatic tension as a result of the surface tension forces across the curved meniscus, according to the well-known laws of capillarity. Under the conditions of thermodynamic equilibrium, the difference in gas and pore water pressure across the curved water–air meniscus in a capillary (with negative curvature) is given by Kelvin's equation as

$$\Delta P_w = \left(\frac{\rho RT}{M}\right) \ln h \tag{A.1}$$

where ΔP_w is the pressure difference between air and pore water pressure due to capillarity and is a negative quantity. Equation (A.1) shows that the capillary tension is given by the relative humidity, h, of the ambient atmosphere with which the pore water presenting a curved meniscus is in equilibrium. Also, ρ is the pore water bulk density. Usually, the air pressure in the concrete pore space is constant and can be assumed to be equal to the atmospheric pressure. Thus ΔP_w represents the deficiency of pressure in the pore water relative to atmospheric pressure and can be expressed as P for brevity, denoting the relative pore water pressure. The negative pressure difference, P, creates an attractive force between the surface of the pore walls. This attraction causes a net compressive stress over a small control volume. If there are no externally applied stresses (free unrestrained sample) then effective stress acting on the solid skeleton would depend on both the magnitudes of the tension and the area where it is applied. In most simplistic terms, the effective stress can be expressed as a product of the relative pore water pressure, P, and a suitable area factor to account for the redistribution of stresses inside the pores as the pores get emptied. Under these conditions, effective stress σ' in the x, y and z directions is equal and can be simply derived from the thermodynamic equilibrium condition given in equation (A.1) across vapour and liquid water interfaces as

$$\sigma' = -S_f P = -S_f \left(\frac{\rho RT}{M} \right) \ln h \tag{A.2}$$

where S_f is the area factor for effectiveness of stress distribution. As an approximation, S_f can be taken as the total liquid water content per unit mortar volume. It has to be remembered here that the water content of a microstructure at a certain relative humidity will be decided by the hystersis isotherms obtained from thermodynamic considerations (Chapter 4).

A.1.2 Unrestrained drying shrinkage strain

If the medium is assumed to be isotropic, corresponding strains in the x, y and z directions would also be the same. The stress–strain relationship which describes the microdeformation of cement paste due to capillary stress might show a nonlinearity. At this stage, however, it is difficult to take into account the nonlinear behaviour of CSH crystals at the microlevel. Therefore, as a simplification, if a linear stress–strain law is assumed, equation (A.3) is obtained as the *free unrestrained drying shrinkage strain* ε in all directions

$$\varepsilon = -\frac{S_f}{E} \left(\frac{\rho RT}{M} \right) \ln h \tag{A.3}$$

where E is the elasticity modulus of mortar and represents various aspects of the microscopic and probably time-dependent nonlinear deformation behaviour in a convenient bulk macroscopic parameter. Note that E used in equation (A.3) is usually 2–4 times smaller than the instantaneous elasticity modulus of mortar depending on the rate of drying and the microstructural conditioning conditions. Equation (A.3) has been used to compute the shrinkage strains while analysing the drying behaviour. The remaining terms in the above equation have their usual meanings.

Although E in equation (A.3) should be known beforehand, the coupled computational system, described in Chapter 6 can compute this quantity dynamically considering the level of maturity of the concrete and the strength development model described in Chapter 3.

A.1.3 Applicability of the model

The methodology described above provides a simple and basic analytical framework for the prediction of drying shrinkage strain in concrete, once the moisture distributions and several other parameters in equations (A.2) and (A.3) are obtained. It must be noted, however, that this methodology is qualitative in nature since rational and reliable methods for the prediction of parameters, such as E, for arbitrary mix proportions have not yet been established. In contrast to this methodology, several empirical equations exist that have been used for the estimation of the shrinkage of concrete. Such equations may be easier to use but are restricted in range and versatility. For example, the influence of different material properties dependent on mix proportions, curing conditions and concrete age, and a time-dependent variation in environmental conditions, etc., cannot be considered rationally for the evaluation of drying shrinkage strains in such schemes. Moreover, the evaluation of stress and strain distributions within structures is not possible in such schemes.

It must be emphasized that the important aspect of the method of shrinkage strain computations described here is not characterized by the assumptions regarding underlying models, but the methodology itself. Although the overall approach can be termed qualitative in nature at this stage of our research, it is hoped that combining this evaluation system with the integrated hydration, structure formation and moisture transport models will enable us to consider rationally the influence of material properties and arbitrary environmental conditions on drying shrinkage strains. Later on in this section, we will show the application of this method to simulate and study the influence of material level macroscopic parameters, such as water-to-powder ratio and concrete age on the development of drying and autogenous shrinkage strains and dewatering (moisture loss due to drying). However, before that, let us refer to the experimental data of drying shrinkage strain of HPC (high-performance concrete) mortars and concrete which were obtained for various test conditions and specimens shown in Table A.2. The experimental details and results are discussed next.

A.2 Outline of drying shrinkage experiments

The specimens considered here are among the 34 mortar and concrete specimens that were subjected to drying under various conditions. Experimental conditions that varied were in the following range:

- water-to-cement ratio by weight: 31.6 ~ 83.9 (%);
- volumetric ratio of aggregate: 21 ~ 45.5 (%);
- age at testing (curing duration): 16 (h) ~ 34 (day)
- specimen size: $4 \times 4 \times 16$ (cm^3) and $10 \times 10 \times 40$ (cm^3);
- environmental conditions: vacuum dry and 60% RH.

A.2.1 Experimental outline

The mix proportions of various mortars is shown in Table A.1. The specimens characterized by their size and drying conditions are shown in Table A.2. As noted before, the mix proportions were determined principally according to the HPC mix-design method. After casting and up to testing, specimens were shielded from both wetting and drying by polyethylene sheeting. After the stipulated curing period, the length change of the specimens was measured across the contact tips that were attached to the two side surfaces of the specimens using epoxy type adhesive. Weight measurements were also taken with time. In the vacuum-drying test, the specimens were dried in a glass desiccator of 20 litre capacity and connected to a vacuum pump (exhaust capacity ~50 litre min). This testing method

Table A.1 Mix-proportions of HPC mortars

Mix proportion	*V_w/V_p (vol %)*	*W/C (wt %)*	*V_{agg} (vol %)*	*Unit weight (kg/m^3)* W	*MC*	*I*	*S*	*AD.*
MS100M	100	33.5	45.5	247	738	40[a]	1191	11.67
MS60L740M	100	55.8	45.5	247	443	24[a], 265[b]	1191	10.98
MS40L460M	100	83.9	45.5	247	295	16[a], 397[c]	1191	10.62
S5P10	100	35.0	35.0	325	1030	–	917	10.30
S7P10	100	31.6	49.0	255	808	–	1284	8.08
S3P10	100	31.6	21.0	395	1252	–	550	12.52

V_w/V_p, volumetric ratio of water and powder materials (MC, L);
W/C, water-to-cement ratio by weight;
V_{agg}, volumetric ratio of sand and gravel;
MC, moderate heat cement, specific gravity 3.17 (g/cm^3);
L, limestone powder specific gravity ~2.67 (g/cm^3) Blaine values, a = 18000, b = 7000, c = 4000;
S, river sand, specific gravity 2.62 (g/cm^3), water absorption 0.6%, FM 6.85, max solid volume 0.61;
Ad., superplasticizer.

Table A.2 Test series and specimen details

Name of specimen	*Mix proportion*	*Size (cm)*	*Testing age (day)*	*Drying condition*
MS16	MS100M	4 × 4 × 16	0.7	Vacuum
MS2	MS100M	4 × 4 × 16	2	Vacuum
MS7	MS100M	4 × 4 × 16	7	Vacuum
L77	MS60L740M	4 × 4 × 16	7	Vacuum
L47	MS40L460M	4 × 4 × 16	7	Vacuum
S5P10	S5P10	4 × 4 × 16	2	Vacuum
S7P10	S7P10	4 × 4 × 16	2	Vacuum
S3P10	S3P10	4 × 4 × 16	2	Vacuum
S5P10S	S5P10	4 × 4 × 16	34	60% RH
S5P10L	S5P10	10 × 10 × 40	34	60% RH

magnified the drying shrinkage behaviour due to the rapid drying under very low humidity conditions. Similar methods of vacuum drying have also been widely reported in the literature. After testing, the specimens were dried in an oven at a temperature of 110 °C for more than two days. Afterwards, weight measurements were taken.

A.3 Experimental results and discussion

The experimental results of dewatering and shrinkage strain with time as obtained for the test series of Table A.2 are shown in Figures A.1–A.4. These figures show the influence of water-to-cement ratio, sand volumetric content, concrete age and specimen size on the drying shrinkage and dewatering (weight loss) behaviours. Since the influence of moisture content on the shrinkage strain has long been recognized, dewatering of the specimens due to drying was also measured. It is believed that a knowledge of the drying rate and dewatering versus drying shrinkage relationships would help in the direct estimation of the development of drying shrinkage strains.

Figure A1.1 shows the results for three mortar specimens with different water-to-cement ratios. The cement content was varied by limestone powder replacement. As can be expected, the resistance to moisture transport and perhaps the stiffness of the mortar composite decreases with an increase in the water-to-cement ratio. Also, the total porosity of the hardened matrix increases with the water-to-powder ratio. This results in a larger and rapid moisture loss for higher water-to-cement ratio mortars. At the same time, due to a higher fraction of larger size pore voids present in the higher water-to-powder mortars, the shrinkage strains associated with the corresponding

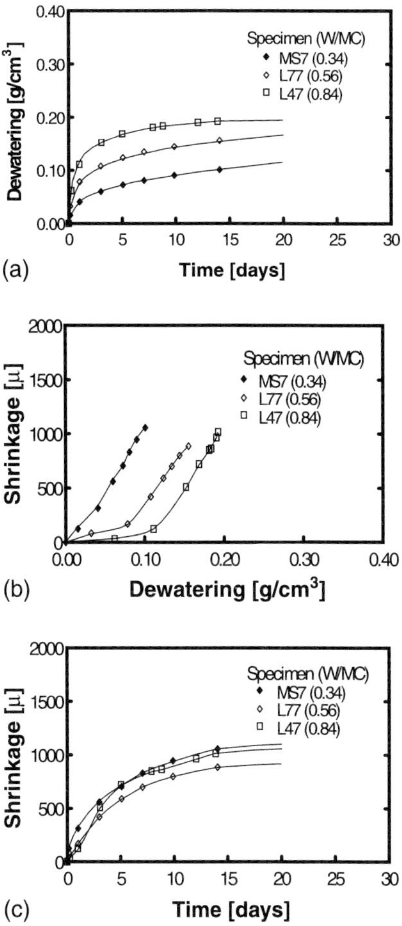

Figure A.1 (a) Experimental measurements of dewatering of mortar specimen versus time, for different water-to-cement ratios. (b) Experimental measurements of dewatering versus drying shrinkage strain. (c) Experimental measurements of drying shrinkage strain versus time, for different water-to-cement ratio cases.

weight loss or dewatering will become smaller (due to a smaller magnitude of capillary stresses). It is interesting to note that a combination of the above two factors (Figures A.1(a,b)) leads to almost identical development of the drying shrinkage strain with time for different water-to-powder ratio mortars (Figure (A.1(c)). However, this observation should not be generalized to concrete or mortars of arbitrary water-to-powder ratio as the properties of the constituent powder materials and admixtures will influence the development of the microstructure. Instead, the discussion above serves to illustrate the possible relationships that might exist between the pore microstructure, water content and the ensuing shrinkage due to a loss of moisture from the porous microstructure.

Figure A.2 shows the influence of sand content in the mix proportion on the weight loss and drying shrinkage relationships with time. If the volumetric sand content and the water-to-powder ratio of the mortar is small, continuous contact of particles across the specimen does not take place and transition zones do not influence the average pore structure

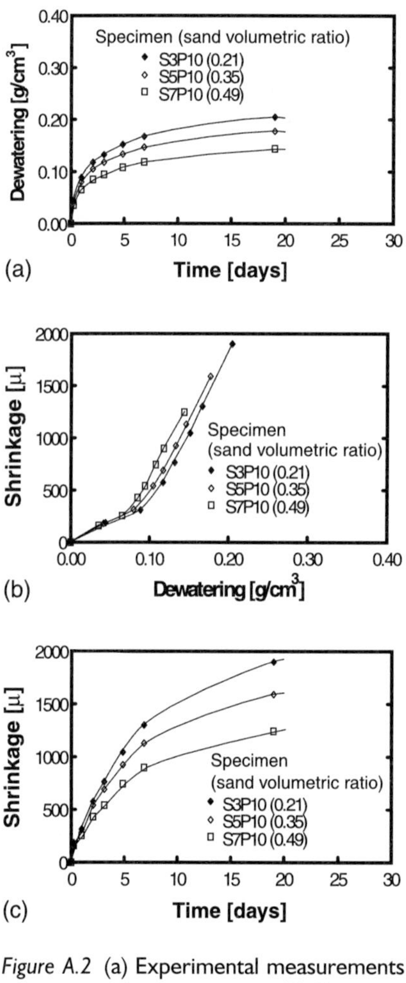

Figure A.2 (a) Experimental measurements of dewatering versus time, for different sand volumetric ratios. (b) Experimental measurements of drying shrinkage strain versus dewatering, for different sand volume ratios. (c) Experimental measurements of drying shrinkage strain versus time, for different sand volumetric ratios.

significantly. Therefore, it can be expected that the dewatering with time relationship for different sand volumetric contents would be simply proportional to the matrix volume fraction in the mortar.

This is indeed observed in the experiments (Figure A.2(a)). Furthermore, as the elastic modulus of the mortar is not significantly influenced by the low sand volume fractions, the shrinkage and dewatering relationships would be almost identical, based upon the concept of origin and redistribution of shrinkage stress from capillary stresses of the micropore structure. Figures A.2(b,c), in fact, show the simple proportional nature of the development of shrinkage strains and dewatering–shrinkage strain relationships for varying sand volumetric contents when the water-to-powder ratio of the mortar is small enough (around 30% by weight). If transition-zone effects become significant, we should expect an earlier onset and development of the shrinkage due to a faster loss of moisture from the hardened matrix. The initial slope of the shrinkage–dewatering curve would be much flatter in such a case. In fact, an initial flatter shrinkage–dewatering curve represents a coarser microstructure and generally leads to higher shrinkage.

Apart from the mix proportion composition, the type of ageing also influences the material properties, such as strength and micropore structure of the cementitious matrix. In general, longer periods of curing will produce a denser and uniform pore structure. Figure A.3 shows the influence of different types of curing conditions on the subsequent dewatering and development of shrinkage strains with time. The specimen cured only for 16 hours shows a large amount of dewatering in a very short time indicating a very coarse microstructure. Very short periods of curing produce a low-strength matrix, too weak to sustain even the low stresses due to capillary surface tension thereby leading to quite large shrinkage strains (Figure A.3(c)). Longer periods of curing lead to an increase in the strength but at the same time the microstructure also becomes denser. Thus, the benefit of the strength gain might be somewhat offset by the increase in shrinkage stresses, with the resultant effect of almost similar shrinkage strains with time. In general, the shrinkage strains in well-aged mortars would be always a bit smaller than, or at most similar to, the shrinkage of corresponding lesser-aged specimens. However, compared to the requirements of lower shrinkage strains, the requirement of longer periods of curing is usually dictated by the strength considerations.

Figure A.4 shows the influence of specimen size on the development of drying shrinkage strains and weight loss with time. Again it is interesting to note that the moisture weight loss due to drying and drying shrinkage exhibit an almost one-to-one relationship despite the difference in the sizes of the specimens.

This result shows the fundamental nature of the relationship between the moisture content in the hardened cementitious matrix and the corresponding shrinkage. Generally, this relationship will be unique for a cementitious matrix only for a particular kind of ageing and drying–wetting history to

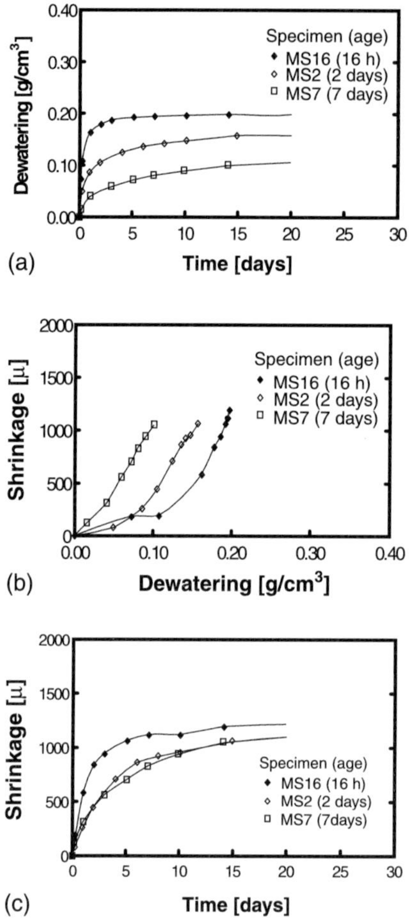

Figure A.3 (a) Experimental measurements of dewatering versus time, for differently aged specimens. (b) Experimental measurements of drying shrinkage strain versus dewatering, for differently aged specimens. (c) Experimental measurements of drying shrinkage strain versus time, for differently aged specimens.

which it is exposed. For example, young mortar and concrete will exhibit a shrinkage–dewatering relationship that is irreversible and exhibits hysteresis due to the combined effects of alterations and rearrangement of microstructure and hysteresis in the drying and wetting isotherms. For mature hardened pastes, this relationship probably exhibits hysteresis but is reversible at the end-points, e.g. at complete saturation. The origins of hysteresis in this case being the behaviour of the absorption isotherms of mature pastes. Similar results have also been reported in the literature.

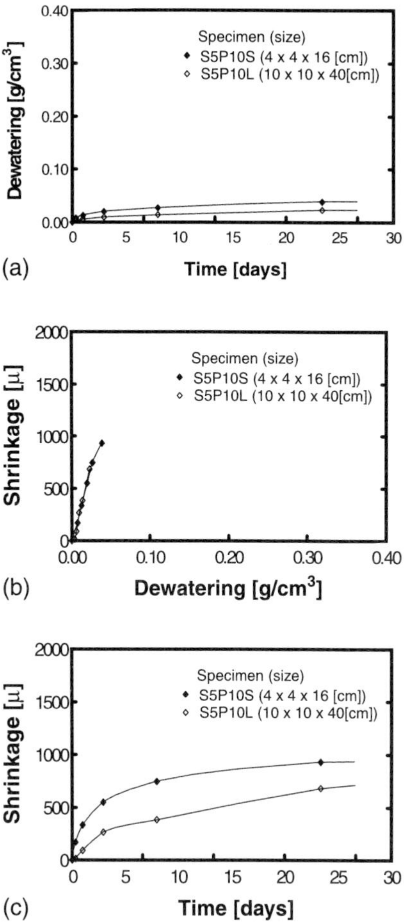

Figure A.4 (a) Experimental measurements of dewatering versus time, for differently sized specimens. (b) Experimental measurements of drying shrinkage strain versus dewatering, for differently sized specimens. (c) Experimental measurements of drying shrinkage strain versus time, for differently sized specimens.

A.4 Computational simulation

The experimental results discussed in the previous section give a general idea of the nature of the relationships of micropore structure with the water content and associated shrinkage of the hardened cementitious matrix. The analytical model of shrinkage strain discussed earlier can be used to simulate the influence of macroscale parameters, such as water-to-powder ratio and curing conditions, on the dewatering, drying and autogenous shrinkage behaviour in the similar integrated scheme of durability evaluations as outlined in Chapter 6. This is done by coupling the shrinkage strain model with the integrated hydration, microstructure formation and moisture transport theory. To treat both autogenous and drying shrinkage behaviour and their coupling in the same scheme, various material properties such as the degree of hydration, pore-water pressure development and distribution, relative humidity in micropores, microstructure characteristics, and deformability against capillary stress, should be predicted in space and time domains. For a given mix proportion, structure configuration and the initial and boundary conditions, the computational program DuCOM discussed in Chapter 6 can obtain these properties. In this computational system, it is not necessary to distinguish between the autogenous and drying shrinkage behaviour, since by calculating the water content and pore structures in the concrete under specified conditions, the volume change due to shrinkage can be predicted from equation (A.3).

Figures A.5 and A.6 show the various simulated results as obtained by the coupled computational program DuCOM described in Chapter 6. The capillary stress redistribution factor, S_f, in equation (A.2) was taken as the total water content of the mortar. Furthermore, the elasticity modulus, E (kgf/cm^2), was computed from the following empirical equation

$$E = 1.333 \times 10^4 f_c^{1/3} \tag{A.4}$$

where f_c (kgf/cm^2) is the compressive strength at any stage of hydration as obtained in equation (3.35). The magnitude of E as obtained from the above equation is about $\frac{1}{2} - \frac{1}{4}$ of the ordinary value for the static elastic modulus of concrete. The reason for this discrepancy might be the assumption of a uniform stress field due to capillary tension or perhaps some intrinsic differences in the mechanisms of deformation at micro- and macroscale due to capillary stress and that due to applied stress. Despite the fact that the model for E in equation (A.4) is quite approximate and may not cover the range of concrete encountered in practice, it would serve our purpose here for qualitative assessments of various material parameters and ageing conditions on the shrinkage and dewatering behaviours. Needless to say, research on the strength development mechanisms from the viewpoint of a physical phenomenon needs to be actively pursued.

Although, the computed results (Figures A.5 and A.6) are qualitative in nature (due to the qualitative nature of E), it could be observed that the trends as well as the range of computed results for various cases in Figure A.5 and Figure A.6 are similar to the experimental values. This is important considering the fact that no adjustment is made to the various material submodels on a case-by-case basis.

The mortar mix proportions and boundary conditions applied are the only input to the simulation system. This fact emphasizes the rationality of the overall simulation system. The coupled computational program DuCOM that is based on thermodynamic considerations can also be applied to understand the phenomenon of autogenous shrinkage. The problem of autogenous shrinkage has hitherto been confined to the realms of empirical relationships that are limited to small ranges of the conditions of cement type, curing temperatures or mix proportions of real concrete. In the integrated scheme discussed in this book, the phenomenon of

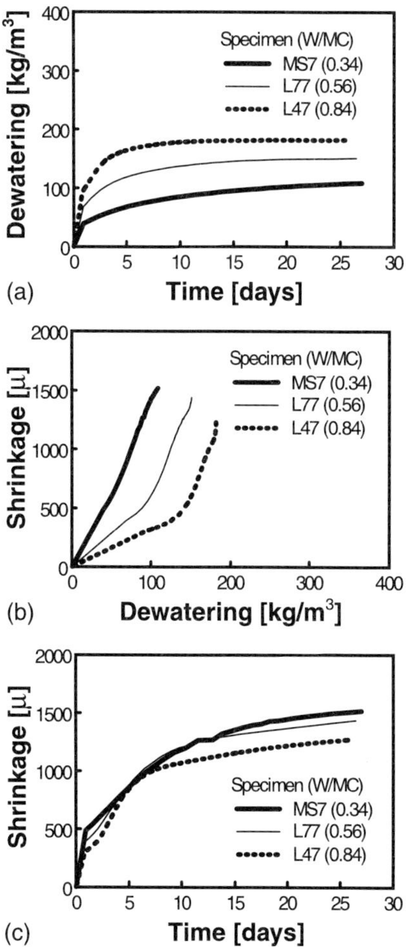

Figure A.5 Computed results for different water-to-cement ratios: (a) time–dewatering relationship; (b) dewatering–shrinkage relationship; (c) time–shrinkage relationship.

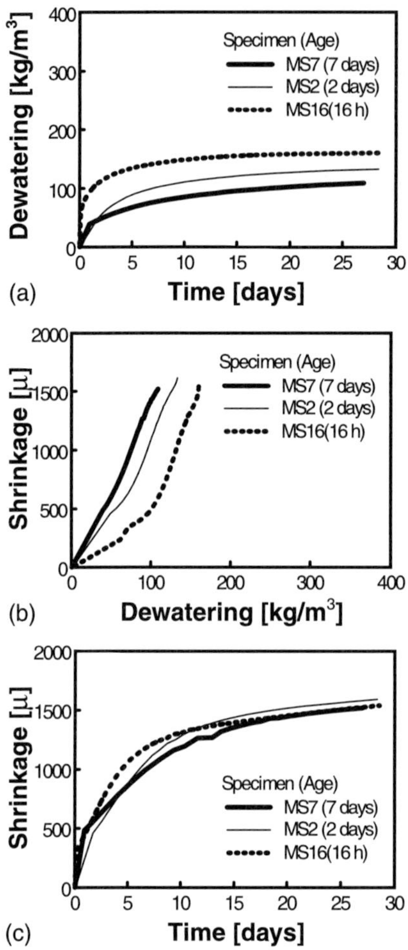

Figure A.6 Computed results for differently aged specimens: (a) time–dewatering relationship; (b) dewatering–shrinkage relationship; (c) time–shrinkage relationship

autogenous shrinkage can be explored in the same framework without requiring extraneous adjustment factors. The source of internal shrinkage stresses is primarily due to the pore water surface tension, especially at higher pore saturation. Therefore, the knowledge of representative pore distribution and pore humidity, as obtained from DuCOM, can be directly used in estimations of the internal tensile stresses at arbitrary stages of hydration even when there is no external moisture loss.

The conditions of thermodynamic compatibility provide us with an automatic balance between the moisture consumed in hydration reactions, the dynamic micropore structure, temperature and the free moisture content in the microstructure. Since, this information is computed at each stage of the early development, autogenous shrinkage computations can be performed in the same framework by applying equation (A.3) for each step over the entire space domain of the specimen. Figure A.7 gives the analytical results of autogenous shrinkage for a few model cement paste specimens. In this framework, factors such as curing and material properties are directly integrated into the unified structure of various physico-chemical and mechanical models. Therefore, durability evaluation is possible from the stage of the casting of concrete without any special treatment for conventional parameters, such as the duration and type of curing. Figures A.8(a)–(c) show comparisons of the experimental and analytical results of drying and autogenous shrinkage of mortars and concrete as obtained from the finite-element program DuCOM for various specimen sizes, mix proportions and curing conditions. Experimental results are represented by marker symbols and lines represent the analytical results. It can be observed that both the drying and autogenous shrinkage strains can be satisfactorily simulated. In these simulations, the compressive strength f_c (MPa, equation (A.4)) is obtained from the following empirical equation.

$$f_c = 78.5\exp(-8.0V_p) \tag{A.5}$$

where V_p is the capillary porosity computed by DuCOM corresponding to pores having radii above 50 nm. In the finite-element simulations, similar material conditions and curing were used to those in the experiments. The analysis can implicitly consider the effects of autogenous shrinkage even

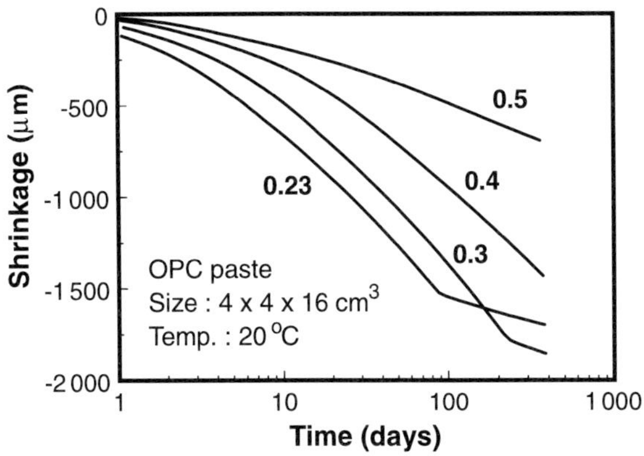

Figure A.7 Analytical autogenous shrinkage strains for cement paste specimens of various W/C ratios (weight fractions).

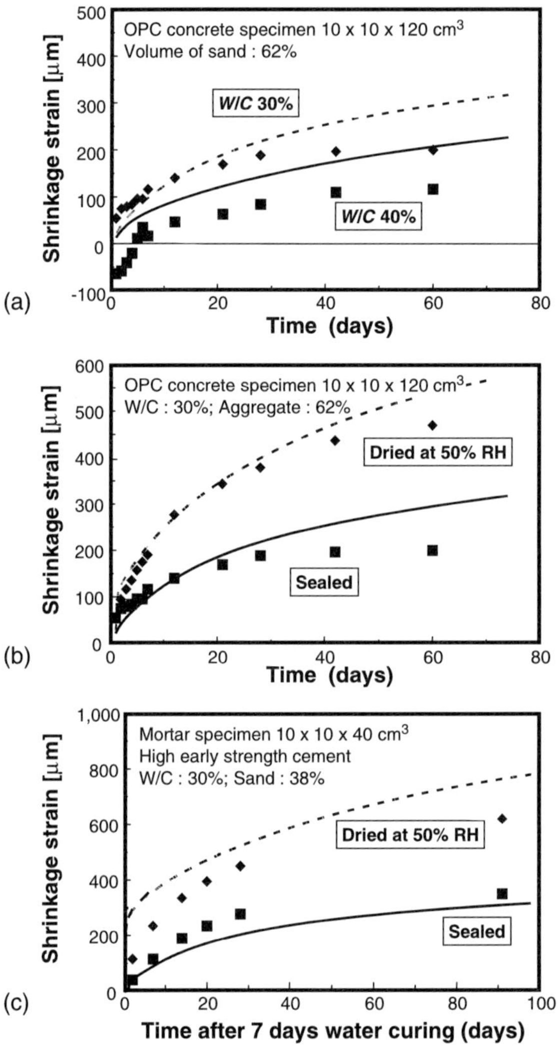

Figure A.8 (a) Prediction of autogenous shrinkage strains of concrete of different W/C ratios. Specimens were kept sealed over the entire testing period. Lines show analytical results. (b) Prediction of autogenous and drying shrinkage strains of concrete specimen. Drying was performed one day after casting. Lines show analytical results. (c) Prediction of autogenous and dry shrinkage strains of dry mortar. Drying was started 7 days after casting. Lines show analytical results.

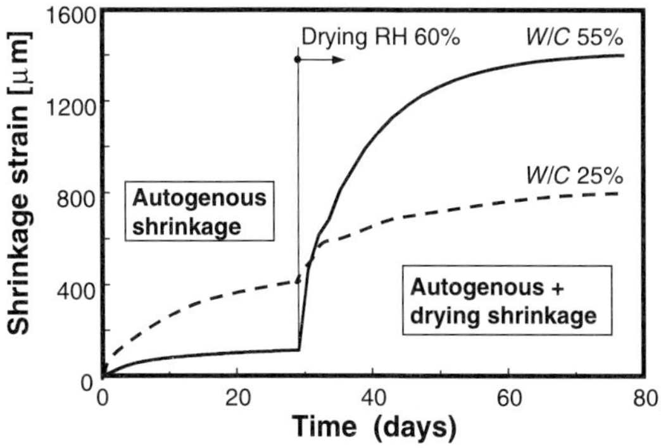

Figure A.9 Numerical simulation of autogenous and drying shrinkage behaviour for different W/C ratios. Drying was started 28 days after casting.

under drying conditions when hydration could still be significant. Of course, the delay of hydration due to a loss of free water is coherently evaluated in the analysis.

Figure A.9 shows the results of a numerical sensitivity simulation of the behaviour of autogenous and drying shrinkage behaviour for different water-to-cement ratio concrete. It has been widely reported that the shrinkage behaviours are quite different in the ordinary concrete and low water-to-cement ratio concrete. In the case of ordinary concrete, the contribution of the autogenous shrinkage to the total volume change is relatively small compared to the drying shrinkage contribution due to a coarser microstructure and lower reduction in the pore humidity due to selfdesiccation (as ample pore water is available). On the other hand, for low water-to-cement ratio concrete, autogenous shrinkage is quite significant compared to the drying shrinkage. This situation is very clearly simulated by DuCOM.

Appendix B: DuCOM on the Internet and its Basic Specifications

As described in Chapter 6, DuCOM is a finite-element computer program that can solve the heat and mass transport problems of porous media. It is the result of ongoing research being carried out at the concrete laboratory of the University of Tokyo, Japan. The generic differential equation that can be solved by DuCOM can be stated as

$$\alpha_i \frac{\partial \theta_i}{\partial t} - \operatorname{div}\left(\sum_{j=1,N} D_i^j \nabla \theta_j \right) + \beta_i \theta_i + Q_i = 0 \tag{B.1}$$

where θ_i is an independent variable. DuCOM is capable of solving this differential equation for any number of independent variables simultaneously (1 to N). This has been achieved by implementing an alternate solving scheme where the latest iteration level values of an independent variable are obtained by using previous iteration level values of other variables in the same time step.

B.1 Material modelling in DuCOM

For heat and moisture transport modelling in concrete, equation (B.1) can be resolved to the simpler form as given in equation (6.1). The material-specific parameters in equation (B.1) are obtained by considering the constitutive laws of powder material hydration, microstructure formation and path-dependent moisture transport based on the microstructure. The material properties are held external to the main solver and can be maintained and developed independently of the core finite-element solver. These material subroutines are basically accessed only during the formation of the core stiffness matrix. Also, most of the material models have been currently implemented as is, without many simplifications. This ensures a high degree of accuracy in the results; however, from the computational efficiency point of view it might not be an optimum solution. The specific simplifications for speedier computations as and where adopted have been noted in the main text of this book. Readers interested in a top-down approach to constitutive material modelling implemented in DuCOM are advised to refer to the material modelling summary table in Chapter 6. This table includes direct pointers to the relevant equations and is a good starting point for the details of the constitutive material models of moisture and heat transport in concrete.

B.2 DuCOM on the Internet

The last few years have seen an exponential growth in the use and application of Internet to exchange information. The ease of use of the World Wide Web (WWW) and the power of online publishing has resulted in a spurt of activity. New WWW sites offering all kinds of contents are posted every day. In accord with the general spirit of the Internet of free information for everyone by everyone, we have created a site dedicated to the online demonstration of DuCOM. It contains most of the information contained in this book in an abstract form: a working demonstration of DuCOM itself; free downloadable binary files of the program and much more. This section is devoted to a discussion about the DuCOM demonstration site. However, it is strongly suggested that for the most up-to-date information you visit the site.

It is likely that a search on DuCOM would land you somewhere on the DuCOM demonstration site, however the recommended entry point is located at `http://concrete.t.u-tokyo.ac.jp/en/demos/ducom/index.html/`. The suggested browsers for this site are currently Netscape 4.0 or Internet Explorer 4.0 or above. The main entry page is shown in Figure B.1.

B.2.1 Content

There are basically three areas which can be explored at the site. We will discuss these one by one. We expect that the analysis area will be of immediate interest to most people as it gives an opportunity to create online virtual concrete and analyse the influence of mixcompositions and environment on several properties of concrete, such as strength, temperature, moisture content, porosity, drying shrinkage strains, etc. Figure B.2 shows a snapshot of the analysis form that is available under the analysis subsection of the DuCOM demonstration page.

There are three main sections in the input form. The first one defines the basic mix proportion of the concrete (or mortar). The second section defines the casting and curing conditions to be simulated. It will be noticed that the online simulation only works for a one-dimensional slab structure. The ambient humidity and temperature are assumed to be constant throughout the period of simulation after the formwork has been removed. Finally, in the last section, the points of interest in the slab structure can be selected by

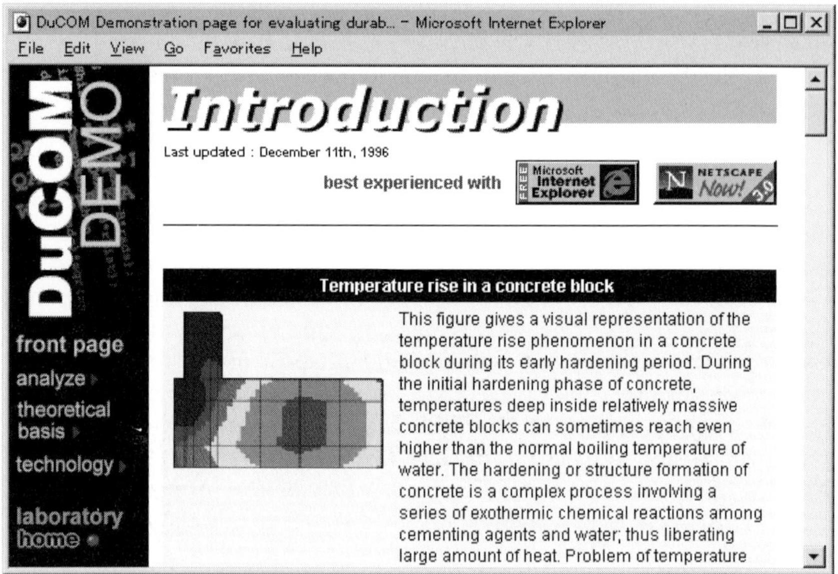

Figure B.1 The World Wide Web DuCOM homepage.

using the two list boxes provided. Thus the simulation results would be presented for these two points only. It would help in understanding and comparing the concrete material properties along the depth from the face which is exposed to the environment. A part of the typical simulation result appears in Figure B.3.

The simulation engine behind the above demonstration is DuCOM, which is executed from a wrapper that is actually a CGI (common gateway interface) script. The options for the web demonstration have been kept at bare minimum, owing to the short response time required for the web. On its own DuCOM can handle any shape and size of concrete structures for any history of boundary conditions applied over the life cycle of a typical structure. Current web technology and inherent difficulties in providing a proper user interface prevent us from implementing the full fledged version of DuCOM on the web. However, executable binary files of DuCOM for HP-UX and Windows 95 are available for download to those who have the proper computational tools at their disposal. Following are some of the technical specifications about the program itself at the time of writing.

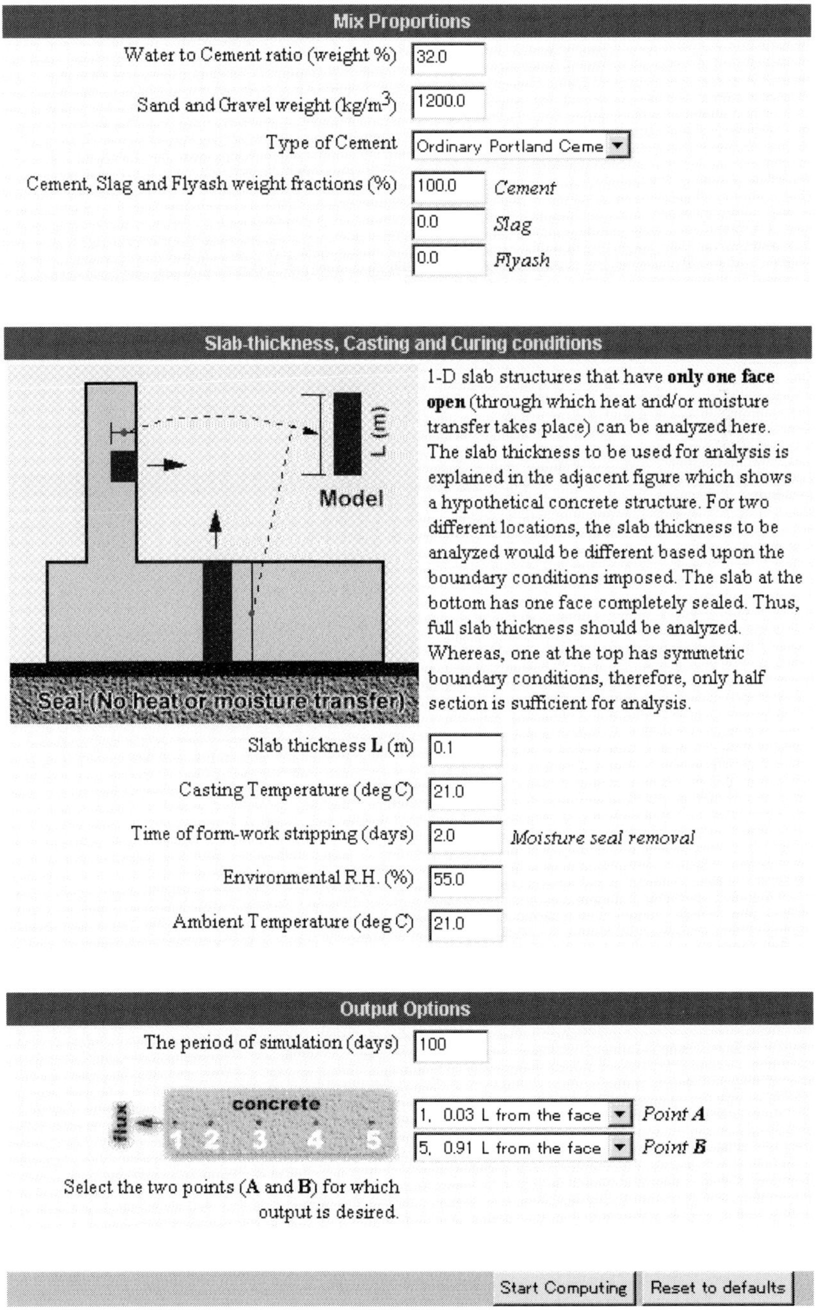

Mix Proportions

Water to Cement ratio (weight %) 32.0

Sand and Gravel weight (kg/m^3) 1200.0

Type of Cement: Ordinary Portland Ceme

Cement, Slag and Flyash weight fractions (%): 100.0 *Cement*; 0.0 *Slag*; 0.0 *Flyash*

Slab-thickness, Casting and Curing conditions

1-D slab structures that have **only one face open** (through which heat and/or moisture transfer takes place) can be analyzed here. The slab thickness to be used for analysis is explained in the adjacent figure which shows a hypothetical concrete structure. For two different locations, the slab thickness to be analyzed would be different based upon the boundary conditions imposed. The slab at the bottom has one face completely sealed. Thus, full slab thickness should be analyzed. Whereas, one at the top has symmetric boundary conditions, therefore, only half section is sufficient for analysis.

Slab thickness **L** (m) 0.1

Casting Temperature (deg C) 21.0

Time of form-work stripping (days) 2.0 *Moisture seal removal*

Environmental R.H. (%) 55.0

Ambient Temperature (deg C) 21.0

Output Options

The period of simulation (days) 100

1, 0.03 L from the face — *Point **A***

5, 0.91 L from the face — *Point **B***

Select the two points (**A** and **B**) for which output is desired.

Start Computing | Reset to defaults

Figure B.2 The input form at the online DuCOM demonstration site.

DuCOM

version 2.2

INPUT DATA	
WC ratio (weight %)	**32.0**
Sand + Aggregate (kg/m^3)	**1200**
Cement Type	**Ordinary Portland Cement**
Cement fraction (%)	**100.0**
SLAG wt. fraction (%)	**0.0**
FLYASH wt. fraction (%)	**0.0**
Slab thickness (m)	**0.10**
Casting Temperature (deg C)	**21.0**
Form stripped at (days)	**2.0**
Ambient Humidity	**55.0**
Ambient Temperature (deg C)	**21.0**
Simulation period (days)	**100**
Point A at (m)	**0.003**
Point B at (m)	**0.091**

Note : All of the following figures correspond to points A and B only. — A — B

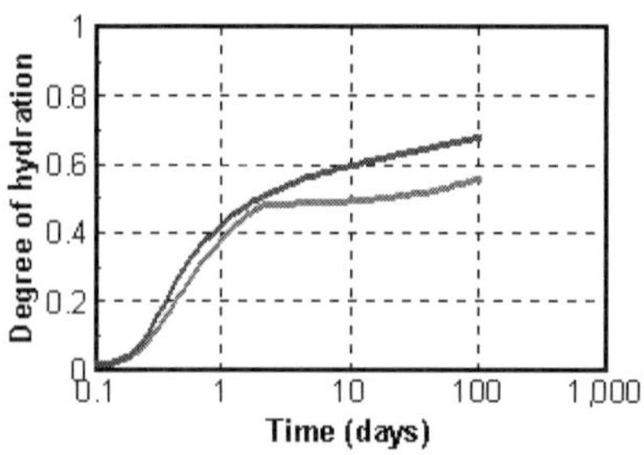

Fig. 1: Average degree of hydration with time

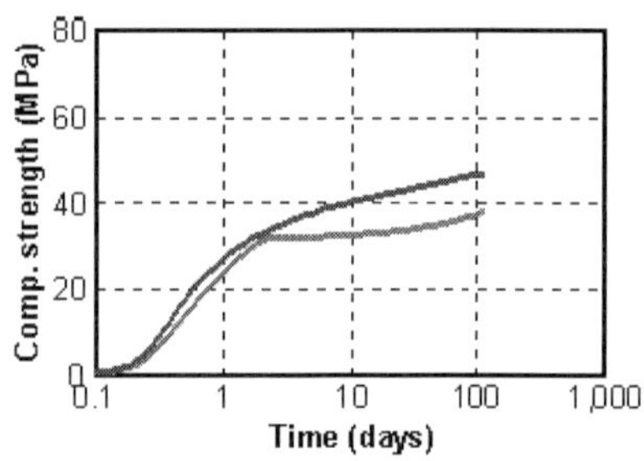

Fig. 2: Development of compressive strength

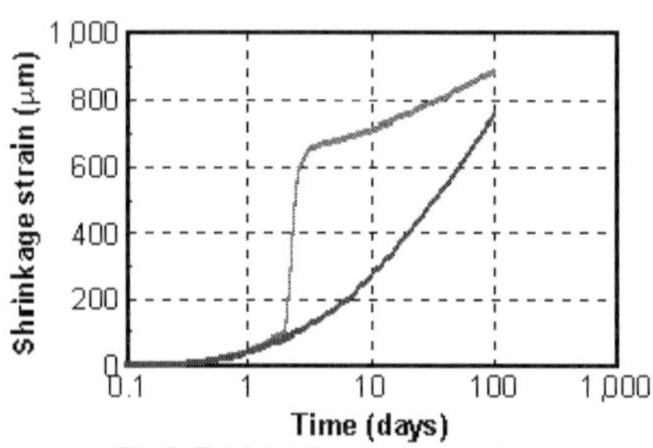

Fig. 3: Total shrinkage strains with time

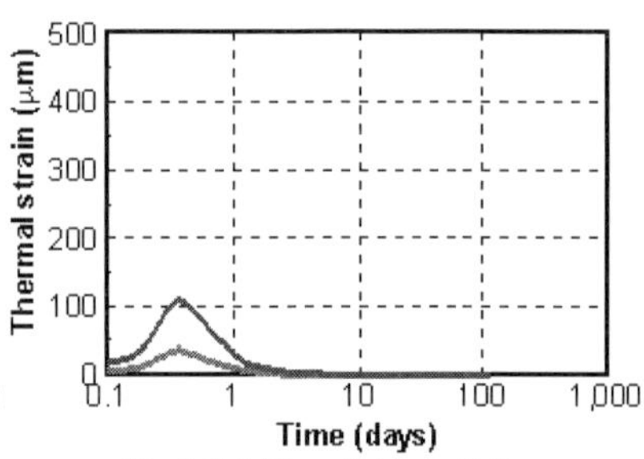

Fig. 4: Total thermal strains with time

Figure B.3 Simulation results obtained online by using the DuCOM demonstration.

B.2.2 Key specifications

- Number of independent variables: unlimited (only *P* and *T* used now);
- Element types: all isoparametric — 4, 8 and 20 node;
- Core storage: skyline indexing;
- Core solver: modified Choleski's (forward and backward) substitutions;
- Core memory management: static (defined at the compile time);
- Boundary condition: absolute value (fixed DOF), convective flux;
- Time stepping and error control: dynamic. (primitive at current stage);
- User interface: command line and I/O files (text + binary);
- Supported OS: HP-UX; port underway for PC platforms;
- Support for primitive real time collaborative data sharing and job control by external process (UNIX version only).

Future versions of DuCOM will include the coupling of several other deterioration and transport phenomena, such as chloride and oxygen diffusion, corrosion of reinforcement, carbonation, etc. Furthermore, efforts are underway to integrate the material durability simulation tools, such as DuCOM with their structural analysis counterpart, such as COM3 so that a homogeneous system of concrete service life-span simulation can be achieved. This would aid in examing several key questions related to the performance-based durability analysis and design methods and their

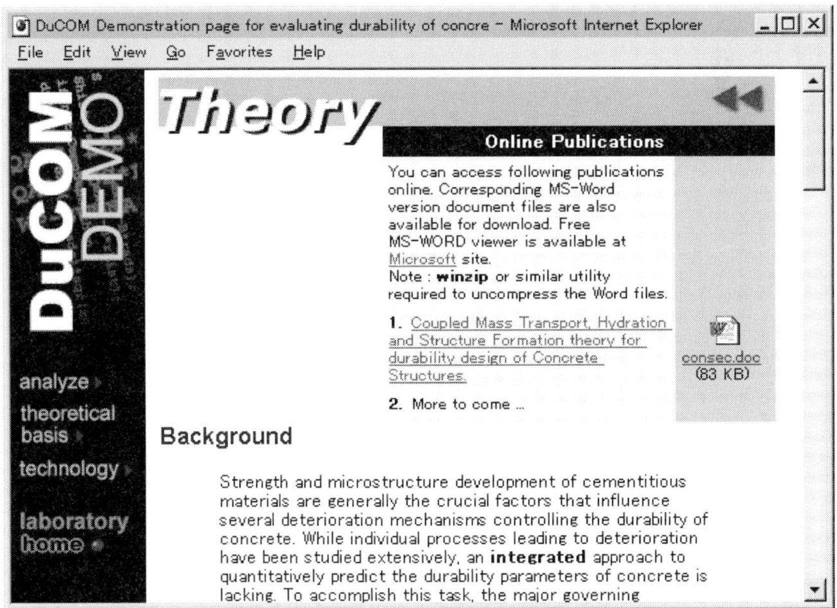

Figure B.4 Reference area of DuCOM demonstration.

feasibility. In the authors' view, such an integrated approach is more rational and certainly feasible provided concerted efforts are undertaken in this direction. For the latest developments in this area and the current versions of DuCOM, one can access the technology section of the DuCOM demonstration page. The outline content of this book as well as several other publications related to the concrete durability research at the Tokyo concrete laboratory can be accessed under the theory page. It also contains several publications in original high quality format for download. Figure B.4 gives a snapshot of the theory page devoted to discussions on the material modelling research of concrete.

Certainly, the Internet is a dynamic area. A huge amount of information appears and disappears everyday from the World Wide Web. We have provided a list of URLs in Table B.1 that might be of interest to the concrete material research community. Due to the dynamic nature of the Web, however, we do not guarantee the availability of the sites at all times.

Table B.1 Concrete research related web sites and their URLs

Web Site	*URL*
Concrete Laboratory, The University of Tokyo. High performance concrete, nonlinear mechanics of concrete and durability design of concrete structures	http://concrete.t.u-tokyo.ac.jp/
Japan Concrete Institute, A list of publications; event schedules, etc.	http://ux01.so-net.or.jp/~jci/
American Society of Civil Engineers, Broadbase information, contains a searchable index of all the ASCE publications	http://www.asce.org
High Performance Construction Materials and Systems, A comprehensive bibilography on concrete material research, online publications and programs; who's who of material research	http://titan.cbt.nist.gov
Concrete engineering discussion list (should be joined by email)	gopher://nisp.nd.ac.uk/11/lists-a-e/engineering-concrete
Shilstone companies, A meta-link page of civil engineering related sites	http://www.shilstone.com/public/online.htm
RILEM, The International Union of Testing and Research Laboratories for Materials and Structures	http://web.ens-cachan.fr/~rilem/

It is hoped that the DuCOM demonstration site will serve to illustrate some of the key technologies that have been developed and adopted in its design. To some extent, it also might counter the wide-spread myth that the physical theory based material models are complex and unsuitable for real-life simulations of material behaviour, especially of something as complex as hardening concrete. Improved understanding of the cementitious materials at microscale, their thermodynamic representation and availability of cheap enhanced computational power should only aid the development of rational simulation tools, such as DuCOM in future.

Appendix C: Multi-component Heat Generation Subroutine for Cement in Concrete

A FORTRAN77 subprogram list for computing the heat of hydration rate of cement in concrete is explained in this appendix. The code name is HEAT. The core of the multicomponent cement heat of hydration model is summarized in Chapter 3, and the theoretical background and details are discussed in Chapter 7. This subprogram is one of essential parts of DuCOM and can be easily implemented in any finite element analysis program for heat conduction and thermal crack risk evaluation for massive concrete structures.

This subprogram is freeware for academic and scientific purposes. By using a WWW browser, anybody can take the program source file from the authors' home page. Those who are interested can visit the authors' home page at `http://concrete.t.u-tokyo.ac.jp/`. When a user utilizes this program for research and study, the authors would appreciate his/her report on how the program works and contributes to engineering problems and welcome users' suggestions for improvements. When this program is installed in other software, some message about the inclusion of HEAT should also be implemented. Claims and questions should be directed to the feedback located at the DuCOM home page.

C.1 Specification of input and output data

The source program list is divided into several segments and each segment will be explained in some detail.

The first segment serves as a preprocessor to initialize variables and to define parameters. The output variables to be computed by this code are heat generation rate of concrete per unit volume denoted by QR (kcal $\mathrm{day}^{-1}\,\mathrm{m}^{-3}$). This variable will be used for the heat conduction analysis. Simultaneously, heat generation of each mineral compound consisting of cement during the time interval is calculated in terms of QGETA, QGETF (ettringite), QGMNA, QGMNF (monosulphate), QG3A (aluminate), QGAF (ferrite), QG3S (erite), QG2S (berite), QGSG (slag) and QGFA (fly ash: kcal/kg). TIME denotes the time in days at the previous step and TIME1 is the updated one concerned.

The above-stated parameters are computed with temperature, denoted by TEMP (degree), and accumulated heat of each mineral component as path-dependent parameters by QAXX (xx-component: kcal/kg). These parameters are input ones. When this program is used within the step-by-step integration scheme in time, path-dependent parameters denoted by QAXX have to be revised outside this program, such that QAXX = QAXX + QGXX, and stored in memory.

The free water-to-powder ratio by weight indicated by FREEW and $CaOH_2$ solution in the pore water by DCAOH2 are computed by assuming no water migration. This means perfect isolation on mass transport, and this condition is generally acceptable for thermal stress computation of massive concrete structures at early age. When water migration is to be considered by DuCOM, the rate of water consumption associated with the heat of hydration of cement has to be specified as a parameter in the statement of SUBROUTINE definition.

```
      SUBROUTINE HEAT(QR,QGEN,QACM,TIME,TIME1,TEMP,THYS,MID,ISTP
     +                ,QAETA,QAETF,QAMNA,QAMNF,QA3A,QAAF,QA3S,QA2S
     +                ,QASG,QAFA,QGETA,QGETF,QGMNA,QGMNF,QG3A,QGAF
     +                ,QG3S,QG2S,QGSG,QGFA,FRCSET,FRCSMN,RQCSET,RQCSMN
     +               ,FREEW,CAOH2,RDCE3A,RDCEAF,RDCE3S,RDCE2S,FISG,FIFA
     +               ,RDSG,RDFA,RBG3A,RBGAF,RBG3S,RBG2S,RBGSG,RBGFA
     +               ,AEPCA,AEPCF,AEPC3,AEPC2,DCAOH2)
C
C
C          Welcome to MULTI-COMPONENT HYDRATION MODEL
C                  FOR CEMENT IN CONCRETE STRUCTURES!
C
C               ((((((( SEPTEMBER 30, 1996 )))))))
C
C This subroutine can be installed in heat conduction analysis program.
C Concerning theoretical background and details, refer to Chapter 7.
C
C INPUT information : temperature,
C                     accumulated heat of each mineral component
C OUTPUT variables  : heat generation rate of each mineral component
C
C ___________________________________________________________________
C
C   (( developed and implemented in FEM program by DR.T.KISHI ))
C
C ===================================================================
C
C   Variable Specifications and Notations defined in this module
C
C     QR   : HEAT GENERATION RATE (kcal/day/m3 : CONCRETE)
C     QACM : ACCUMULATED HEAT GENERATION (kcal/kg : cement)
C            at time=TIME0 (previous time)
C     QGEN : HEAT GENERATION DURING STEP (kcal/kg : cement)
C            (TIME1 - TIME0)
C     HT** : HEAT GENERATION RATE (kcal/kg/hr)
C [*] HS** : SPECIFIED HEAT GENERATION RATE at T=BTEMP (kcal/kg/hr)
C [*] ZO** : ACTIVATION ENERGY as a function
```

```
C     BTEMP: SPECIFIED TEMPERATURE (degree : centigrade)
C     TEMP  : UP-DATE TEMPERATURE (degree : centigrade)
C     WC    : UNIT WEIGHT OF CEMENT IN CONCRETE (kg/m3)
C     MID   : MATERIAL INDEX defined in each element [ELEM DATA]
C
C ------------------------------------------------------------------
C
C     PPC : PSG : PFA = weight percentage of portland, slag and fly ash
C    (PPC+PSG+PFA=100)
C    P3A:P3S:P4AF:P2S:PPCS2H = weight percentage of C3A, C3S, C4AF, C2S
C                                              and CSH2(monosulfate)
C    SGCS2H = weight percentage of monosulfate in SLAG
C    FACS2H = weight percentage of monosulfate in FLY ASH
C    WP      = water to powder ratio by weight
C    RBLN:RBLNSG:RBLNFA = rate parameter related to Blaine Values
C    QSGMX  = final heat generation of slag (kcal/kg)
C    QFAMX  = final heat generation of fly ash (kcal/kg)
C    RWMONO:RSGW1:RFAW1 = weight percentage of consumed water (%)
C                         (monosulfate, slag and fly ash)
C    RSGCA:RFACA = weight ratio of consumed Ca(OH)2 when reaction of
C    slag and fly ash proceeds.
C
C ------------------------------------------------------------------
C
     DIMENSION THYS(1),TAF(90),T2S(55),TTS(80),TFA(55)
C
C       MATERIAL CONSTANTS FOR CEMENT AND CONCRETE
C       ========================================
    DATA WC/400.0/
    DATA P3A/10.4/,P3S/47.2/,P4AF/9.40/,P2S/27.0/,PPCS2H/3.87/
    DATA PPC/100.0/,BLN/3380/
    DATA PSG/0.00/,BLNSG/3300/,SGCS2H/0.00/,WSGMOL/0.0/
    DATA PFA/0.00/,BLNFA/3280/,FACS2H/0.00/,WFAMOL/0.0/
    DATA PLS/0.00/,BLNLS/7000/
    DATA WP/39.2/,QSP/0.250/,QSPAD/0.00/,CHARSP/5.00/
    DATA QSGMX/110.0/,RSGW1/0.30/,RSGCA/0.22/
    DATA QFAMX/50.00/,RFAW1/0.10/,RFACA/1.00/
    DATA RH3AMN/0.67/,RHAFMN/1.00/
    DATA SLLDED/0.015/
    DATA ALPHA/1.00/
C
C
    QR=0.0
    IF(MID.EQ.2) RETURN
    SLL=TIME1-TIME
C
C       ((((((((powder fineness factor on the rate of hydration))))))))
C
     RBLN =BLN /3380
     RBLN3A=RBLN
     RBLNAF=RBLN
     RBLN3S=RBLN
     RBLN2S=RBLN
     RBLNSG=BLNSG/4330
     RBLNFA=BLNFA/3280
     RBLNLS=BLNLS/7000
```

```
C
   RSGMX=QSGMX/110.0
   RFAMX=QFAMX/50.0
C
   QLV3S=(QA3S/120)*100
   QLV2S=(QA2S/62 )*100
C
   QSPDED=(0.080*P3A+0.020*P4AF)/100*PPC*RBLN+0.005*PSG*RBLNSG
  \                                         +0.025*PFA*RBLNFA
   ESP=QSP*CHARSP-QSPDED
    IF(ESP.LT.0.0) THEN
      ESP=0.0
      QSPDED=QSP*CHARSP
    END IF
C
   ESP=ESP+QSPAD*CHARSP+0.020*PFA*RBLNFA
    IF(ESP.LT.0.0) ESP=0.0
C
   SUP=(0.020*P3S+0.010*P2S)/100*PPC*RBLN
  \                         +0.005*PSG*RBLNSG+0.150*PLS*RBLNLS
   RSP=ESP/SUP
C
   RFL=1-EXP(-5000.0*((P3S+P2S)/100*PPC/100)**10.0)
C
   RCMAL=(1-(EXP(-0.48*(P3S/P2S)**1.4))*RFL)*(1.0+0.4*RFL)+0.1*RFL
     RCM3A=1.0
     RCMAF=1.0
     RCM3S=RCMAL
     RCM2S=RCMAL
     RCMSG=1.0
     RCMFA=1.0
C
   IDED=0
   IEFW=0
     RBG3A=1.0
     RBGAF=1.0
     RBG3S=1.0
     RBG2S=1.0
     RBGSG=1.0
     RBGFA=1.0
C
     IF (ISTP.NE.1) GO TO 200
C
     CAOH2=0.0
     CS2H=PPCS2H*PPC/100+SGCS2H*PSG/100+FACS2H*PFA/100
     FRCSET=CS2H
     FRCSMN=CS2H
C
C  ((((((( initialization of heat generation accumulator ))))))))
C
     QAETA=0.0
     QAETF=0.0
     QAMNA=0.0
     QAMNF=0.0
     QA3A =0.0
     QA3S =0.0
     QAAF =0.0
     QA2S =0.0
   QASG =0.0
   QAFA =0.0
```

```
C
      IDED=1
      SLL=SLLDED
C
C ----------------- continued later
C
```

Material data and the properties of cement and concrete are defined in the DATA statement. Pure ordinary Portland cement concrete is set as default. Of course, it is possible to input the material constants from a data file or through subroutine parameters given by the main frame to include in this program. The following are definitions of the constants.

- Concrete mixture data:
  WC = unit cement content of concrete (kg/m^3), WP = initial water-to-powder ratio by weight (percentage)
- Cement chemical component data:
  P3A:P3S:P4AF:P2S:PPCS2H = weight percentage of OPC minerals and mono-sulphate, PPC:PSG:PFA:PLS = weight percentage of OPC, slag, fly ash and non-reactive materials
- Physical and chemical constants of raw materials:
  BLN, BLNSG, BLNFA, BLNLS = Blaine values of OPC and pozzolans (cm^2/g), SGCS2H, FACS2H = weight percentage of mono-sulphate in slag and fly ash (%), QSP, QSPAD, CHARSP = constants for the effect of superplasticizer dosage, QSGMX, QFAMX = final accumulated heat generation of slag and fly ash (kcal/kg), RSGW1, RFAW1 = weight percentage of consumed water of slag and fly ash (%), RSGCA, RFACA = weight ratio of consumed $Ca(OH)_2$ when reaction of slag and fly ash proceeds, RH3AMN, RHAFMN = mono-sulphate conversion factors, SLLDED = defaulted time duration at initial stage (day), ALPHA = model constant for cluster.

C.2 Heat of hydration computation of mineral compounds

In the second segment, the heat generation of ettringite supplied from aluminate and ferrite is computed. This term is closely related to the temperature computation of concrete structures at very early age.

In this program, variables HTXX and HSXX (kcal $kg^{-1} h^{-1}$) indicate the heat generation rate of each mineral component-XX and their reference heat generation rate which corresponds to the reference temperature denoted by BTMXX (degree centigrade). The reference heat generation rate is a function of the accumulated heat which represents the path dependence of the hydration process, and the sensitivity of temperature on the rate of hydration is formulated based on activation energy by denoted by ZXX. These values are not user definition constants but default ones.

```
 200 CONTINUE
C
C ==========================================
C ============  ETTRINGITE MODEL  ===========
C ==========================================
C
C  ===== C3A ETTRINGITE (GYPSUM-2-HYDRATE) =====
C
C
    DATA BTMETA/20.0/,ZETA/-6500/
C
    TEMETA=TEMP
C
     IF(QAETA.LT.3.94) THEN
        HSETA=TQHS(QAETA,0.0,80.00,3.94,20.00)
      ELSE IF(QAETA.LT.31.52) THEN
        HSETA=TQHS(QAETA,3.94,20.00,31.52,8.00)
      ELSE IF(QAETA.LT.78.8) THEN
        HSETA=TQHS(QAETA,31.52,8.00,78.8,4.00)
      ELSE IF(QAETA.LT.197.0) THEN
        HSETA=TQHS(QAETA,78.8,4.00,197.0,2.000)
      ELSE IF(QAETA.LT.394.0) THEN
        HSETA=TQHS(QAETA,197.0,2.000,394.0,0.0)
      ELSE
        HSETA=0.0
     END IF
C
    IF(HSETA.LT.0.0) HSETA=0.0
C
    HTETA=HSETA*RBLN3A
   \           *EXP(ZETA*(1.0/(TEMETA+273.0) - 1.0/(BTMETA+273.0)))
C
    QGETA=HTETA*SLL*24.0
    IF(QGETA.LE.0.0) GO TO 500
C
    QETAAD=QAETA+QGETA
    IF(QETAAD.GT.394.0) THEN
      RRETA=(394.0-QAETA)/QGETA
      HTETA=RRETA*HTETA
      QGETA=RRETA*QGETA
    ENDIF
C
 500 DHTETA=HTETA*P3A/100*PPC/100
C
C
C  ===== C4AF ETTRINGITE (GYPSUM-2-HYDRATE) =====
C
C
    DATA BTMETF/20.0/,ZETF/-4000/
C
    TEMETF=TEMP
C
     IF(QAETF.LT.1.0) THEN
       HSETF=TQHS(QAETF,0.0,13.54,1.0,3.384)
      ELSE IF(QAETF.LT.8.0) THEN
       HSETF=TQHS(QAETF,1.0,3.384,8.0,1.354)
      ELSE IF(QAETF.LT.20.0) THEN
       HSETF=TQHS(QAETF,8.0,1.354,20.0,0.677)
```

```
      ELSE IF(QAETF.LT.50.0) THEN
       HSETF=TQHS(QAETF,20.0,0.677,50.0,0.338)
      ELSE IF(QAETF.LT.100.0) THEN
       HSETF=TQHS(QAETF,50.0,0.338,100.0,0.0)
      ELSE
       HSETF=0.0
     END IF
C
    IF(HSETF.LT.0.0) HSETF=0.0
C
    HTETF=HSETF*RBLNAF
   \      *EXP(ZETF*(1.0/(TEMETF+273.0) - 1.0/(BTMETF+273.0)))
C
    QGETF=HTETF*SLL*24.0
    IF(QGETF.LE.0.0) GO TO 600
C
    QETFAD=QAETF+QGETF
    IF(QETFAD.GT.100.0) THEN
     RRETF=(100.0-QAETF)/QGETF
     HTETF=RRETF*HTETF
     QGETF=RRETF*QGETF
    ENDIF
C
  600 DHTETF=HTETF*P4AF/100*PPC/100
C
C
C ===== TOTAL ETTRINGITE =====
C
C
    RQETA=QGETA/394.0*P3A*PPC/100/270.2*(172.182*3)
    RQETF=QGETF/100.0*P4AF*PPC/100/(485.92/2)*(172.182*3)
   RQCSET=RQETA+RQETF
C
    IF(RQCSET.LE.0.0) GO TO 800
C
    RCSET=FRCSET/RQCSET
    IF(RCSET.GT.1.0) RCSET=1.0
    IF(RCSET.LT.0.0) RCSET=0.0
C
    GO TO 900
C
  800 RCSET=0.0
C
  900 QGETA =QGETA *RCSET
      QGETF =QGETF *RCSET
      DHTETF=DHTETF*RCSET
      DHTETA=DHTETA*RCSET
      RQCSET=RQCSET*RCSET
C
C ---------------------continued later
C
```

C.3 Interaction of hydration process among constituent minerals

The hydration of the cement constituent minerals proceeds concurrently in the thermodynamic environment of concrete. So, the hydration of some minerals may affect the simultaneous hydration process of other minerals. This means that the overall reaction of cement is not a simple summation of each pure hydrating heat generation process, but is of mutually interrelated complexity.

In this third segment, these interactive factors are first computed with respect to the free water denoted by FREEW and FRW and cluster thickness of each mineral denoted by THCKXX. The content of free water represents the residual space in which newly produced hydrate can be dissolved and crystallized with gel. Finally, the interactive control parameters FIXX among each component are evaluated and used for modifying the reference heat generation rate. For computing the free water, the consumed water trapped inside the hydrated crystal and gel, indicated by SWXX, is calculated, based on the mineral chemical reaction equation.

```
C
C    ============================================
C    ========== HYDRATION HEAT MODEL =============
C    ============================================
C
 1000 SWMNA=QAMNA/207*P3A/100*PPC*(0.6668+0.15)
      SWMNF=QAMNF/100*P4AF/100*PPC*(0.7414+0.15)
      SW3A =(QA3A-QAMNA)/207*P3A/100*PPC*(0.4001+0.15)
      SWAF =(QAAF-QAMNF)/100*P4AF/100*PPC*(0.3707+0.15)
      SW3S =QA3S/120*P3S/100*PPC*(0.2367+0.15)
      SW2S =QA2S/62*P2S/100*PPC*(0.2092+0.15)
      SWSG =QASG/(QSGMX/RSGMX)*PSG*(RSGW1+0.15)
      SWFA =QAFA/(QFAMX/RFAMX)*PFA*(RFAW1+0.15)
C
      FREEW=WP-SWMNA-SWMNF-SW3A-SWAF-SW3S-SW2S-SWSG-SWFA
C
      FREEWO=FREEW/((PPC+PSG+PFA)/(PPC+PSG+PFA+PLS))
C
C
C     ========================================================
C     COMPUTING THICKNESS OF CEMENT HYDRATE CLUSTER AROUND POWDERS
C     ========================================================
C
      TEI3A=100-QA3A/207*100
       IF(TEI3A.LT.0) TEI3A=0
      TEIAF=100-QAAF/100*100
       IF(TEIAF.LT.0) TEIAF=0
      TEI3S=100-QA3S/120*100
       IF(TEI3S.LT.0) TEI3S=0
      TEI2S=100-QA2S/62*100
       IF(TEI2S.LT.0) TEI2S=0
      TEISG=100-QASG/(QSGMX/RSGMX)*100
       IF(TEISG.LT.0) TEISG=0
      TEIFA=100-QAFA/(QFAMX/RFAMX)*100
       IF(TEIFA.LT.0) TEIFA=0
```

```
C
      THCK3A=(100-(10000*TEI3A)**0.333333)
      THCKAF=(100-(10000*TEIAF)**0.333333)
      THCK3S=(100-(10000*TEI3S)**0.333333)
      THCK2S=(100-(10000*TEI2S)**0.333333)
      THCKSG=(100-(10000*TEISG)**0.333333)
      THCKFA=(100-(10000*TEIFA)**0.333333)
       IF(THCK3A.LT.1.0) THCK3A=1.0
       IF(THCKAF.LT.1.0) THCKAF=1.0
       IF(THCK3S.LT.1.0) THCK3S=1.0
       IF(THCK2S.LT.1.0) THCK2S=1.0
       IF(THCKSG.LT.1.0) THCKSG=1.0
       IF(THCKFA.LT.1.0) THCKFA=1.0
C
 1200 CONTINUE
C
      IF(IEFW.EQ.0) THEN
        FRW=FREEWO
       ELSE
        FRW=FREEWN
      END IF
C
      PI3A=FRW/(THCK3A**ALPHA)/RBLN3A**0.5
      PIAF=FRW/(THCKAF**ALPHA)/RBLNAF**0.5
      PI3S=FRW/(THCK3S**ALPHA)/RBLN3S**0.5
      PI2S=FRW/(THCK2S**ALPHA)/RBLN2S**0.5
      PISG=FRW/(THCKSG**ALPHA)/RBLNSG**0.5
      PIFA=FRW/(THCKFA**ALPHA)/RBLNFA**0.5
C
      FI3A=1-EXP(-5.0*PI3A**2.4)
       IF(PI3A.LT.0.0) FI3A=0.0
      FIAF=1-EXP(-5.0*PIAF**2.4)
       IF(PIAF.LT.0.0) FIAF=0.0
      FI3S=1-EXP(-5.0*PI3S**2.4)
       IF(PI3S.LT.0.0) FI3S=0.0
      FI2S=1-EXP(-5.0*PI2S**2.4)
       IF(PI2S.LT.0.0) FI2S=0.0
      FISG=1-EXP(-5.0*PISG**2.4)
       IF(PISG.LT.0.0) FISG=0.0
      FIFA=1-EXP(-5.0*PIFA**2.4)
       IF(PIFA.LT.0.0) FIFA=0.0
C
C    ----------------------continue later
C
```

C.4 OPC compound hydration routine

By using the modification factors and interaction index, the heat of hydration rate of each constituent mineral of ordinary Portland cement is calculated as below. These hydrate reactions are the major ones for micropore structural formation, solidification and the development of strength. The basic scheme of computation is similar to that of the ettringite model stated above.

The reference heat rate of minerals is here defined by a piecewise linear function with respect to the accumulated heat. The function TQHS gives intermediate values between specified points. In future, the relation of heat rate and the accumulated heat should be formulated with smooth analytical functions of continuity. The accumulated heat of each mineral normalized by the final heat generation at which the heat generation terminates is regarded as an indicator of the degree of hydration.

```
C
C
C  ===== C3A (MONOSULFATE & HYDRATION) =====
C
     DATA BTM3A/20.0/,ZC3A/-6500/
C
     TEM3A=TEMP
C
        IF(QA3A.LT.2.070) THEN
            HS3A=TQHS(QA3A,0.0,1.3800,2.070,1.3800)
         ELSE IF(QA3A.LT.6.21) THEN
            HS3A=TQHS(QA3A,2.070,1.3800,6.21,5.000)
         ELSE IF(QA3A.LT.16.56) THEN
            HS3A=TQHS(QA3A,6.21,5.000,16.56,5.000)
         ELSE IF(QA3A.LT.41.40) THEN
           HS3A=TQHS(QA3A,16.56,5.000,41.40,3.105)
         ELSE IF(QA3A.LT.103.5) THEN
           HS3A=TQHS(QA3A,41.40,3.105,103.5,1.656)
         ELSE IF(QA3A.LT.207.0) THEN
           HS3A=TQHS(QA3A,103.5,1.656,207.0,0.0)
         ELSE
           HS3A=0.0
      END IF
C
     IF(HS3A.LT.0.0) HS3A=0.0
C
     HT3A=HS3A*RBLN3A
    \          *EXP(ZC3A*(1.0/(TEM3A+273.0) - 1.0/(BTM3A+273.0)))
C
     IF(QA3A.LT.2.07) THEN
       RBG3A=EXP(-2.0*RSP)
     END IF
C
     IF(FRCSMN.GT.0.0) RCM3A=1.0
C
     IF(RCM3A.GT.1.0) GO TO 1500
     IF(RCM3A.LT.FI3A) THEN
       FI3A =1.0
      ELSE
       RCM3A=1.0
     END IF
C
  1500 RDCE3A=FI3A*RCM3A
C
     QG3A=HT3A*(1-RCSET)*RDCE3A*RBG3A*SLL*24.0
     IF(QG3A.LE.0.0) GO TO 2000
```

```
C
    Q3AD=QA3A+QG3A
    IF(Q3AD.GT.207) THEN
      RR3A=(207-QA3A)/QG3A
      HT3A=RR3A*HT3A
      QG3A=RR3A*QG3A
    ENDIF
C
 2000 DHT3A=P3A/100*(HT3A*(1-RCSET)*RDCE3A*RBG3A)*PPC/100
C
C
C ===== C4AF (MONOSULFATE & HYDRATION) =====
C
    DATA BTMAF/20.0/,ZC4AF/-4000/
C
    TEMAF=TEMP
C
     IF(QAAF.LT.1.0) THEN
        HSAF=TQHS(QAAF,0.0,0.6667,1.0,0.6667)
      ELSE IF(QAAF.LT.3.0) THEN
        HSAF=TQHS(QAAF,1.0,0.6667,3.0,1.610)
      ELSE IF(QAAF.LT.8.0) THEN
        HSAF=TQHS(QAAF,3.0,1.610,8.0,1.610)
      ELSE IF(QAAF.LT.20.0) THEN
        HSAF=TQHS(QAAF,8.0,1.610,20.0,1.0)
      ELSE IF(QAAF.LT.50.0) THEN
        HSAF=TQHS(QAAF,20.0,1.0,50.0,0.533)
      ELSE IF(QAAF.LT.100.0) THEN
        HSAF=TQHS(QAAF,50.0,0.533,100.0,0.0)
      ELSE
        HSAF=0.0
     END IF
C
    IF(HSAF.LT.0.0) HSAF=0.0
C
    HTAF=HSAF*RBLNAF
  \         *EXP(ZC4AF*(1.0/(TEMAF+273.0) - 1.0/(BTMAF+273.0)))
C
    IF(QAAF.LT.1.0) THEN
      RBGAF=EXP(-2.0*RSP)
    END IF
C
    IF(FRCSMN.GT.0.0) RCMAF=1.0
C
    IF(RCMAF.GT.1.0) GO TO 2300
    IF(RCMAF.LT.FIAF) THEN
      FIAF=1.0
    ELSE
      RCMAF=1.0
    END IF
C
 2300 RDCEAF=FIAF*RCMAF
C
    QGAF=HTAF*(1-RCSET)*RDCEAF*RBGAF*SLL*24.0
    IF(QGAF.LE.0.0) GO TO 2400
```

```
C
    QAFD=QAAF+QGAF
    IF(QAFD.GT.100) THEN
      RRAF=(100-QAAF)/QGAF
      HTAF=RRAF*HTAF
      QGAF=RRAF*QGAF
    ENDIF
C
 2400 DHTAF=P4AF/100*(HTAF*(1-RCSET)*RDCEAF*RBGAF)*PPC/100
C
C
C ====== TOTAL MONOSULFATE & HYDRATION ======
C
    QGMNA=QG3A/RH3AMN
    QGMNF=QGAF/RHAFMN
C
    RQMNA =QGMNA/207*P3A*PPC/100/270.2*172.182
    RQMNF =QGMNF/100*P4AF*PPC/100/242.99*172.182
    RQCSMN=RQMNA+RQMNF
C
    IF(RQCSMN.LE.0.0) GO TO 2500
C
    RCSMN=FRCSMN/RQCSMN
    IF(RCSMN.GT.1.0) RCSMN=1.0
    IF(RCSMN.LT.0.0) RCSMN=0.0
C
    GO TO 2600
C
 2500 RCSMN=0.0
C
 2600 QGMNA=QGMNA*RCSMN
      QGMNF=QGMNF*RCSMN
      QG3A =QG3A*(RCSMN/RH3AMN+(1-RCSMN))
      QGAF =QGAF*(RCSMN/RHAFMN+(1-RCSMN))
      RQCSMN=RQCSMN*RCSMN
C
C
C ============ C3S =============
C
    DATA BTM3S/20.0/
C
 3000 TEM3S=TEMP
C
     IF(QA3S.LT.1.2) THEN
        HS3S=TQHS(QA3S,0.0,0.8,1.2,0.8)
      ELSE IF(QA3S.LT.7.5) THEN
        HS3S=TQHS(QA3S,1.2,0.8,7.5,3.0)
      ELSE IF(QA3S.LT.18.0) THEN
        HS3S=TQHS(QA3S,7.5,3.0,18.0,3.0)
      ELSE IF(QA3S.LT.30.0) THEN
        HS3S=TQHS(QA3S,18.0,3.0,30.0,1.5)
      ELSE IF(QA3S.LT.60.0) THEN
        HS3S=TQHS(QA3S,30.0,1.5,60.0,0.8)
      ELSE IF(QA3S.LT.120.0) THEN
        HS3S=TQHS(QA3S,60.00,0.80,120.0,0.0)
      ELSE
        HS3S=0.0
     END IF
```

```
C
      IF(HS3S.LT.0.0) HS3S=0.0
C
      IF(QA3S.LT.1.2) THEN
         ZC3S=TQZ(QA3S,0.-5000,0,1.2,-5000)
       ELSE IF(QA3S.LT.7.5) THEN
         ZC3S=TQZ(QA3S,1.2,-5000,7.5,-6000)
       ELSE IF(QA3S.LT.18.0) THEN
         ZC3S=TQZ(QA3S,7.5,-6000,18.0,-6000)
       ELSE IF(QA3S.LT.30.0) THEN
         ZC3S=TQZ(QA3S,18.0,-6000,30.0,-5000)
       ELSE
         ZC3S=-5000
      END IF
C
      IF(QLV3S.LT.15.0) THEN
        RBLN3S=1.0
       ELSE IF(QLV3S.LT.25.0) THEN
        RBLN3S=((RBLN3S-1.0)/(25.0-15.0))*(QLV3S-15.0)+1.0
      END IF
C
      HT3S=HS3S*RBLN3S
     \         *EXP(ZC3S*(1.0/(TEM3S+273.0) - 1.0/(BTM3S+273.0)))
C
      IF(QA3S.LT.1.20) THEN
        RBG3S=EXP(-2.0*RSP)
      END IF
C
      IF(QLV3S.LT.15.0) THEN
        RCM3S=1.0
       ELSE IF(QLV3S.LT.25.0) THEN
        RCM3S=((RCM3S-1.0)/(25.0-15.0))*(QLV3S-15.0)+1.0
      END IF
C
      IF(RCM3S.GT.1.0) GO TO 3500
      IF(RCM3S.LT.FI3S) THEN
        FI3S=1.0
       ELSE
        RCM3S=1.0
      END IF
C
 3500 RDCE3S=FI3S*RCM3S
C
      QG3S=HT3S*RDCE3S*RBG3S*SLL*24.0
      IF(QG3S.LE.0.0) GO TO 3600
C
      Q3SD=QA3S+QG3S
      IF(Q3SD.GT.120) THEN
        RR3S=(120-QA3S)/QG3S
        HT3S=RR3S*HT3S
        QG3S=RR3S*QG3S
      ENDIF
C
 3600 DHT3S=P3S/100*(HT3S*RDCE3S*RBG3S)*PPC/100
C
```

```
C
C ==================== C2S ==================
C
    DATA BTM2S/20.0/
C
    TEM2S=TEMP
C
     IF(QA2S.LT.0.62) THEN
        HS2S=TQHS(QA2S,0.0,0.4133,0.62,0.4133)
      ELSE IF(QA2S.LT.4.65) THEN
        HS2S=TQHS(QA2S,0.62,0.4133,4.65,0.861)
      ELSE IF(QA2S.LT.11.2) THEN
        HS2S=TQHS(QA2S,4.65,0.861,11.2,0.861)
      ELSE IF(QA2S.LT.18.6) THEN
        HS2S=TQHS(QA2S,11.2,0.861,18.6,0.3513)
      ELSE IF(QA2S.LT.31.0) THEN
        HS2S=TQHS(QA2S,18.6,0.3513,31.0,0.2067)
      ELSE IF(QA2S.LT.62.0) THEN
        HS2S=TQHS(QA2S,31.0,0.2067,62.0,0.0)
      ELSE
        HS2S=0.0
     END IF
C
    IF(HS2S.LT.0.0) HS2S=0.0
C
    IF(QA2S.LT.0.62) THEN
       ZC2S=TQZ(QA2S,0.0,-2500,0.62,-2500)
     ELSE IF(QA2S.LT.4.65) THEN
       ZC2S=TQZ(QA2S,0.62,-2500,4.65,-5000)
     ELSE IF(QA2S.LT.11.2) THEN
       ZC2S=TQZ(QA2S,4.65,-5000,11.2,-5000)
     ELSE IF(QA2S.LT.18.6) THEN
       ZC2S=TQZ(QA2S,11.2,-5000,18.6,-2500)
     ELSE
       ZC2S=-2500
    END IF
C
    IF(QLV2S.LT.18.0) THEN
       RBLN2S=1.0
     ELSE IF(QLV2S.LT.30.0) THEN
       RBLN2S=((RBLN2S-1.0)/(30.0-18.0))*(QLV2S-18.0)+1.0
    END IF
C
    HT2S=HS2S*RBLN2S
   \        *EXP(ZC2S*(1.0/(TEM2S+273.0) - 1.0/(BTM2S+273.0)))
C
    IF(QA2S.LT.0.62) THEN
       RBG2S=EXP(-2.0*RSP)
    END IF
C
    IF(QLV2S.LT.18.0) THEN
       RCM2S=1.0
     ELSE IF(QLV2S.LT.30.0) THEN
      RCM2S=((RCM2S-1.0)/(30.0-18.0))*(QLV2S-18.0)+1.0
    END IF
```

```
C
      IF(RCM2S.GT.1.0) GO TO 4000
      IF(RCM2S.LT.FI2S) THEN
         FI2S=1.0
       ELSE
         RCM2S=1.0
      END IF
C
 4000 RDCE2S=FI2S*RCM2S
C
C
      QG2S=HT2S*RDCE2S*RBG2S*SLL*24.0
      IF(QG2S.LE.0.0) GO TO 4100
C
      Q2SD=QA2S+QG2S
      IF(Q2SD.GT.62) THEN
         RR2S=(62-QA2S)/QG2S
         HT2S=RR2S*HT2S
         QG2S=RR2S*QG2S
      ENDIF
C
C
 4100 DHT2S=P2S/100*(HT2S*RDCE2S*RBG2S)*PPC/100
C
C     ---------------------continued later
C
```

C.5 Pozzolan hydration routine (slag and fly ash)

Pozzolans are effective for designing low heat-cementitious powder used for massive concrete. In this program, slag and fly ash reactions are treated separately from the OPC routine, but some interactions are taken into account through the free water content and calcium hydroxide concentration in the pore water. Furthermore, mutual interaction between slag and fly ash is considered with respect to consumption of pore water.

```
C
C
C     ================= SLAG ===================
C
      DATA BTMSG/20.0/,ZSG/-4500/
C
      TEMSG=TEMP
C
        IF(QASG.LT.1.10) THEN
           HSSG=TQHS(QASG,0.0,0.733,1.10,0.733)
         ELSE IF(QASG.LT.6.875) THEN
           HSSG=TQHS(QASG,1.10,0.733,6.875,2.750)
         ELSE IF(QASG.LT.16.5) THEN
           HSSG=TQHS(QASG,6.875,2.750,16.5,2.750)
         ELSE IF(QASG.LT.27.5) THEN
           HSSG=TQHS(QASG,16.5,2.750,27.5,0.4125)
        ELSE IF(QASG.LT.55.0) THEN
           HSSG=TQHS(QASG,27.5,0.4125,55.0,0.220)
```

```
       ELSE IF(QASG.LT.110.0) THEN
         HSSG=TQHS(QASG,55.0,0.220,110.0,0.0)
       ELSE
         HSSG=0.0
      END IF
C
     IF(HSSG.LT.0.0) HSSG=0.0
C
     QLVSG=(QASG/(QSGMX/RSGMX))*100
     IF(QLVSG.LT.15.0) THEN
        RBLNSG=1.0
      ELSE IF(QLVSG.LT.25.0) THEN
        RBLNSG=((RBLNSG-1.0)/(25.0-15.0))*(QLVSG-15.0)+1.0
     END IF
C
     HTSG=HSSG*RBLNSG
    \         *EXP(ZSG*(1.0/(TEMSG+273.0) - 1.0/(BTMSG+273.0)))
C
     IF(QASG.LT.1.1) THEN
        RBGSG=EXP(-2.0*RSP)
     END IF
C
C
C ================== FLY ASH ======================
C
     DATA BTMFA/20.0/,ZFA/-8000/
C
     TEMFA=TEMP
C
      IF(QAFA.LT.0.50) THEN
         HSFA=TQHS(QAFA,0.0,0.005,0.50,0.005)
       ELSE IF(QAFA.LT.2.50) THEN
         HSFA=TQHS(QAFA,0.50,0.005,2.50,0.0625)
       ELSE IF(QAFA.LT.9.00) THEN
         HSFA=TQHS(QAFA,2.50,0.0625,9.00,0.0625)
       ELSE IF(QAFA.LT.15.0) THEN
         HSFA=TQHS(QAFA,9.00,0.0625,15.0,0.04688)
       ELSE IF(QAFA.LT.25.0) THEN
         HSFA=TQHS(QAFA,15.0,0.04688,25.0,0.02188)
       ELSE IF(QAFA.LT.42.5) THEN
         HSFA=TQHS(QAFA,25.0,0.02188,42.5,0.00531)
       ELSE IF(QAFA.LT.50.0) THEN
         HSFA=TQHS(QAFA,42.5,0.00531,50.0,0.0)
       ELSE
         HSFA=0.0
      END IF
C
     IF(HSFA.LT.0.0) HSFA=0.0
C
     HTFA=HSFA*RBLNFA
    \         *EXP(ZFA*(1.0/(TEMFA+273.0) - 1.0/(BTMFA+273.0)))
C
     IF(QAFA.LT.1.5) THEN
        RBGFA=EXP(-0.0*RSP)
     END IF
C
 4200 CONTINUE
C
```

```
C
C ================ SLAG & FLY ASH ===================
C
     RQCASG=((HTSG*RBGSG*SLL*24.0)/(QSGMX/RSGMX)*PSG)*RSGCA
     RQCAFA=((HTFA*RBGFA*SLL*24.0)/(QFAMX/RFAMX)*PFA)*RFACA
     RQCAOH=RQCASG+RQCAFA
C
     IF(RQCAOH.GT.0.0) THEN
        RCOH=CAOH2/RQCAOH
      ELSE
        RCOH=10.0
      END IF
C
     IF(RCOH.LT.0.0) RCOH=0.0
C
C    ----------------------------------------------------
C    ---------------------- SLAG -----------------------
C    ----------------------------------------------------
     RCOHSG=RCOH
     RSGCAV=RSGCA
C
     RSGRED=1-EXP(-2.0*RCOHSG**5.0)
     IF(RQCASG.EQ.0.0) RSGRED=1.0
C
     IF(RSGRED.LT.FISG) THEN
        RDSG=RSGRED
      ELSE
        RDSG=FISG
     END IF
C
     IF(RDSG.LT.RBGSG) THEN
        RBGSG=1.0
      ELSE
        RDSG=1.0
     END IF
C
     QGSG=HTSG*RDSG*RBGSG*SLL*24.0
     IF(QGSG.LE.0.0) GO TO 4600
C
     QSGD=QASG+QGSG
     IF(QSGD.GT.QSGMX) THEN
        RRSG=(QSGMX-QASG)/QGSG
        HTSG=RRSG*HTSG
        QGSG=RRSG*QGSG
     ENDIF
C
 4600 DHTSG=(HTSG*RSGMX)*RDSG*RBGSG*PSG/100
      IF(PSG.EQ.0.0) QGSG=0.0
C
C    -----------------------------------------------------
C    --------------------- FLY ASH ---------------------
C    -----------------------------------------------------
     RCOHFA=RCOH
     RFACAV=RFACA
C
     RFARED=1-EXP(-2.0*RCOHFA**5.0)
     IF(RQCAFA.EQ.0.0) RFARED=1.0
```

```
C
      IF(RFARED.LT.FIFA) THEN
         RDFA=RFARED
       ELSE
         RDFA=FIFA
      END IF
C
      IF(RDFA.LT.RBGFA) THEN
         RBGFA=1.0
       ELSE
         RDFA=1.0
      END IF
C
      QGFA=HTFA*RDFA*RBGFA*SLL*24.0
      DHTFA=(HTFA*RFAMX)*RDFA*RBGFA*PFA/100
      IF(PFA.EQ.0.0) QGFA=0.0
C  --------------------------------------------------
C  ----------------- EFFECTIVE WATER -----------------
C  --------------------------------------------------
      IF(IEFW.EQ.1) GO TO 4800
C
      IF(RSGRED.GE.FISG) THEN
         RSGEFW=1.0
       ELSE IF(FISG.GT.0.0) THEN
         RSGEFW=RSGRED/FISG
       ELSE
         RSGEFW=1.0
      END IF
      IF(RSGEFW.LT.0.0) RSGEFW=0.0
C
      IF(RFARED.GE.FIFA) THEN
         RFAEFW=1.0
       ELSE IF(FIFA.GT.0.0) THEN
         RFAEFW=RFARED/FIFA
       ELSE
        RFAEFW=1.0
      END IF
      IF(RFAEFW.LT.0.0) RFAEFW=0.0
C
      FREEWN=FREEW/((PPC+RSGEFW*PSG+RFAEFW*PFA)/(PPC+PSG+PFA+PLS))
C
      IEFW=1
      GO TO 1200
C -----------------------------------------------------------
C ---------- CALCULATION OF CALCIUM HYDRO-OXIDE LEFT ----------
C -----------------------------------------------------------
C
 4800 DCA3S=QG3S/120*P3S/100*PPC*0.4868
      DCA2S=QG2S/62*P2S/100*PPC*0.2151
      DCAAF=QGAF*(1-RCSMN)/100*P4AF/100*PPC*0.3049
      DCASG=QGSG/(QSGMX/RSGMX)*PSG*RSGCAV
      DCAFA=QGFA/(QFAMX/RFAMX)*PFA*RFACAV
C
      DCAOH2=DCA3S+DCA2S-DCAAF-DCASG-DCAFA
C
C     ---------------------continued later
C
```

C.6 Total heat of hydration of OPC and mixed cement

Finally, the hydration rate of each mineral component, including pozzolans, is accumulated as for the cement heat generation in concrete as a whole. For the cement-based formulation, the dimension of time is the hour, but the final output variables are expressed in days since the thermal stress analysis is conducted over several months.

The effect of superplasticizer and other chemical admixture agents is taken into account in terms of RBGXX. The retarded hydration by the chemical admixture is an observed fact but its chemico-physical mechanism is still under investigation. Then, this program adopts a simple formulation as a tentative modelling for the chemical admixture.

```
C
C ===========================================================
C ================ TOTAL HEAT GENERATION ==================
C ===========================================================
C
      IF(IDED.NE.1) GO TO 5000
      QAETA=QGETA
      QAETF=QGETF
      QAMNA=QGMNA
      QAMNF=QGMNF
      QA3A =QG3A
      QAAF =QGAF
      QA3S =QG3S
      QA2S =QG2S
      QASG =QGSG
      QAFA =QGFA
      IDED =0
      SLL=TIME1-TIME
      FRCSET=FRCSET-RQCSET
      FRCSMN=FRCSMN-RQCSMN
      GO TO 200
C
 5000 HTETRG=DHTETA+DHTETF
         HTCL=DHT3A+DHT3S+DHTAF+DHT2S
C
      HT=HTETRG+HTCL+DHTSG+DHTFA
      QGEN=HT*SLL*24.0
      QR=HT*WC*24.0
C
 5300 CONTINUE
C
 5500 RETURN
      END
C
C
      FUNCTION TQHS(QA,VQ1,VS1,VQ2,VS2)
C
C     ----------FUNCTION FOR HEAT GENERATION RATE----------
C
        TQHS=((VS2-VS1)/(VQ2-VQ1))*(QA-VQ1)+VS1
      RETURN
      END
```

```
C
C
      FUNCTION TQZ(QA,VQ1,NZ1,VQ2,NZ2)
C
C       ----------FUNCTION FOR ACTIVATION ENERGY---------
C
        TQZ=((NZ2-NZ1)/(VQ2-VQ1))*(QA-VQ1)+NZ1
      RETURN
      END
C
C
```

C.7 Adiabatic temperature rise

The subroutine HEAT is multi-purpose, and the hydration of cement in concrete can be simulated under any temperature history. The adiabatic temperature rise condition can be also covered by this program. In this thermally isolated condition, the temperature and heat generation are not independent but closely associated.

For computing the adiabatic temperature rise, the program HEAT is first called and the heat generation during the specified time interval is obtained. This heat energy is fully used for elevating the temperature of the concrete. The elevated temperatures become the revised input data for the next heat generation of cement. This routine is cycled step by step. Since this scheme is an explicit one, the time interval should be small enough to obtain convergence of the computation.

Appendix D: BET adsorption model

The model of isotherms of moisture retention discussed in Chapter 4 uses a modified BET model (named after S. Brunauer, P.H. Emmet and E. Teller) of water adsorption to describe the combined condensation–adsorption phenomenon of pore water in a cementitious microstructure. The relationship given by equation (4.24) is also known as the BET model for adsorption over a plane surface. Of course, to consider adsorption in micropores, a modification to the original model is required so as to incorporate the effect of the shape of the pores. In this section an abbreviated and basic derivation of the BET model is given and the assumptions involved are discussed.

A schematic representation of adsorption of water molecules on a planar surface is given in Figure D.1. Here V_1, V_2, etc., represent the amount of water adsorbed in each layer. A fully adsorbed layer has a volume given by V_m. At any instant, the situation as depicted in Figure D.1 is not static. The molecules are in a dynamic situation, with some leaving the adsorbed layer, while others get caught up. Under constant conditions of temperature and humidity, the situation is quasiconstant, and the number of molecules leaving a surface equals the number of molecules being caught up in a particular layer. The original BET theory makes some basic assumptions to

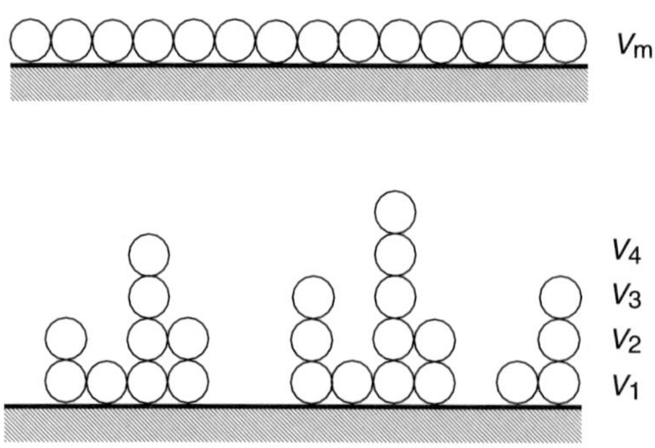

Figure D.1 Adsorption of water molecules on a plane surface.

describe the volume of molecules adsorbed in each layer considering this dynamic situation.

These basic assumptions can be stated as:

- The number of molecules leaving a layer is proportional to the directly exposed area of the layer (the area not covered by the next layer above).
- The number of molecules being attached in a layer is proportional to the the noncovered area of the preceding layer.
- The number of molecules being attached to a layer is proportional to the number of molecules in the gas phase, i.e. to the relative humidity h.

From these assumptions the following relationships can be obtained for layers 1 to n, where k_1, k_2 are the constants of proportionality.

$$V_1 - V_2 = k_1(V_m - V_1)h \tag{D.1}$$

$$V_2 - V_3 = k_2(V_1 - V_2)h \tag{D.2}$$

$$V_n - V_{n+1} = k_n(V_{n-1} - V_n)h \tag{D.3}$$

The adsorption of layers 2 to n takes place on a layer of water molecules, whereas layer 1 is adsorbed on a solid surface. It can be therefore reasonably be assumed that constant k is the same for layers 2 to n except for layer 1, i.e.

$$k_n = k_2 \qquad n \geqslant 2 \tag{D.4}$$

Applying equation (D.4) to equations (D.2) and (D.3) and the addition of all layers 2 to n gives

$$V_2 - V_{n+1} = k_2(V_1 - V_n)h \tag{D.5}$$

For n tending to infinity, V_n approaches zero, therefore

$$V_2 = k_2 V_1 h \tag{D.6}$$

Addition of layers 3 to n, and so on up to layers n, in the same way gives

$$V_3 = k_2 V_2 h = V_1(k_2 h)^2 \qquad V_n = V_1(k_2 h)^{n-1} \tag{D.7}$$

The total amount of adsorbed water V_a is the sum of the water adsorbed in each layer. Summation of V_1 to V_n, where n is tending to infinity, gives

$$V_a = V_1[1 + k_2 h + (k_2 h)^2 + \cdots] = \frac{V_1}{1 - k_2 h} \tag{D.8}$$

From equations (D.8) and (D.1) we get

$$V_1 = \frac{k_1 V_m h}{1 - k_2 h + k_1 h} \tag{D.9}$$

and therefore, (from equations (D.9) and (D.8))

$$V_a = \frac{k_1 V_m h}{(1 - k_2 h)(1 - k_2 h + k_1 h)} \tag{D.10}$$

When condensation over the plane surface occurs at a particular relative humidity, h_m, all the layers are fully filled. In other words, $V_1 = V_2$ for $h = h_m$. Applying this condition to equation (D.6) yields

$$k_2 = 1/h_m \tag{D.11}$$

Therefore, equation (D.10) can be rewritten as

$$V_a = \frac{k_1 V_m h}{(1 - h/h_m)(1 - h/h_m + k_1 h)} \tag{D.12}$$

For a planar surface, BET theory assumes $h_m = 1$, corresponding to pure water condensation over a plane surface. Application of this condition to equation (D.12) gives the classical BET theory. In the modified adsorption theory used in Chapter 4, the relative humidity, h_m, at which complete condensation of a micropore occurs is obtained by considering the curvature of the water surface as given by the Kelvin equation. As a result, h_m would be dependent on the pore size and be different for pores of different size. This issue is discussed more fully in Chapter 4. The average number of completely filled adsorbed layers at a particular relative humidity is given by the ratio of total adsorbed water, V_a, to the volume of water in one completely filled layer, V_m. The product of this ratio and the thickness of one layer of water molecule (about 0.35 nm) would give the thickness of adsorbed water over a surface. The constant k_1 is a material constant and usually obtained experimentally. For silicate materials, $k_1 = 15$ has been reported to be a reasonable value. Thus, thickness of the adsorbed layer of water, t_a, over a surface of silicate material in nanometres can be readily obtained from equation (D.12) as

$$t_a = \frac{k_1 V_m h}{(1 - h/h_m)(1 - h/h_m + k_1 h)} \tag{D.13}$$

This is one of the basic relationships used in Chapter 4 to obtain the sorption isotherms of a cementitious microstructure under virgin drying or wetting conditions.

Index